HANDBOOK OF HUMAN FACTORS TESTING AND EVALUATION

Second Edition

HANDBOOK OF HUMAN FACTORS TESTING AND EVALUATION

Second Edition

Edited by

Samuel G. Charlton
Waikato University

Thomas G. O'Brien
O'Brien and Associates

 LAWRENCE ERLBAUM ASSOCIATES, PUBLISHERS
2002   Mahwah, New Jersey   London

Lawrence Erlbaum Associates, Inc., Publishers
10 Industrial Avenue
Mahwah, New Jersey 07430

Cover design by Kathryn Houghtaling Lacey

Library of Congress Cataloging-in-Publication Data

Handbook of human factors testing & evaluation / editors: Samuel G. Charlton,
Thomas G. O'Brien.–2nd ed.
p.   cm.
Includes bibliographical references and indexes.
ISBN 0-8058-3290-4 (alk. paper) — ISBN 0-8058-3291-2 (pbk. : alk. paper)
1. Human engineering–Handbooks, manuals, etc.   2. Anthropometry–Handbooks,
manuals, etc. I. Charlton, Samuel G. II. O'Brien, Thomas G.
TA166 .H276   2002
620.8'2—dc21   2001033265
                                                              CIP

Books published by Lawrence Erlbaum Associates are printed on acid-free paper,
and their bindings are chosen for strength and durability.

Printed in the United States of America
10   9   8   7   6   5   4   3   2   1

To my love Jean, who has stuck with me through two hemispheres, and to Colin, Nina, and Julie for letting me tie up the computer.

—SGC

To my beautiful and understanding wife, Patty, and to my wonderful children, TJ and Karen, who still ask, "What is it you do?"

—TGO

Contents

Preface

In the first edition of this handbook we posed the question "Why a book on human factors test and evaluation?" Our answer was that we felt that human factors test and evaluation (HFTE) is one of the most important applications of an avowedly applied discipline. HFTE is the principal means of ensuring that systems and products will be effective and usable before they are placed in the hands of the public, and yet, it remains more or less invisible to those outside the profession.

Implicit in that statement was the presupposition that HFTE was somehow more pragmatic, less theoretical, and less appreciated than the brand of human factors and ergonomics practiced by researchers in academic institutions. Faced with the need to "make it work" in the field environment, one could not afford the luxury of laboratory methods and true experimental designs, nor could one typically reason from general theoretical principles to the specific cases at hand. An oft-repeated refrain of HFTE professionals is that HFTE is an art; no standard formula for conducting HFTE exists any more than a standard system exists.

Have we changed our view in these past 5 years? Maybe. Certainly it has been tempered somewhat. Is HFTE an entity distinct from human factors and ergonomics generally? Our answer is still yes, but perhaps the distinction is not as great as it once was (or as once was believed). HFTE practitioners are sometimes characterized as preferring the methods sections of journal articles, looking for measures that might be appropriate for some future test, blithely skipping the literature review and discussion sections.

While this portrayal may have some truth to it, is it an indication of an inclination towards pragmatism at the expense of theoretical curiosity? There are certainly HFTE practitioners with strong interests in theoretical issues and who appreciate a philosophical conundrum as much as an academic. By the same token, since the publication of the first edition, we have become more aware of academic researchers that view test and evaluation as a fertile ground for uncovering issues germane to theory construction and laboratory research.

Is HFTE an art? If one views art as a creative interplay between environment, materials, and message and an artist as one skilled in a set of tools for conveying the message afforded by the materials in their environment, our answer is yes. Is a generic formula for doing HFTE possible? Perhaps not. But we do believe that there are some generalizations to be drawn. If we believed wholeheartedly that every HFTE effort were truly unique, there would be little point in attempting a volume such as this. Although one may not be able to design an HFE test from first principles laid out in the laboratory, there are some robust test and evaluation principles that can be seen by reviewing many of the examples of HFTE that have gone before. Thus, one of the principal reasons for embarking on this project was to showcase the many creative techniques and intriguing results obtained by HFTE practitioners.

One of the peculiarities of HFTE is that the enterprise is not often well-communicated to the larger human factors and ergonomics community, or even between HFTE practitioners. Methods and results are often documented only in closely held technical reports and are only rarely suitable for publication in mainstream journals and books. One of our goals in preparing this handbook was to introduce readers to many of the tools of the HFTE trade, outline some of the major issues confronting HFTE, and offer as many examples and case studies as practical in the hopes that practitioners could add to their knowledge of what works and draw their own generalizations and abstractions. We also hope that an inductive enterprise of this sort might provide a service to academics and theoreticians as well. A theory of HFTE? Not quite, but maybe some insights into issues and approaches meriting further examination in the controlled confines of the research laboratory.

Which brings us to the present, and now begs the questions "Why another edition? What's new?" One of the goals of offering a second edition is to expand and emphasize the applications chapters, providing contemporary examples of HFTE enterprises across a range of systems and environments. We have updated our coverage of the standard tools and techniques used in HFTE as well; however, based on feedback from readers of the first edition of the handbook, we feel more strongly than ever that it is the case studies and examples that serve to lift HFTE's veil of obscurity.

The revelations acquired from the first edition were not just for those new to HFTE but for established practitioners as well. Thus, we have doubled the number of applications chapters and introduced many more examples and illustrations throughout the other chapters.

As with the first edition, the orientation of this edition has been towards breadth of coverage, at the inevitable expense of in-depth treatment of a few issues or techniques. Experts in a particular field may be disappointed by the amount of coverage given to their specialty, but we are of the view that HFTE is somewhat better served by a rich general knowledge rather than narrow specialization. Once again, we hope that the book contains something of interest for every reader. Experienced testers will find much that is familiar, but we also hope they find a few new tools, creative approaches, and perhaps a rekindled enthusiasm. Newcomers to HFTE will discover the diversity of issues, methods, and creative approaches that makes up our field. The tools and techniques of HFTE should not remain arcane knowledge in the sole possession of a few practitioners. It is our hope that this book has been written in such a way that individuals outside the profession will learn the intrinsic value and pleasure in ensuring safe, efficient, and effective operation and increased user satisfaction through HFTE.

—*Samuel G. Charlton*
—*Thomas G. O'Brien*

Contributors

Brett D. Alley is a human factors researcher with Transport Engineering Research NZ. In 1999 he completed his master's degree in human factors psychology supported by the Road Safety Trust Research Scholarship and a University of Waikato Masters Scholarship. His areas of specialization include visual processing, driver perception, and cognitive processing. He has particular experience in measuring driver perception and behavior using driving simulator technology, including projects investigating the impact of cell phone use on driver attention and behavior, the influence of roadside advertising signage on driver attention, and driver behavior at passing lanes.

Stephen D. Armstrong began his career with a human factors engineering degree from the United States Air Force Academy and has over 13 years of experience in human factors engineering (HFE), including human–computer interaction, psychometrics, and user-centered design. During his career, he has contributed HFE expertise to the U.S. Air Force's B-2 stealth bomber program and missile defense projects, the commercial aviation industry, and various other defense and civilian efforts. He has served as a member of the faculty at State Fair Community College in Missouri and speaks regularly at national conferences on human factors and user-centered design issues. Currently, Mr. Armstrong holds positions as manager of human factors engineering for Genomica Corporation and senior consultant for Colorado Ergonomics, a human factors consulting firm.

Liz Ashby is a human factors researcher of the Centre for Occupational Human Factors and Ergonomics (COHFE). Her interests lie in injury prevention and ergonomics in forestry and related industries. Liz joined COHFE after having worked in England with an ergonomics consultancy, where her role included the development and delivery of manual handling risk assessment competency training, and she was also involved with associated ergonomics consulting. Liz has qualifications in physiotherapy (Graduate Associateship in Physiotherapy) and Ergonomics (MSc, with distinction).

Peter H. Baas has been involved in road transport research for over 18 years. His special interest is in road safety, especially vehicle performance and the interactions among the driver, the vehicle, and the road. He played a major role in the New Zealand House of Representatives Transport Committee inquiry into truck crashes as its technical adviser, including undertaking the inquiry's investigation into truck driver fatigue. Peter was formerly the coordinator of the Industrial Research Ltd. Transport Research Centre and is now the managing director and a principal researcher at Transport Engineering Research New Zealand Limited.

Dr. Herbert H. Bell is a branch chief and senior research psychologist with the Air Force Research Laboratory, Warfighter Training Division, in Mesa, Arizona. His research interests focus on applied research involving the acquisition and maintenance of combat mission skills. He is currently investigating the relation between specific training experiences and mission competency. Dr. Bell received his PhD in experimental psychology from Vanderbilt University.

Dr. Tim A. Bentley is manager and programme leader of the Centre for Occupational Human Factors and Ergonomics (COHFE). Tim joined COHFE following a successful year with Massey University in Albany leading a multidisciplinary program of research on safety in New Zealand's adventure tourism industry. Tim's recent UK employment history includes extensive safety research with the Royal Mail and a study considering training needs in the British health sector. Tim has qualifications in psychology (BSc, Honors), ergonomics (MSc) and health and safety ergonomics (PhD), which are reflected in his broad range of work interests from the fields of injury prevention, occupational psychology, and ergonomics.

Dr. Michael A. Biferno is the manager of human factors and ergonomics technology for the Boeing Company Phantom Works and is a Boeing technical fellow. He has led the development of new human factors technologies that have been transitioned to programs in the commercial, military, and space divisions to support vehicle design, production, operations, and maintenance. Dr. Biferno is Chairman of the Society of Automotive Engineers (SAE) G-13 Committee on Human Modeling Technology and Standards.

William C. Brewer has worked in the user interface development field since 1984, participating in the development of aircraft cockpits, flight instrument displays, and mobile command and control shelters, as well as participating in several human–computer interface development efforts. Curt Brewer has designed and executed many usability tests both for the purpose of validating these interfaces and to make comparisons between multiple design alternatives. The results of these efforts have been published at the Military Operations Research Society Symposium and in numerous corporate technical reports. Mr. Brewer holds a bachelor's degree in experimental psychology from the University of California, Santa Barbara, and has studied human factors psychology at California State University, Northridge. He is currently employed as a system engineer at TRW, Inc. in Colorado Springs, Colorado.

Dr. Lyn S. Canham has worked for the last 11 years at the Air Force Operational Test and Evaluation Center at Kirtland AFB in New Mexico. Prior to that, she received her PhD in cognitive psychology from the University of New Mexico. Her work is focused on measurement and modeling of human performance and human design making in military command and control environments.

Dr. Antonio B. Carvalhais is a human factors program analyst with the United States Coast Guard's Office of Safety and Environmental Health, where he serves as the primary human factors resource to major system acquisitions, sponsors and conducts research to explore how operational policies and procedures impact human alertness and performance, and develops policy and support guidance for human factors-related issues in Coast Guard activities. Tony serves as the Coast Guard representative on various committees to address human factors and ergonomic issues in the Department of Transportation (DOT) and Department of Defense (DOD). Tony has authored numerous scientific papers and book chapters on human factors, ergonomics, shiftwork, fatigue, and human alertness.

Dr. Samuel G. Charlton received his PhD in experimental psychology in 1983 from the University of New Mexico. He has held positions at the University of Albuquerque, the University of New Mexico, the BDM Corporation, the Veterans Administration Medical Center, and the Air Force Operational Test and Evaluation Center (AFOTEC). While serving as the human factors technical director at AFOTEC, he was principal investigator of operator fatigue, workload, and human performance issues for airborne, space, and ground-based systems. In 1995 Samuel moved to New Zealand and took up a position as a lecturer in cognitive psychology. He is currently a senior lecturer in psychology at Waikato University and principal human factors scientist for Transport Engineering Research New Zealand Limited.

Dr. Jerry M. Childs serves as director of performance engineering with TRW. He has 25 years of experience in the management and conduct of performance improvement programs within both government and private sectors. His technical skills are in the areas of training systems development and evaluation, human factors, and performance assessment. His clients include the DOD, DOE, Federal Aviation Administration (FAA), NASA, Federal Express, General Motors, Ford, Caterpillar, Bechtel Energy, Hughes, Sprint, Exxon, and US West. Dr. Childs is a member of the American Society for Training and Development, the Aviation Industry CBT Committee, Department of Energy Training Managers Quality Panel, and Human Factors and Ergonomics Society. He served on Southwest Airlines' Training Advisory Committee and has authored more than 75 professional publications and presentations. He holds a PhD in engineering psychology from Texas Tech University.

Thomas W. Dennison is currently employed by Sikorsky Aircraft as a senior crew station design engineer on the U.S. Army RAH-66 Comanche helicopter program. He has held that position for 3 years and was previously the lead crew station design engineer for the S-92 Civil Helicopter, which first flew successfully in December 1998. He worked for Honeywell's Avionics Division and the U.S. Army in various positions, performing basic research, applied research, and test and evaluation. He earned a master of arts in experimental psychology from Marquette University and a bachelor of arts in psychology from State University of New York College at Buffalo.

Dr. Peter J. M. D. Essens works for the Netherlands Organization for Applied Scientific Research in the area of system ergonomics and team-based human machine systems. Since 1985, he has been involved in studies of current and future command and control in the military and civil domain. In particular, his interest lies in the interaction among individual, team and organization. He is developing an approach to human-based design of complex systems that should link early analysis and the future way of work. Dr. Essens is a member of the European Association of Cognitive Ergonomics and is currently chairman of the NATO Exploratory Group on Team Effectiveness.

Marty Gage is one of four principals at SonicRim, a research firm which helps companies learn to involve users in their process. Prior to founding SonicRim, Marty worked at Fitch, an international design firm, for 10 years. During this time he conducted user research to support branding and product development. Marty has a bachelor's in psychology from Hendrix College and a master's in human factors psychology from Wright State University.

Dr. Valerie J. Gawron is a level 5 (world-class engineer) at Veridian Engineering and an expert in computer-aided engineering, controls and displays, environmental stressors, human performance measurement, simulation, situational awareness, test and evaluation, training, Uninhabited Aerial Vehicles, and workload. She received a PhD in engineering psychology from the University of Illinois and master's degrees in psychology, industrial engineering, and business administration from the University of New York. She is a fellow of the Human Factors and Ergonomics Society and an associate fellow of the American Institute of Aeronautics and Astronautics. She was chair of the Human Factors and Ergonomics Society Test and Evaluation Technical Group from 1996 through 2000 and a member of the U.S. Air Force Scientific Advisory Board in the same period. Her book, *Human Performance Measures Handbook*, was published in July of 2000 by Lawrence Erlbaum Associates.

Dr. Douglas H. Harris is chairman and principal scientist of Anacapa Sciences, Inc., a company formed in 1969 to improve human performance in complex systems and organizations. His principal contributions have been in the areas of inspection, investigation, and maintenance performance. His publications include *Human Factors in Quality Assurance* and *Organizational Linkages: Understanding the Productivity Paradox*. He is a past president of the Human Factors and Ergonomics Society and a past chair of the Committee on Human Factors of the National Research Council.

James C. Higgins is the group leader of the Systems Engineering and Safety Analysis Group at Brookhaven National Lab (BNL). His general areas of expertise include nuclear facility operations, testing, control-room design review, and PRA. He holds a bachelor's in naval engineering from the U.S. Naval Academy and a master's in mathematics from the U.S. Naval Postgraduate School. In the U.S. Navy, he served as an engineering division officer and department head on two nuclear-powered submarines. For the U.S. Nuclear Regulatory Commission (NRC), he served as a region-based reactor inspector and as a senior resident inspector. At BNL, he provided support for onsite reviews, analysis, and research for the NRC and Department of Energy on various programs including sensitivity of risk to human errors, upgrades to alarm systems and local control stations, and advanced reactor control room reviews.

Darcy L. Hilby has been with the Boeing Commercial Airplane Group since 1988. Her work at Boeing includes flight deck ergonomics and certification for all Boeing models produced in the Puget Sound region. She was responsible for the anthropometric and ergonomic analysis of the newly designed 777 flight deck as well as helping develop the Boeing Human Model. Prior to joining Boeing, Ms. Hilby spent 3 years at Westinghouse as an engineer on an automated fuel-processing project. Ms. Hilby holds a master's degree from the University of Oregon in the specialty area of biomechanics.

Dr. Alan R. Jacobsen, currently a technical fellow of The Boeing Company, received his PhD in experimental psychology from the State University of New York at Stony Brook with an emphasis in visual perception. For the past 13 years, he has been working at Boeing in the area of flight deck and avionics design. He was the lead engineer for the flight deck group responsible for the overall design and integration of displays and lighting on the 777 flight deck, as well as the new generation 737 airplane. He was also responsible for human factors requirements and evaluations of the new flat-panel liquid crystal displays that were introduced into commercial aviation to a significant degree on the 777 airplane. His participation on the liquid crystal display program spanned the entire gamut from basic initial research and development (IR&D) to production implementation and certification. Dr. Jacobsen has served on numerous industry standards committees and chaired the Society for Automotive Engineering subcommittee that wrote the Aerospace Recommended Practice guidelines for the effective utilization of color in avionics displays. For the past several years, he has also lectured at the University of California at Los Angeles (UCLA) on human factors issues in flight deck design. Prior to joining Boeing, Dr. Jacobsen was employed as a research psychologist at the Naval Submarine Research Laboratory at Groton, Connecticut, where his work focused on the application and evaluation of color in new generation sonar displays.

Brian D. Kelly received a bachelor of science in aerospace engineering from the University of Southern California in 1978 and a master of science in aeronautics and astronautics from Stanford University in 1979. He joined Boeing in 1979 at the beginning of the 757 and 767 programs in flight deck integration, assisting in the development and certification of the crew interfaces for the Engine Indication and Crew Alerting System (EICAS) and other airplane systems, including piloted validation testing of the new crew-alerting system. Subsequent assignments involved development of guidance, alerting systems, and displays, including piloted development testing of several candidate display concepts in support of the Boeing-led Windshear Task Force, development and testing of primary flight control interfaces and displays concepts on the 7J7 program, and participation on industry committees associated with the definition of crew interfaces and procedures for the Traffic Alert and Collision Avoidance System (TCAS). As a manager in flight crew operations integration from 1991 to 1998, Mr. Kelly oversaw several aspects of the development and certification of the 777 flight deck and subsequently other models. In late 1998, he was accepted as a Boeing associate technical fellow and has recently served on the RTCA Certification Task Force Human Factors Subcommittee. Mr. Kelly hold three patents and is author or coauthor of several papers and presentations.

Rebecca E. Luther is a human factors researcher with Transport Engineering Research New Zealand. She completed her master's degree in psychology at the University of Waikato in 1999, specializing in human factors and organizational psychology. Her thesis focused on improving the training of information-processing skills (situation awareness) in air traffic controllers. Her particular areas of interest include human information processing (situation awareness, cognitive workload, and memory) and driver training and education.

Dr. Thomas B. Malone received his PhD from Fordham University in experimental psychology in 1964. For the past 20 years, he has been president of Carlow International of Falls Church, Virginia, and Carlow Associates Ltd. of Dublin, Ireland. He is a past president of the Human Factors and Ergonomics Society (HFES), of which he is also a fellow. He currently serves as chair of the HFES Systems Development Technical Group. His primary professional interests currently lie in integrating human factors into systems engineering, enhancing the interaction between humans and automated systems, reducing the incidence and impact of human error in complex systems, such as medical systems, and reducing manning levels and enhancing human performance on ships and offshore systems.

Dr. David Meister received his PhD in experimental psychology from the University of Kansas in 1951. Since then he has worked in government and industry and as a contract researcher. He is a former president of the Human Factors Society (1974) and the author of more than a dozen textbooks on human factors in design and measurement. He is a fellow of the Human Factors and Ergonomics Society and the International Ergonomics Association, and he has received awards from both of these societies, as well as the American Psychological Association.

Dr. Randy J. Mumaw has a PhD in cognitive psychology from the University of Pittsburgh and did his graduate research at the Learning Research and Development Center in Pittsburgh. He has been with the Boeing Commercial Airplane Group for 2 years, working in the areas of flightcrew understanding of flightdeck automation and flightcrew performance. Prior to joining Boeing, Dr. Mumaw spent 7 years at Westinghouse, where he worked on nuclear power plant control-room design and evaluation. He had a major role in developing the operator performance component of the Verification and Validation (V&V) plan for an advanced technology control room (AP600). Also, at Westinghouse, he conducted both simulator and observational studies to investigate operator monitoring, decision making, procedure use, and emergency management. Prior to working at Westinghouse, Dr. Mumaw worked on projects involving analysis of skills and training design in areas such as air traffic control, electronics troubleshooting, and military aviation. Primary areas of interest continue to be interface design, automation design, decision making, analysis of expertise, human error, and training design.

Dr. Jean E. Newman is an experimental psychologist specializing in the study of human cognition and language understanding. She received her PhD in cognitive psychology from the University of Toronto, Canada, in 1981. Dr. Newman has over 20 years of research experience that includes expertise in research design and statistical analysis and the study of human information processing and memory. She was an associate professor of linguistics and psychology at the University of New Mexico, where she also served as the chairperson of the Department of Linguistics from 1991 until joining the staff of the University of Waikato Department of Psychology as a lecturer in 1995. Some of her current work includes the analysis of risk perception in road-user interactions.

Thomas G. O'Brien is an engineering psychologist and owner of O'Brien & Associates. He received his master's degree from Towson University in 1980. His first job was with the U.S. Army Human Engineering Laboratory, Aberdeen Proving Ground in Maryland, where he provided human factors advice to the U.S. Army Materiel Systems Analysis Activity. From 1984 to 1986 he worked for the BDM Corporation in Albuquerque, New Mexico, conducting human factors operational tests for the U.S. Air Force Operational Test and Evaluation Center. In 1986 he went to work for Battelle Memorial Institute, where he performed human factors assessments of foreign threat combat systems. In 1988, he established his consultancy in human factors engineering and generalized testing and evaluation. Currently, Mr. O'Brien's consultancy applies usability engineering conventions to help various military and industrial clients develop efficient and effective user interfaces for complex software-controlled logistics and e-business operations. End clients include Delta Airlines, the Federal Aviation Administration, the U.S. Air Force Space Command, and others. A former member of the U.S. Navy special forces, Mr. O'Brien dabbles in combat ergonomics, a method he devised for testing combat systems without the usual government constraints.

Dr. John M. O'Hara is currently a scientist at Brookhaven National Laboratory (BNL) and is the human factors research manager. He has served as principal investigator of numerous research programs in human systems integration, addressing the effects of technology on crew performance in complex systems and system safety. The results of the research are currently being used to help develop guidance for the design of the human factors aspects of complex systems, such as power plants and merchant ships. Prior to BNL, he served as principal investigator in Grumman's Space Division for various NASA programs in automation and robotics, extravehicular activities systems, and shuttle and space station workstation designs. Prior to Grumman, John was research manager of the Computer Aided Operations Research Facility, where he headed studies on bridge system technology, performance measurement, the use of simulation for investigating restricted waterway navigation by large vessels, and risk assessment and management using an integrated systems approach.

Aernout Oudenhuijzen was born in Neede in the Netherlands and is currently employed at the Netherlands Organization for Applied Scientific Research (TNO), Human Factors (HF). He is a scientist of physical ergonomics in the department of work space ergonomics. He is involved in human modeling technology. He began using these models while obtaining his degree in industrial engineering at the Technical University of the city of Delft. He worked for Fokker Aircraft from 1991 until 1996, where he was responsible for human factors of the F50, F60, F70, and F100 aircraft and was involved in the development of new products. At TNO HF, he has worked on the design and the development of several workspaces; car interiors, truck cabins, cockpits, command and control centers, office buildings, and armored vehicles. Since 1993 he has served on the SAE-G13 Human Modeling Standards and has been the lead for the international verification and validation project for human modeling systems.

Richard J. Parker is a researcher with the Centre for Occupational Human Factors and Ergonomics (COHFE) and was formerly a human factors researcher with the New Zealand Logging Industry Research Organisation (LIRO). He has practical experience in the logging industry, working as a faller and breaker out for 2 years. His academic background includes a bachelor of science in zoology and postgraduate diplomas in agricultural science (animal physiology and behavior) and ergonomics. He is currently completing a master's in ergonomics, investigating safety issues of people working in close proximity to mobile forestry machines. He has been involved in studies on spiked boots, high-visibility clothing, logging hazards, physiological workload, harvester ergonomics, and protective footwear for loggers.

Dr. Brian Peacock graduated from Loughborough University in 1968 after studying ergonomics and cybernetics. He obtained his PhD from Birmingham University in 1972. He worked in academia from 1972 to 1986 with appointments in Birmingham University, Hong Kong University, Monash University, Dalhousie University, and the University of Oklahoma before moving to General Motors (GM), where he is manager of the GM Manufacturing Ergonomics Laboratory. He is now Discipline Coordinating Scientist at NASA's National Space Biomedical Research Institute. He has published numerous technical articles and two books on automotive ergonomics and statistical distributions. He has been vice president of the Board of Certification in Professional Ergonomics, director of the Institute of Industrial Engineers Ergonomics Division, chairman of the Human Factors and Ergonomics Society (HFES) Industrial Ergonomics Technical Group, and he is currently a member of the HFES Executive Council.

Dr. Anita M. Rothblum earned her PhD in experimental psychology at the University of Rochester in 1982. She began her human factors career at AT&T Bell Laboratories as a systems engineer and human factors consultant on the design and evaluation of telephone and computer products. Currently, Dr. Rothblum is with the U.S. Coast Guard Research and Development Center in Groton, Connecticut. She has managed human factors research in the areas of casualty investigations, fatigue, the effects of automation on job design and training, and crew size modeling.

Dr. Elizabeth B. N. Sanders is the president of SonicRim, where she explores generative search and research activities with collaborators in many different industries. She is best known for introducing participatory design practices to the consumer product marketplace. She also serves as an adjunct faculty member in both the Industrial Design Department and the Industrial and Systems Engineering Department at Ohio State University in Columbus, Ohio. Liz has undergraduate degrees in psychology and anthropology and a PhD in experimental and quantitative psychology.

Richard K. Steinberg is a senior scientist at Schafer Associates in Management of Software Development, Usability and Human Performance Engineering. He has held positions as the National Missile Defense Command and Control Display Design technical lead and display prototype and usability testing lead for the U.S. Army's Theater High Altitude Area Defense (THAAD) Battle Management Command, Control, and Communications (BMC3) and Radar Displays. While serving as the technical lead for these programs, he designed and developed rapid prototypes for usability testing. He has designed displays and written software for the Army's Exo-atmospheric Re-entry Interceptor Subsystem Mission and Launch control displays. Mr. Steinberg has published numerous articles on human–computer interaction and holds a bachelor's degree in mechanical engineering from the Auburn University.

William F. Stubler is a research engineer at Brookhaven National Laboratory (BNL). His specific areas of interest include development and application of human factors guidelines for evaluation of advanced control rooms and control-room upgrades and investigation of issues that affect operator performance in advanced control rooms, including computer-based display, control, and alarm systems. Prior to joining BNL, Bill was a human factors engineer with Westinghouse Electric Company. He conducted research addressing topics such as the verification and validation of advanced control rooms, advanced human–computer interaction concepts for nuclear and fossil power plant operation, and outer-air battle management.

Colin T. William is a researcher at SonicRim. He has participated in generative and explorative projects on user experiences in diverse areas. Colin has an undergraduate degree in psychology and religion/philosophy and a master's degree in cognitive/developmental psychology. He is completing a PhD in cognitive psychology at Emory University and serves as an adjunct faculty member at Columbus State Community College in Columbus, Ohio.

Dr. Glenn F. Wilson is currently a senior research psychologist at the U.S. Air Force Research Laboratory's Human Effectiveness Directorate. He works with Air Force test and evaluation teams in their evaluation of new and modified Air Force systems. His research interests include using psychophysiological methods to understand operator functional state. To study physiological responding in complex real-world environments, he has recorded data from crew members in a number of different types of aircraft. His current focus is on the use of operator functional-state information in adaptive aiding systems to improve overall system performance. Dr. Wilson has written numerous articles, chapters, and technical reports dealing with the application of psychophysiology to human factors issues.

Part I

TOOLS AND TECHNIQUES

INTRODUCTION

We have devoted the first half of this second edition to the description of the methods and approaches that have been used in human factors test and evaluation (HFTE). Some of the chapters are updated versions from the first edition of the handbook, others are new treatments of subjects covered previously, and others are new additions to our coverage of HFTE tools and techniques. Undoubtedly, we will have missed some topics; that is an inevitability in any treatment of our diverse discipline. Although we may not be able to provide an exhaustive account of every HFTE technique, nor treat every topic in the depth deserved, we have sought to strike a balance between the scope of our coverage and the detail provided. In other words, while these chapters will not suffice in making someone an expert in HFTE, there will be enough information for a reader to grasp the essentials of the methods. It is hoped that by the end of this section, even the casual reader will become a more knowledgeable consumer of HFTE.

Our coverage of these methods begins with a brief historical overview of HFTE. As with many other technology driven enterprises, HFTE has a legacy to the machinery of armed conflict. Our first chapter describes these historical roots and how HFTE evolved from the somewhat informal practice of field testing weaponry to the development of more systematic time and motion studies in industrial settings. Our history goes on to describe how the practice of HFTE as a separate discipline was initiated by the establishment of government-directed laboratories and acquisition procedures, and it concludes by covering the recent emergence of modeling, simulation, and usability testing methods.

The current processes and procedures of HFTE are described in detail in the second chapter. We describe the formalized system development process used by government agencies, the corresponding practices used by private industries, and the newer practice of integrated product development. The position of HFTE has shifted from the end of the design and engineering chain to become more thoroughly integrated with the system development process; HFTE activities occur in the earliest stages of concept exploration through to production. This chapter discusses the activities and issues confronting human factors testers at each stage of the system development process.

Although it can rightly be said that there is no standard formula for planning and conducting a human factors test, the SITE methodology presented in the third chapter outlines one approach that has been used to introduce students and new human factors professionals to the testing discipline. The approach is essentially an organizational aid for selecting human factors test measures and placing test results in the context of product development and performance. The chapter provides several examples of how the approach has been implemented in both student projects and full system tests.

Chapter 4 covers test support documentation and builds on the planning and reporting theme, providing the reader with a guide through the host of human factors standards and HFTE reporting requirements. From the design, performance, and process standards used in the early stages of test planning through the test plans themselves, and ultimately to the myriad of test reporting formats, this chapter provides many descriptions and examples that will be helpful to both new and experienced testers.

The fifth chapter addresses one of the classic human factors test methods: human performance testing. This chapter begins by describing the theory and discussing 50 years of history associated with performance testing. The chapter provides the reader with detailed guidance for designing and assessing performance tests, with special attention given to attaining adequate test reliability. The final part of the chapter addresses how to administer performance tests, particularly in the critical area of establishing standardized testing conditions.

Another classic HFTE methodology is the measurement of cognitive states. Chapter 6 explores some of the techniques for measuring cognitive states, particularly cognitive workload and situation awareness. Here we examine the theoretical and definitional issues that underlie the measurement of these elusive but important determinants of human behavior. The chapter also outlines several criteria that can be used to compare the various measures of workload and situation awareness currently in use. Descriptions of the most frequently used workload and situation measures are provided, along with advice on how to collect and analyze data with each of them.

The subject of chapter 7 is psychophysiological measurement, another classic HFTE methodology. The author provides an in-depth look at an extensive range of these measures, organized according to the type of system application where they can best be employed. Because the use of psychophysiological measures is still relatively new to many areas of HFTE, the chapter contains thoughtful guidance on the types of equipment available, the practicalities involved in using the measures for best effect, and a brief bibliography for additional reading. A thorough review of the latest HFTE literature, as well as a look at what the future holds for psychophysiological methods, will bring the reader current with the latest developments in this challenging field.

In the chapter describing the purposes, processes, and products of measurement in manufacturing ergonomics, the author introduces us to a variety of ergonomics measurement and analysis techniques and provides an overview of the design principles of the ergonomics measurement tools themselves. For example, the issues of resolution and precision of measurement, decision thresholds, and interactions in complex stress-exposure measures, are all clearly described. Finally, the chapter covers the key intervention opportunities for ergonomists in manufacturing settings.

The increasing emphasis on performing HFTE as early as possible has meant testers have had to turn to methods such as mock-ups, models, and simulations to evaluate designs, even before the first article is produced. Chapter 9 provides the reader with a wealth of information on these tools, from the old standards, such as plywood mock-ups, to the latest in computer-based manikins and operator performance models. The authors also describe various evaluation support tools that can help the tester select the most appropriate tool for a given test. The chapter goes on to describe various embedded testing techniques and the latest trends and developments in HFTE tools.

The final two chapters in this section present two commonly used, but frequently unpopular, HFTE techniques: questionnaires and environmental measurement. These techniques feature to some degree in nearly every human factors test, but the ease with which they can be poorly performed, along with the perceived difficulties associated with doing them properly, means that these methods are rarely employed to their full potential. These two chapters are designed to acquaint the reader with the necessary basics of how to properly use these techniques. It is our hope that removing many of the misconceptions surrounding questionnaires and environmental measurement will lead to both greater care in their application and to a better appreciation of their potential benefits to HFTE.

Chapter 1

Human Factors Testing and Evaluation: An Historical Perspective

Thomas G. O'Brien
O'Brien & Associates

David Meister
Human Factors Consultant

History gives us an appreciation of where we are today and what it took to get us here. Like anything else, HFTE has a history, and it is important, as well as interesting, to understand and appreciate the early techniques that led to modern practices.

Where did the formal practice of HFTE begin? For argument's sake, let's say it began about the same time as the discipline of human factors engineering (HFE). Although there are arguably some differences in definitions, for the purpose of discussion, we consider *engineering psychology* and *ergonomics* synonymous with HFE, and later, we encompass all of these under the general term *human factors*, to include, in addition, specialties such as training, manpower, personnel, and safety. Human factors was begot from a number of other disciplines, as Meister points out in his history of human factors and ergonomics (Meister, 1999). For this reason, to assign a chronological beginning would be difficult, if not impossible. Indeed, the origin of human factors has been a matter of recent debate. Meister provides a broad historical explanation of the discipline, outlining the morphology of human factors and ergonomics in various cultures and providing a great deal of insight into its early days as the natural offspring of experimental psychology and the progenitor of what might be the next phase of evolution—usability engineering. But what of HFTE?

Like a two-sided coin, HFTE compliments the design aspect of human factors. Without it, we might never know if our system or component design meets our intended technical and operational objectives. Moreover, without HFTE, it would be difficult to impart measurable improvements to our design. And so, like all methodologists, we should define HFTE. Our simple definition is "Human factors testing and evaluation is a set of methodologies to characterize, measure, assess, and evaluate the technical merit and operational effectiveness and suitability of any human-system interface." A more detailed definition would include training, manpower and personnel, health hazards, and perhaps others.

Measurement is as integral to human factors as it is to every scientific and engineering discipline. Indeed, it would be correct to say that HFTE preceded the development of the discipline because it is only when one can measure phenomena that one can organize these phenomena into a discipline. Ordinarily, we associate human factors with World War II; however, to get a more complete understanding of the underlying precepts and causes leading to our contemporary notion of HFTE, we must reach back to the days prior to World War II and even World War I. For example, in the American Civil War, organized efforts emerged to test how well soldiers and sailors performed with new weapons and accoutrements. Indeed, throughout written history, makers of weapons and apparel tested their wares for designs that best suited the owner in form, fit, and function. For example, a suit of armor was no good to the knight unless it fit him.

In the first edition of this book (O'Brien & Charlton, 1996), we described a general historical model of HFTE comprising premodern and modern periods. In this edition, we refine our argument to make the point that HFTE has roots, deep and penetrating, reaching well before the 19th century. The examples we cite are mostly military because, in the early days, the military aspect of technology was more important to the movers and shakers than it was for domestic use. As with the advancement of military applications, there is a corresponding, later concern for human factors, particularly in medicine. Today, the application of HFTE is as well-founded in the industrial environments as it is in the military.

THE PREMODERN ERA

Prior to the Second World War, the only test of human fit to the machine was an informal process of trial and error. The human either could or could not function with the machine. If the human could not, another human was selected until the right match was made. The operator was viewed as a convenient and dispensable element whose importance was limited to operating the machine. While this selection model (Brogan, Hedge, McElroy, & Katznelson, 1981) worked well with mass-produced weapons and equipment, we might argue that other, more sophisticated methods were engaged for individual components, for example, body armor.

Although scant written evidence survives to support the notion that HFTE was anything more than trial and error prior to World War II, there is evidence of the use of tools to measure bodily parts, for example, calipers to measure the cranium for fitting helmets. Also, chroniclers of the Middle Ages tell of inventors who took the time to record the effects of different equipment designs on the ability of humans to perform tasks. Recording the pattern of arrows striking a target set at measured distances from archers helped 13th century developers determine the optimal length and curvature for that weapon, considering an average man's height, reach, and draw strength. Today, we would call this human performance testing. The results of such testing ultimately contributed to the invention of the crossbow, which was more powerful. Some crossbows required a windlass (i.e., a hand-crank mechanism for drawing back the crossbow string into firing position). This mechanical device helped the soldier in preparing the weapon for firing. Notably, like our example of the crossbow, the goal of modern HFTE is to improve the efficiency and effectiveness of systems.

Later, possibly as early as the 17th century, British gun crews were timed in the loading and firing of deck guns. Early infantry were also timed in loading and firing their weapons. To shorten the time it took to deliver a volley, new firelock designs were introduced. For example, the basic infantry firearm evolved from the highly unwieldy matchlock to the less-unwieldy wheel lock and then to the somewhat unwieldy flintlock. Finally, it graduated to the percussion lock before being replaced altogether by the modern rifled arm. Each time, the soldier was able to load and fire faster and with greater accuracy, as evidenced in human factorslike testing.

Naval gun designers saw that crews performed the task of loading and firing much faster if the guns were fitted with improved sighting devices and with carriages that allowed the weapon to be hauled easier in and out of battery. Based on observations during battle, naval commanders saw that by painting gun decks red, the sight of bloodshed by injured crewmen during battle had a less deleterious effect on gun-crew performance when compared with crews on decks that were not painted red. That is, the crewmen became somewhat desensitized to the color red and tended not to become distraught at the sight of so much blood spilled during battle. Commanders also quickly realized that a well-trained crew would perform better than one that was not trained or one that was poorly trained. Thus, training became an important aspect of shipboard routine as it is now for all military systems. Later, HFTE specialists would include training as an important aspect of evaluation criteria. Organization and tactics soon drew the attention of military planners who now had to consider not only the technological benefits of newer weapons but also, in a general context, the capabilities and limitations of soldiers and seamen who operated and maintained those weapons. In this latter respect, as weapons became more accurate at greater distances, infantry formations went from the tightly packed phalanx formation of advancing on the enemy to a more flexible and effective dash-and-take-cover maneuver, as seen in World War I. Regardless of the period, commanders would often note during battles the effects of various tactics, roughly approximating what contemporary HFTE specialists might do when measuring the effects of infantry deployed against various threats.

These are a few examples of how the human factor was taken into account well before the advent of sophisticated machines and formalized testing. However, it was not until the American Civil War that we clearly see a methodical effort to include the human in systems testing.

The American Civil War

The American Civil War saw a leap in technology, which resulted in a greater degree of sophistication in the design, operation, and maintenance of systems and equipment. The United States Patent Office took special interest in whether mass-produced uniforms and accoutrements fit Union infantrymen or whether weapons possessing new loading and firing procedures could be effectively utilized by the common soldier. Although human factors engineering was still years away from being recognized as a profession, design engineers were beginning to recognize that they had to involve humans in their tests to avoid certain problems in their design and to verify that certain inventions performed as advertised.

Literature of the Civil War period is replete with examples of inventions to reduce the number of humans needed to operate a weapon or to improve efficiency and effectiveness. (Safety probably remained less of a concern than it would become in the 20th century.) Breech-loading rifles and machine guns, for example, saw very limited service from 1861 to 1865, not because inventors didn't recognize the importance of producing rapid-fire weapons. Rather, the constraint lay in the metallurgical properties of guns, black powder propellants, and other technological factors. To demonstrate the effectiveness of new weapons, tests were conducted by the U.S. Patent Office under controlled conditions. With U.S. Patent Office employees at the trigger, tests were conducted throughout the war. Using our imagination, we might view these people as early precursors to human factors test specialists.

To prove the efficacy of new inventions, soldiers were often provided the opportunity to sample new weapons in the field, which meant in battle. For example, some Union regiments were armed with the new Spencer repeating rifle. The Spencer, manufactured both as a rifle and carbine, was the most widely used breechloader of the war (Coggins, 1983). Although other repeating rifles were manufactured, the Spencer was the most sought after because of its ability to hold eight rounds (seven in its tube magazine and one in the chamber); ease of use (it could be fired as fast as you could work the lever); and a revolutionary type of ammunition, much like today's rim fire .22 cartridge, only much larger (.52 caliber was the preferred bore size). During tests, Spencer himself fired as many as 21 rounds/minute. This was compared to the average muzzle-loader, which could be fired at the laborious rate of 4 rounds/minute at best but more typically, 3, using the standard 12- and 4-count firing procedure of the period (*U.S. Infantry Tactics*, 1861). Additional human performance tests were conducted in handling and durability. Personnel of the Ordnance Department bespoke the simplicity and compactness in construction and thought they were less liable to malfunction than any other breech-loading arm in use. Yet the true value of the repeater was demonstrated in battle. Today we might refer to this as operational testing. Confederate soldiers who had witnessed the effect of Union regiments armed with the Spencer repeater attested to the weapon's devastating effect. In his diary, one Confederate soldier described an attack by his battalion on a Union position armed with Spencer repeaters. "The head of the column, as it was pushed on by those behind, appeared to melt away, or sink into the earth. For although continually moving, it got no nearer" (Greener, 1870).

In the period between the American Civil War and World War I, testing by U.S. War Department and U.S. Patent Office personnel continued for both military and civilian-industrial patent submissions and included informal measurement and observational techniques, in which humans played a part.

Turn of the 20th Century

By the turn of the century, machines had become much more sophisticated. Consequently, the demands on human operation required more than simply fitting the human to the machine. Perhaps it was during this period that inventors gave serious thought to fitting the machine to the human.

Around 1900, Simon Lake tested operators during submarine trials for psychophysiological attributes. He measured their ability to withstand unpleasant and dangerous environmental conditions. Included were oxygen deficit, exposure to noxious and toxic gases, seasickness, and cramped quarters. His testing perspective remained solely as one that considered the human as a constraining variable on the capability of the submarine, not as one that considered ways to improve the safety, efficiency, and effectiveness of the machine.

Shortly thereafter, a new perspective in testing the human–system interface appeared. Just prior to World War I, Taylorism, or the scientific study of the worker, emerged as a means to increase the efficiency of humans in the workplace (Taylor, 1919). In 1898, F. W. Taylor restructured an ingot-loading task at the Bethlehem Steel Plant just outside of Baltimore, Maryland (Adams, 1989). Taylor demonstrated through a formalized method of data collection and statistical analyses that by proper selection, training, and work–rest cycles, an individual could substantially increase his work output. Later, Frank and Lilian Gilbreth, students of Taylor's, laid the foundation for what has become known as time-and-motion study in their analyses of bricklaying and surgical procedures. They featured measurement of work elements based on molecular body movements, a technique still important to modern industrial engineering (Adams, 1989).

World War I

Human performance testing continued as technology advanced. The First World War saw new aircraft tested and selected, mostly on the basis of pilot opinion. Fighter aces, such as Guynemer and von Richtoffen, flew experimental aircraft, pushing the envelope to see how far was too far; manufacturers relied on these individuals to tell them how well the planes handled in acrobatic maneuvers, and, consequently, decisions were made whether or not to proceed with prototype development. Such tests were not actually tests of pilot performance but primarily of the aircraft, with the pilot as a necessary part of the test. An important new development in HFTE emerged; little had changed since the Civil War, except that now, as systems became increasingly sophisticated, a new dimension in personnel selection had to be considered: human intelligence. The selection of personnel on the basis of intelligence and the relationship between individual differences in intelligence and various types of work became an area of intensive investigation during the years surrounding World War I. The necessity for drafting large numbers of men into the armed forces and fitting them into various specialized niches required the development of both intelligence and special aptitude tests. Although the World War I experience contributed little to human factors test methods as we know them today, it did force a new look at differences between individuals and how those differences could affect the efficiency of performance.

World War I also saw the beginning of aeromedical research and an accompanying need for related test and measurement methods. At the beginning of World War I, almost anyone with the courage to fly was permitted to try his luck. This nonselective approach quickly resulted in an increasing number of fatalities. The British noted, for example, that 90%

of all fatal accidents were the result of incapacity, or individual deficiencies (U.S. Army, 1941). This observation led to the establishment in several countries of laboratories designed to study problems connected with flying. Some effort was made at the conclusion of hostilities to correlate personal qualities with skill in aerial combat (Dockery & Isaacs, 1921), which was repeated after World War II but with similar negative results.

Essentially, World War I launched human factors researchers into a direction that would encourage the development of modern HFTE methods. Most American psychologists who rushed to the colors in World War I were involved in personnel selection and testing (i.e., the development and administration of intelligence tests). A few, however, such as Raymond Dodge (Alluisi, 1992), anticipated human factors activities in World War II by developing training aids (e.g., naval gunnery trainers) and measuring performance (e.g., how long soldiers could wear a gas mask—not very long, it turned out).

Between the World Wars

By the end of the First World War, two laboratories had been established, one at Brooks Air Force Base, Texas, and the other at Wright Field, Ohio. These laboratories performed a great deal of research dealing with human perception, motor behavior, and complex reaction time. The research focused attention on the human factors problems associated with new military aircraft. Studies on the effects of high altitude, acceleration, anthropometry, and pilot and aircrew performance were conducted as early as 1935 (Dempsey, 1985).

During this period, some noteworthy research on human performance was also conducted in the civilian sector. Fascination with the automobile led to a great deal of research on driving (Forbes, 1939). This interest has continued to the present. A very significant part of the human factors measurement literature is related to driving performance, second only to the continuing measurement interest in the human factors of flying. In the industrial area, the most significant testing research was conducted at the Hawthorne Plant of the Western Electric Company from 1924 to 1933 (Hanson, 1983). Here, the effects of illumination on worker productivity were examined. Although the results showed no difference between the control and experimental groups, it did demonstrate a new phenomenon quite by accident: the Hawthorne effect showed that motivational forces, such as subjects knowing they are being observed, can influence performance in powerful ways.

Although our examples in this period have been drawn from predominantly North American activities, significant human factors work was performed concurrently in Britain and on the European and Asian continents, particularly in Russia. Until the conclusion of World War II, human performance measurement and human factors existed as part of psychology. Thus, when examining human performance measurement between the World Wars, one must consider the psychological orientation in which researchers of the period functioned. In North America, research on driving, flying, and in industry was performed within the then-dominant school of psychology—behaviorism—and emphasized the stimulus and response aspects of performance. Although the concept of the mind had been summarily disposed of in North America by Watson's behavioristic approach, it was still much a part of the research performed in the Union of Soviet Socialist Republics (USSR) by Anokhin (1935), Bernstein (1935), and Leontov (1977). Later, we shall return to this work and to what eventually became activity theory, but for now we take note of Russian psychology of that period as it related to the political system then in power.

The Marxist ideology that governed the USSR from 1917 to 1991 opposed most psychological research as representative of bourgeois thinking. During the repression of the 1930s, many psychologists were imprisoned, sent into exile, and even executed. This inevitably retarded psychology in Russia until the 1960s, when human factors in the USSR was rehabilitated, mostly because of an accelerated space program. For a major industrialized country, the number of those who considered themselves, and were considered by others, as human factors people has never numbered more than about 300. This is in contrast to the United States, in which 5,000 are members of the Human Factors and Ergonomics Society.

THE MODERN ERA

World War II

With World War II came another exponential leap in technology, resulting in aircraft that flew higher and faster and required complex cognitive and physical skills, radar systems that presented more information than a single operator could comprehend, submarines that went deeper and could stay below the surface longer than ever, sonar that detected submarines, and so on. In addition to the increased physical and mental abilities needed to operate these sophisticated machines, HFTE would likewise have to make a substantial leap.

Personnel selection continued to play a vital role; new air-crew selection methods led to the Air-Crew Classification Test Battery (Taylor & Alluisi, 1993), which was a paper-and-pencil test. Various apparatuses were also used to measure coordination and decision time. Depth perception, for example, was measured by having the subject align two markers using two cable pulleys in a box.

Additionally, large numbers of personnel were required to operate and maintain these systems. Prior to this time, when technological progress was somewhat more gradual and the number of skilled personnel needed was relatively small, it was possible to be highly selective of operators. This luxury was not an option in World War II. More than ever, machines had to be designed to fit the capabilities and limitations of operators. Consequently, a systematic program of human-machine interface testing and research emerged.

Arguably the best known of the World War II American researchers was Paul Fitts, who worked with collaborators on aircraft controls and displays (Fitts & Jones, 1947). Researchers who entered the military were experimental psychologists who adapted their laboratory techniques to applied problems. For example, early studies of signal discrimination as it applied to sonar were directed at auditory perceptual capabilities (National Research Council, 1949). Similar research was performed to determine the visual capabilities needed to distinguish targets on radar. In this research attention was focused on what could be done to make controls and aural and visual displays more accommodating to the capabilities and limitations of the operator.

At Wright Field's Army Air Corps Aeromedical Research Unit, human factors specialists conducted a wide variety of other research as well: tests of human tolerance limits for high-altitude bailout, automatic parachute-opening devices, cabin pressurization, pressure breathing apparatus, and protective clothing for use at high altitudes. Also, studies examined new G-suit designs, ejection seats, and airborne medical evacuation facilities.

Post-World War II

Fueled by a new threat of expansionism by the Soviet Union, the U.S. government continued to fund human performance research. Individual laboratories that were established during the war expanded. For example, during the war, the University of California Division of War Research established a laboratory in San Diego. Later, this laboratory became the U.S. Navy Electronics Laboratory. Through expansions and reorganizations, it evolved into the Naval Ocean Systems Center and, finally, into its present organization, the Naval Research and Development Center. In 1953 the Department of the Army established its Human Engineering Laboratory, now the Army Research Laboratory's Human Research and Engineering Directorate, located at Aberdeen Proving Ground in Maryland. In fact, each of the services developed human performance research laboratories either during the war or shortly thereafter. The Navy Electronics Laboratory's Human Factors Division consisted of three branches specializing in the psychophysics, human engineering, and training of sonar devices and sonarmen. The first branch determined what human auditory and visual capabilities were needed, the second considered how these could be utilized in machine form, and the third examined how personnel could best be trained to operate these machines. In his excellent book, Parsons (1972) lists 43 individual laboratories and projects and describes in detail the most important studies that were performed. Many of these institutions continue to evolve, providing valuable human factors research and improved HFTE methods to the community.

Nearly all human factors research during and immediately following the war was military sponsored. Paul would move to Wright Field in 1946 to continue his work with sonar systems. Universities were granted large sums to conduct basic and applied research (e.g., the Laboratory of Aviation Psychology at Ohio State University). Other so-called think tanks, such as the System Development Corporation in Los Angeles and the RAND Corporation that split off from it, were established and funded by the military. Whereas during the war research had concentrated on smaller equipment components such as individual controls and displays, the new studies performed by the laboratories embraced larger equipment units, such as an entire workstation or an entire system, which corresponded in point of time to acceptance by the discipline of a new concept: the system concept. To quote Parsons (1972), "The technological development which had the greatest impact in fostering . . . man-machine research seems to have been radar, in conjunction with the production of high-performance military and commercial aircraft" (p. 7). Most of the activity Parsons describes took place from the 1950s to the mid-1960s. Obviously, the work has continued and has expanded, as witnessed by the growth of membership in the HFES from only several hundred in 1957 when the Society was founded as the Human Factors Society to its present membership.

The 1950s to the Present

Prior to the advent of powerful computers, sophisticated simulators were not available. Instead, test equipment consisted of mock-ups and simulated displays. Terrain models were built for laboratory tests, but military sites were increasingly used to provide a more realistic measurement of human performance in a natural environment. For example, the U.S. Army's Behavioral Sciences Research Laboratory (now the Army Research Institute for Behavioral Sciences) established field stations at various U.S. Army installations, such as at Fort Rucker, Alabama, for helicopter-related HFTE and at Fort Knox, Kentucky, for armor-related HFTE. The Army's

Human Engineering Laboratory likewise sent human factors specialists to field sites to monitor and assist in conducting human factors engineering analyses. In the 1950s, the U.S. Air Force established a very extensive research program at several airfields, which, under the name of the Air Force Personnel and Training Research Center (AFPTRC), studied a variety of subjects of interest to the Air Force, including personnel selection, training, equipment development, and personnel proficiency, to name a few.

By the 1950s, many private companies began hiring their own human factors engineering staff. Companies such as Boeing, Lockheed, North American Aviation, and Convair found that human factors specialists were essential to the design and testing of manned systems. Although these were primarily airframe companies, electronics, communications, and automotive firms also hired human factors specialists as employers began to realize the value of human factors expertise. Today, human factors specialists can be found in almost every civilian industry.

Most recently, the U.S. military, as part of an endeavor to incorporate human factors principles in all phases of its system development efforts, has established formal human factors programs. These programs are meant to help ensure that behavioral factors are fully integrated during system development. The best example is the Army's Manpower and Personnel Integration (MANPRINT) program, which includes seven domains: human factors engineering, manpower, personnel, training, system safety, health hazards, and personnel survivability. In addition, the U.S. Department of Energy has a well-established human reliability program; the Nuclear Regulatory Commission has applied formal HFTE processes to upgrades in control-room design; and the Department of Transportation's Federal Aviation Administration has an entire human factors division dedicated to the air industry. Europe and Asia also have similar ergonomic organizations to support military and industrial endeavors.

It would be impractical to attempt to describe all of these programs in an historical review. Rather, the questions are How have these programs affected the way in which specialists approach human factors testing? Would it be more economical or practical to concentrate on technical testing, or should the HFTE specialist focus on field performance? Some would argue that human factors testing under the rigors of a controlled and instrumented test environment, or developmental testing, yields more information about the human-system interface than does field, or operational, testing. Others would argue just the opposite: that operational testing is more important because it provides a more accurate and realistic environment in which the human-system interface can be observed. Perhaps HFTE should strike a balance between the two. Indeed, O'Brien (1983, 1985) has demonstrated that under certain conditions, combined developmental and operational testing is not only possible but also can save money and yield more information relative to a system's human-system interface. As we enter the 21st century, this combined developmental-operational testing philosophy has become generally accepted.

The significance of military-mandated programs, such as MANPRINT, to HFTE is that these programs require more systematic human factors testing than might otherwise be performed by industry and they move testing much more into the field, or the operational environment. This does not necessarily mean that new test methods are required, only that those methods already exist, have been empirically validated, and should be considered first. Also, where current methods are found to be inadequate, improvements should be sought.

THEORETICAL CONTEXT OF HFTE

Measurement always takes place in relation to some theoretical orientation, even though the details of the testing may appear to be far removed from that orientation. In North America and Britain, as well as in Europe and Japan, that orientation has evolved from classic behaviorism (i.e., measurement of stimuli and responses only, with little effort to measure cognitive activity) to an information-processing orientation. This shift has been aided and influenced by the increasing computerization of equipment and the development of equipment for uses in which the major function is itself information processing. The operators of such systems, equipment, and products (hereinafter referred to simply as *systems*) act more as monitors of computer-driven operations. These operations are reflected in displays that represent the outward expression of internal functioning. The operator's function in these systems is more cognitive now than before computerization; the operator must not only monitor the system processes but also diagnose the causes of an actual or impending malfunction. All of this has resulted in an increased emphasis on information processing in testing as well. Because operator information processing is largely covert, data about the effectiveness of that processing must be secured directly from the operator, or at least the operator must aid in the measurement. The development of rapid prototyping, a testing as well as a design methodology, has thus featured subjective techniques, including single and focus group interviews and expressions of preferences using scales as measurement devices, as well as others. The relatively recent focus on what has been called user-centered design, particularly in commercial systems, made it necessary to secure information about the user, and this emphasized a number of subjective techniques, including direct observation; audio/video recording of user performance for later, more detailed analysis; and interviews with subject matter experts. The conditions under which user information is elicited is less structured and objective than in formal testing. It is, moreover, measurement to elicit information rather than to measure effectiveness, but the two functions are associated in rapid prototyping and usability testing. The test environment may also vary from highly structured to very informal.

A new theoretical orientation developed from Russian psychology (Bedny & Meister, 1997) has become popular in European human factors and can be expected to influence HFTE. This orientation emphasizes a very subjectivist viewpoint in such features as the importance of the individual's goals in information processing; images in perception; the nature of the goal as not necessarily fixed but as determined by the human's motivation, at least in part; an extremely detailed analysis of human functions; and a highly molecular measurement technique related to time and motion methods in industrial engineering. It is not yet clear how HFTE methods will be influenced by activity theory, but its subjectivist viewpoint is correlated with the increasing emphasis on the user in design (e.g., rapid prototyping) and in the collection of information about user performance and their design preferences.

Modeling and Simulation

As computing processors and their software become more and more powerful in their capacity to process and store data, and as the worldwide electronic network evolves to ever greater levels of sophistication, the roles of computer-based modeling and simulation in both design and testing become apparent. Few will argue that the use of models and system-level simulations based on modeling data can save time and money; however, the question remains: How valid are the results? More particularly, how useful are those results to HFTE?

Initial attempts based on variations of task analysis to represent the human in computer models and to generalize those results across a heterogeneous population drew limited results, especially when attempting to represent large-scale operations. Employing the model data, it was quickly realized that validating the performance outcome of computer-based simulations was difficult and cost ineffective. Indeed, some would argue that if it was necessary to validate the model through the need for actual, operational testing, then why bother with the model. One answer came in the form of distributed interactive simulations (DIS). With DIS, it is now possible to conduct cost-effective, large-scale simulations with participants from geographically distributed areas. Work, however, continues on models that focused on specific system functionality.

Early unit-task models, such as the Keystroke-Level Model (Card, Moran, & Newell, 1983), attempted to represent human performance at varying levels of task complexity. Beginning with part-task observations of operator physical actions, modelers later attempted to capture the mental tasks of operators to predict the time it took to perform tasks in user–computer interactions. Some models that were developed in the late 1970s and early 1980s (e.g., the Systems Analysis of Integrated Network Tasks [SAINT]) have evolved into highly sophisticated versions and continue to assist designers and testers in modeling and simulating part- and whole-task requirements (O'Brien & Charlton, 1996). Many other computer-based modeling and simulation tools were developed to assist in time-line analysis, link analysis, work-load analysis, and more. Some of these tools include the Simulation for Workload Assessment and Modeling (SIMWAM); Computer-Aided Design and Evaluation Techniques (CADET); and Computerized Biomechanical Man-Model (COMBIMAN). Today, software models and simulations have taken an increasingly important role in HFTE.

Contemporary testing methods are being augmented by models and simulations in which human performance is represented across several dimensions of physical and mental behavior. Although there may always be the need to exercise the system and gather traditional human factors data, computer models could one day replace some of the task-based measures to represent the human-in-the-loop in what would be considered operationally realistic conditions. Measurement is involved in simulation and modeling in several ways. Because these models are supposed to describe some actual or anticipated human performance, data must be collected about that performance, whereby that data, in turn, can be input into a model to exercise it. Such a model must be validated before it can be legitimately employed, and this means that performance must be measured in the operational environment to permit comparison of the actual performance with modeled performance. This may be thought to gild the lily, as it were, because if one has to measure actual performance to validate the model, why model at all? The utility of a model is that of replicating the described performance many times to reduce performance variations; however, the actual performance may be so lengthy that only one or two actual measurements are possible. Specific uses of models and simulations as part of HFTE are discussed in greater depth later in this book.

Usability Testing

The term *usability testing* is mentioned here because, as we progress into the 21st century, people have come to rely more on computers and software-dependent interactive systems. At the time this book is published, most people will have a computer in their home and virtually all businesses will depend on computers for one thing or another. Rubin's *Handbook of Usability Testing* (1994) was among the first to recognize that human–computer interaction had become an issue of such prolific proportions within our society that special testing techniques to address usability were needed. In 1993 John Rieman and Clayton Lewis published the first edition of *Task-Centered User Interface Design* through the University of Colorado, Boulder. The book provided several approaches to evaluating human-computer interfaces. Theo Mandel (1997) and others continue to characterize and measure with greater accuracy and fidelity the usability aspect of human–computer interaction. As technology progresses from mainframe to desktop, home computers, and even wearable computers, we should expect usability testing to play an increasingly important role as a subspecialty of HFTE.

SUMMARY

In this chapter, we have argued that HFTE has roots much deeper than one might think. We divided the history of HFTE into two periods: premodern, the period preceding World War II, and modern, inclusive of and after World War II. We see evidence during the premodern period that some equipment and armor manufacturers included humans during performance tests (e.g., crew performance time and accuracy observations during gun drills onboard 17th century ships). As technology progressed and systems became more complex, so too did the need to ensure that the human was fully integrated into systems design. Sometime around the turn of the 20th century, HFTE began to emerge as a separate discipline. For the first time, methods to measure and evaluate worker productivity were formalized, and personnel selection techniques were established. The beginning of the modern era, World War II, saw the development of sophisticated human performance testing, which was necessary to evaluate large numbers of people over many perceptual and physical dimensions. Today, as human-system interfaces become even more complex, human factors specialists continue to develop new methods and measures. It is possible to anticipate an HFTE future in which the discipline is expanded to include closer relationships with theory, development, modeling and simulation, performance prediction, and design.

REFERENCES

Adams, J. A. (1989). *Human factors engineering*. New York: Macmillan.

Alluisi, E. A. (1992). *Roots and rooters*. Washington, DC: American Psychological Association.

Anokhin, P. K. (1935). *The problem of center and periphery in the physiology of higher nervous activity*. Moscow: Gorky Publishers.

Bedny, G. Z., & Meister, D. (1997). *The Russian theory of activity: Current applications to design and learning*. Mahwah, NJ: Lawrence Erlbaum Associates.

Bernstein, N. A. (1935). The problem of the relationship between coordination and localization. *Archives of Biological Science, 38*(1), 1–34.

Brogan, R., Hedge, J., McElroy, K., & Katznelson, J. (1981). *Human factors engineering*. U.S. Army Human Engineering Laboratory and the Pacific Missile Test Center, Aberdeen Proving Ground, MD.

Card, S., Moran, T., & Newell, A. (1983). *The psychology of human-computer interaction*. Hillsdale, NJ: Lawrence Erlbaum Associates.

Coggins, J. (1983). *Arms and equipment of the Civil War*. New York: Fairfax Press.

Dempsey, C. A. (1985). *Fifty years research on man in flight*. Dayton, OH: Wright-Patterson Air Force Base.

Dockery, F. C., & Isaacs, S. (1921). Psychological research in aviation in Italy, France, and England. *Journal of Comparative Psychology, 1*, 115–148.

Fitts, P. M., & Jones, R. E. (1947). Psychological aspects of instrument display. In *Analysis of 270 "pilot error" experiences in reading and interpreting aircraft instruments* (Report TSEAA-694-12A). Dayton, OH: Aeromedical Laboratory, Air Materiel Command.

Forbes, T. W. (1939). The normal automobile driver as a traffic problem. *Journal of General Psychology, 20*, 471–474.

Greener, W. (1870). *Modern breechloaders*. London: Cassell, Petter, and Galpin.

Hanson, B. L. (1983). A brief history of applied behavioral science at Bell Laboratories. *Bell Systems Technical Journal, 62*, 1571–1590.

Leontov, A. N. (1977). *Activity, consciousness and personality*. Moscow: Political Publishers.

Lewis, C., & Rieman, J. (1993). *Task-centered user interface design*. Boulder: University of Colorado.

Mandel, T. (1997). *The elements of user interactive design*. New York: Wiley.

Meister, D. (1999). *The history of human factors and ergonomics*. Mahwah, NJ: Lawrence Erlbaum Associates.

National Research Council. (1949). *Human factors in undersea warfare*. Washington, DC: Author.

O'Brien, T. G. (1983). *Test protocol for the M1E1 Abrams main battle tank, micro climate control system, prototype qualification test-government, developmental testing-III, desert phase testing*. Aberdeen, MD: U.S. Army Materiel Systems Analysis Activity.

O'Brien, T. G. (1985). Human factors engineering test and evaluation in the U.S. military. *Proceedings of the Human Factors Society 29th Annual Meeting*, 499–502.

O'Brien, T., & Charlton, S. (Eds.). (1996). *Handbook of human factors testing and evaluation*. Mahwah, NJ: Lawrence Erlbaum Associates.

Parsons, H. M. (1972). *Man-machine system experiments*. Baltimore: Johns Hopkins University Press.

Rubin, J. (1994). *Handbook of usability testing*. New York: Wiley.

Taylor, F. W. (1919). *Principles of scientific management*. New York: Harper.

Taylor, H. L., & Alluisi, E. A. (1993). Military psychology. In V. S. Ramachandran (Ed.), *Encyclopedia of human behavior*. San Diego, CA: Academic Press.

U.S. Army. (1941). *Notes on psychology and personality studies in aviation medicine* (Report TM 8-320). Washington, DC: Author.

U.S. infantry tactics for the instruction, exercise, and maneuvers of the United States infantry, including infantry of the line, light infantry, and riflemen. (1861). Philadelphia: Lippincott.

Chapter 2

The Role of Human Factors Testing and Evaluation in Systems Development

Samuel G. Charlton
Waikato University
and Transport Engineering Research New Zealand Ltd.

Thomas G. O'Brien
O'Brien & Associates

In our experience, human factors and ergonomics testing is a discipline quite unlike any other. By virtue of their training, human factors test specialists bring a substantially different perspective to the development of a new system, compared with the rest of the system development community in which they operate. A fundamental tenet of human factors is that the human operator lies at the center of system design, the yardstick by which the form, fit, and function of hardware and software must be gauged. This is in contrast to the typical engineering view of a system as an interrelated collection of hardware and software that is operated by humans. It is this difference in philosophies, placing the human at the center of the design (i.e., user-centered) versus the creation of an elegant engineering solution to a design problem (that operators somehow fail to appreciate), that all too often sets human factors testers on the road to conflict with engineers and other members of a system development effort.

The human factors tester is also quite different from his human factors brethren who work in academia or other research laboratories. The goal of the human factors tester is to develop or refine human-machine systems and the approach is typically one of pragmatic expediency, employing whatever methods and measures are available to get the job done. Human factors research conducted in the laboratory, however, is to a much greater degree guided by the theoretical inclinations and interests of the individual researcher. In this sense, human factors testers are phenomenon-driven or situation-driven, compared with the typically paradigm-driven academic researcher. Further, whereas many academic researchers enjoy the collegial company of like-minded individuals, human factors testers find themselves to be on their own, surrounded by individuals with quite different backgrounds, perspectives, and testing objectives.

The products of human factors and ergonomics testing—technical reports and test reports—also serve to make human factors testing a somewhat insular activity. Because human factors testing does not readily conform to the controlled manipulations and statistical considerations expected of traditional laboratory research, human factors test results are not readily publishable in academic journals and books. Although technical reports do sometimes appear in the reference sections of journal articles, they are difficult to find and obtain. Learning about the existence of an important technical paper is haphazard at best, and obtaining a copy is dependent on government or commercial restrictions and, ultimately, the goodwill of the authors.

All of these conditions have had the effect of keeping human factors testers, and their work, out of the mainstream of human factors. This isolation has a number of important effects on human factors testers, as well as on the field of human factors in general. First, human factors testers typically become staunch advocates for human factors principles as the only spokespeople for their discipline in their workaday world. Second, and somewhat paradoxically, isolation has the effect of making the tester more acutely aware of other perspectives, those of engineers, designers, and users. In spite of this tradition of isolation, we believe there is an untapped wealth of information that human factors testers and test results can offer laboratory researchers as well as other testers. Human factors testing is an underused source of applied research topics, and it offers data with which to verify the external validity of generalizations made in the laboratory.

This chapter attempts to provide some insight into the world of the human factors tester, for both the benefit of others in the testing community as well as researchers outside of it. In the first half of the chapter, we describe how the formal systems development process works and the role human factors testing plays at each stage of that process. The second half of the chapter describes some of the common issues and dilemmas faced by human factors testers during a development effort.

THE SYSTEMS DEVELOPMENT PROCESS

As alluded to above, a system designed exclusively from a hardware and software performance perspective may not be the most efficient, effective, or safest design when considered from the human operator's perspective. Similarly, even the most elegant design solution will not be a commercial success unless potential customers are satisfied with its performance when they attempt to use it. These dual measures of human factors in systems

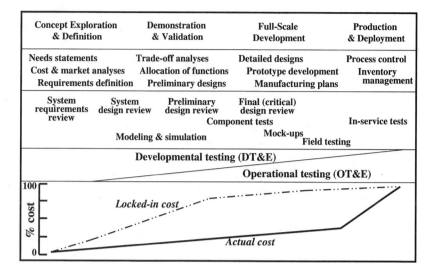

FIG. 2.1. The phases of system development and their associated products, activities, and costs.

design, effectiveness, and suitability (user satisfaction), although not universally acknowledged, have made enough of an economic impression on industry to merit explicit human factors design and testing programs by major manufacturers for a variety of consumer products. Setting aside for a moment the economics of ergonomics, nowhere has HFTE been adopted more uniformly than in the world of military acquisition, where it has been mandated by public law (at least in the United States) and implemented by regulations in each of the armed services. Because of its formalized structure, the military systems acquisition process provides a useful starting point in describing how systems are developed and tested.

Not surprisingly, systems development is a highly structured and regulated process for the acquisition of new defense systems. Shown in Fig. 2.1 are the four phases of the systems development process, as labeled by U.S. defense systems acquisition regulations. Although the terminology may vary, these four phases are common to nearly all systems development activities, commercial products as well as defense systems. Each of the acquisition phases shown in the figure has required HFTE activities and is separated by distinct product milestones.

Concept Exploration and Definition

During this initial phase of system development, the focus is on identification of user needs. This process is usually initiated by identifying shortfalls in current capabilities, deficiencies in existing systems, or new technologies that can be exploited. The U.S. defense industry establishes these needs and desired results in three significant documents: a Mission Needs Statement (MNS), a Cost and Operational Effectiveness Analysis (COEA), and an Operational Requirements Document (ORD). The MNS is based on a mission area analysis that identifies deficiencies in existing defense capabilities or more effective means of performing existing functions based on changes in defense policy, external threats, or new technologies. Following publication of an MNS, the relevant government laboratories and agencies, along with their private industry contractors, work to identify system design alternatives that may meet the mission need. Included in these analyses are the producibility and engineering considerations, logistics support analyses, and manning and operations options. The COEA provides a formalized method for trading off these alternative solutions in terms of cost-benefit analysis. The ORD is used to document the required performance parameters for each concept or system that is proposed to advance to the next phase of system development.

In nondefense industries, the process is less formal but consists of the same essential elements. A commercial need or opportunity is identified based on existing products, new market demands, or new technologies (e.g., "We need a new automobile that meets stricter environmental standards," or "There is a market for a video recorder that is easier to operate."). As with defense systems, alternative product ideas to satisfy the need are generated and compete for the attention and approval of managers and decision makers. One difference in the commercial world is that because of the larger size of the marketplace, and the potential payoff for a successful product, a relatively greater number of bottom-up technological developments are allowed to become full-fledged development efforts. There are, to be sure, a significant number of top-down development programs that are initiated by management or marketing directives, but in contrast to defense systems, they are not the only means of initiating a new product. Finally, as with defense acquisitions, only the most commercially viable or reasonable of the ideas are taken to the next stage in development: analysis of design alternatives in terms of their performance capabilities, market success, and engineering and cost parameters.

Notable from the human factors engineering perspective is that even at this early stage, the requirements analyses and trade-off studies should include the human factors issues associated with each alternative proposed. In the ORD of a new defense system, this is contained in a required human-systems integration section that identifies critical manpower, personnel, training, safety, and human factors engineering issues for the system (DOD Instruction 5000.2, 1991). For commercial products, these issues are often the provenance of one or more specialty engineering teams within the company. For both types of systems, analysis of human factors issues at this early phase is almost entirely analytic because the system is only an idea—nothing has yet been built. The human factors tester must, however, consider the full range of potential human factors and ergonomics issues: the type of mission to be performed or product niche to be filled; possible secondary uses and configurations; critical operations and maintenance tasks; interface characteristics (particularly those that may require extensive cognitive, physical, or sensory skills); factors that may result in frequent or critical human performance errors; and the operator, maintainer, and training manpower requirements. All of these elements must be identified and considered at the outset of the design process (i.e., requirements identification) if they are to have any impact on the ultimate design. Additional commercial analyses not found in military systems may include a market analysis, an analysis of competitors' systems, and an analysis of design preferences voiced by samples of the intended user population. Information from these analyses permits the developer to define the system requirements in enough detail to actually begin the engineering design.

To help define the system requirements described above, the human factors engineer can turn to analyses of existing systems or to a variety of computer models. A completely new system or concept, however, will demand a creative approach by the human factors tester to assess the potential impact of various system alternatives on human performance, maintenance, and training requirements issues. Once the human-systems integration requirements are identified and the decision is made to proceed with the development of a new system, the development effort progresses to the next phase: demonstration and validation.

Concept Demonstration and Validation

In this phase, the actual design effort begins in earnest. Among the major products of this phase are the system specifications and design baseline. These design documents allocate the requirements to the various system components (hardware, software, and personnel). The allocation of functions is a critical step in the design and includes trade-off analyses to arrive at the optimal balance of performance, cost, and reliability. The human factors issues to be considered during these analyses focus on the level of automation desired, analysis of the tasks and skills required for operation, software usability considerations, training requirements, and designing for maintainability.

Here again, computer design tools, simulations (three-dimensional man models), and physical mock-ups are used by human factors engineers to evaluate alternative designs. It is during this phase that many of the most significant human factors engineering decisions are made. As system development progresses from this point forward, any changes to the design become more difficult and much more costly to accomplish. For this reason, the concept demonstration/validation phase culminates in a formal system design review. The system design review allows system engineers to present their designs to product managers and other technical reviewers to ensure that the requirements allocation and design specifications are feasible from a technical and manufacturing perspective and that they will meet the operational requirements identified for the new product or system.

Engineering and Manufacturing Development

This phase, also known as full-scale development, will translate the design approach into the documentation necessary for production, validate the manufacturing and production processes, and demonstrate that the minimum acceptable operational requirements can be met. The culmination of this phase is one or more fully functional prototypes that can be tested with a sample of the intended user population.

In this phase, HFTE assumes a new role. Component or developmental testing focuses on the details of engineering the controls and displays; examining environmental factors, such as lighting, noise levels, and airflow; and investigating anthropometric and safety issues. Human performance is considered throughout, progressing through rigorous function-task allocation, task and time-line analyses, work-flow analyses, and part-task simulations and culminating in full operational tests. This type of testing is sometimes called beta testing in some consumer products industries, particularly in the field of software development.

As alluded to above, there are two distinct types of testing performed throughout the engineering and manufacturing development phase, each with a different human factors emphasis. The first of these is called developmental test and evaluation (DT&E). The principal function of DT&E is to reveal and correct problems through an iterative design-test-redesign process. The specific objectives are to identify potential operational and technological limitations, verify that technical performance objectives and specifications have been met, identify and describe potential design risks, support the identification of cost-performance trade-offs, and certify that the system is ready for operational testing. Developmental tests are conducted on a variety of mock-ups, engineering components, and prototypes as the various configuration items are designed and developed to meet the system's design specifications.

In contrast, the goal of operational test and evaluation (OT&E) is to determine if the new system can meet the requirements and goals identified at the outset during the concept exploration and definition phase. OT&E is conducted on production representative articles under realistic operational conditions and operated by representative users. By this we mean operators and maintainers who have levels of experience and training typical of people who will use the final production system. Specifically, OT&E is structured to determine the operational effectiveness and operational suitability of the system under realistic conditions and to determine if the minimum acceptable operational performance requirements of the system have been satisfied. Most important, OT&E is conducted by independent testers so that the determination of effectiveness and suitability will be made without any undue influence from the system developers.

In practice, DT&E and OT&E activities occur throughout all four phases of system development. DT&E conducted during the concept exploration and definition phase, called DT-1, can include paper studies trading off various system options in terms of price and performance. OT&E activities at this point are typically directed toward ensuring that the operational requirements developed will be reasonable and testable. DT&E conducted during the concept demonstration and validation phase, DT-2, is directed at evaluating design alternatives, and identifying cost-schedule-performance trade-offs. OT&E conducted during this phase typically consists of early operational assessments (EOAs) of any prototypes built, identification of any shortfalls in the design process in addressing the operational requirements, and identification of any adverse trends in the ultimate operational effectiveness and suitability of the system. As described earlier, DT-3 is conducted on developmental components during the engineering and manufacturing development phase to support the determination that design specifications have been met and that a production representative article is ready for OT&E. OT&E is then conducted to determine operational effectiveness and suitability in support of the decision to actually produce the product or system. Finally, as will be described, DT&E and OT&E activities actually extend past the production and fielding of the system as well.

Occasionally, because of cost and schedule constraints, some system development programs will combine developmental and operational test activities to address simultaneously technical and operational issues. That is, rather than scheduling and funding separate DT&E and OT&E, test managers take advantage of the fact that, if designed carefully, some test activities can provide data on both specification-level and operations-level issues. Even in these situations, the data collection and analysis personnel are kept separate to ensure independent evaluations.

Production and Deployment

HFTE continues after the production of the system. Once the system is deployed, or the product released to users, human factors test issues can range from examination of manufacturing processes (i.e., process and production control, packaging, materials handling, shipping, and set-up), to consumer safety and satisfaction. Here, the human factors tester conducts follow-on tests to determine whether there are any residual human engineering discrepancies that may need correcting. Training and personnel requirements receive a final evaluation regarding their applicability and adequacy in meeting the demands of the production system. Moreover, HFTE at this phase is conducted so that human-system interface information can be gathered for the next version or update of the product.

HUMAN FACTORS ISSUES IN SYSTEM DEVELOPMENT

Given the thorough examination of human factors issues in system development, as described in the preceding sections, why is it that we so often encounter poorly designed systems and products? Unfortunately, the realities of cost and schedule are not always kind to the system development process. The process we have described to this point is an ideal, or how it is supposed to work, textbook view of the world. There are three fundamental areas where the process can go awry from the HFTE perspective: requirements issues, testing issues, and evaluation issues.

Requirements Issues

The act of stating human factors requirements is perhaps the single most important activity in the system development process. The requirements drive both the design effort and the nature of the test and evaluation program. Yet requirements definition is as difficult as it is important. The difficulty lies in stating human factors requirements in testable terms with measurable criteria. Many system developers, particularly those without human factors expertise, are hard-pressed to identify what, if any, human factors requirements a new system or product should possess beyond the simple statement that it should be user friendly. With the exception of a few ergonomic aids and niche markets, systems and products do not hold human factors as an intrinsic element or explicit goal; they exist to achieve some other end or perform some other function. Thus, human factors requirements are often given only a cursory treatment, if they are considered at all. Poorly defined requirements lead inexorably to a haphazard system design, incomplete test and evaluation program, and ultimately to a system that is neither effective nor suitable (i.e., satisfying to the users). As human factors professionals, it is up to us to ensure that human factors requirements are identified, in clear and testable terms, at the outset of the development process.

There are two fundamental methods of defining human factors requirements: a characteristic-based approach and a function-based approach. Many human factors testers are familiar only with the former approach, identifying the characteristics associated with good human factors design. Defining human factors requirements in this fashion involves citation of the relevant anthropometric and ergonomic standards in regard to physical sizing of equipment, locations of controls, labeling and display information, and so forth. Requirements stated in this manner translate easily into system design specifications and go a long way to ensuring that the product will be safe, usable, and comfortable.

The alternative method of defining human factors requirements is to state them in terms of their contribution to specific system functions and overall effectiveness (Andre & Charlton, 1994). This form of human factors requirement definition is much less frequently used and is somewhat more difficult to implement. For example, instead of stating that the system shall comply with a particular usability standard, the functional definition would state that the system should be designed to enable a novice user to operate it proficiently within 1 hour, allow no critical system failures to result from operator error, enable a trained operator to complete an average of 10 transactions within 30 minutes, or some similar statement. The key to the function-based method is defining human factors requirements in terms of operator performance and system functions. The difficult bit is working with the system developer to identify acceptable levels of human performance for all of the important system functions.

Let us consider the design and test implications of these two alternatives. A characteristic-based set of human factors requirements will encourage a system design that is sensitive to the needs and limitations of the human operator. The HFTE program will verify that human engineering design standards are met and that the intended user population will be satisfied with the system's ease of use (i.e., the system is suitable to the operator). In contrast, with function-based requirements, the human must be treated as an integral component of the system being designed; the inputs and outputs of the operator in relation to the other system components become explicit factors in the design process. Here, the test program must adopt a greater focus on human performance; evaluation of operability considerations is important only in relation to effective human and system performance.

Which method should be used to define human factors requirements? The answer, not surprisingly, is both. A combination of both approaches offers several significant and complimentary advantages. Characteristic-based requirements will ensure compliance with human factors design standards in the physical layout of consoles and cockpits and enhance their acceptance by users. A function-based set of requirements will add to this a more explicit consideration of the operator tasks and functions and ensure that the resulting human-machine system will be effective as well as suitable.

Testing Issues

A second area where human factors all too often fails the system development process is the omission of sufficiently robust test and evaluation methods. In recent years, there has been a progressive change in the focus and methods of HFTE to include more objective measures (e.g., operator task times and error rates) in an attempt to move away from an over-reliance on subjective test measures, such as questionnaires (Charlton, 1988). There has not, however, been a concomitant strengthening of the basic test design considerations, which often serve to constrain the value of the test results. Test constraints are factors that restrict the conduct of the test or tend to place limitations on how far the evaluator can generalize the results. While many of these factors are outside of the evaluator's control (e.g., bad weather) or difficult to control (e.g., test resource limitations), others, such as the degree of test realism, subject representativeness, and comprehensiveness of test measures selected, can be anticipated and planned for (Bittner, 1992; Kantowitz, 1992; Meister, 1986).

The degree of control the evaluator has over a test is typically a trade-off with the operational realism of the test conditions. By employing a carefully controlled test scenario in a laboratory or simulator environment, the researcher runs the risk of decreasing the representativeness of the test scenario and, thus, the generalizability of the results produced. Nonetheless, there is an achievable balance between control and test fidelity toward which the human factors tester should strive.

The literature on HFTE suggests four general categories of preventable test constraints: inadequate measurement methods, test participant issues, operational fidelity, and test design problems. Measurement methodologies may be inadequate to support a test and evaluation program because of a failure to select a comprehensive set of assessment variables. Further, selection of measures that are insensitive or irrelevant to the system-specific issues at hand may provide false or misleading human factors data on the system design.

In practice, human factors test measures are often selected on the basis of the expertise and experience of the individual tester rather than any systematic analysis of operational requirements, system design characteristics, or appropriateness of candidate measures to the evaluation issues. Although this does not necessarily result in an inadequate test (it is dependent on the capabilities and wisdom of the tester), it does open the door to selecting measures because they are familiar and convenient, at the expense of measures that may be better suited to a particular system test program. Although there is no standard suite of test measures, any more than there is a standard system to be tested, there have been several attempts to develop rules of thumb and approaches for selecting test measures that are comprehensive, valid, and reliable (Bittner, 1992; Charlton, 1996; Kantowitz, 1992; Meister, 1986). (See chapter 3 for one such example.)

The second issue concerns test participants and could include problems such as using test subjects that are not representative of the intended user population in terms of training and experience, including insufficient numbers of test subjects, and including subjects that are negatively predisposed to participate in the test. It is often quite difficult to obtain truly representative participants for the test of a new system or product. For example, in the classic case of aircraft design, highly experienced test pilots are used instead of typical military or civilian aviators for safety reasons and because the expertise these individuals adds to the design and engineering effort. In other cases, time constraints may mitigate against the use of actual members of the intended user population. This is often the case in the design of new training systems; rather than test a new training system by subjecting representative students to a lengthy and unproven course of instruction, subject matter experts and training professionals are often brought in to thoroughly exercise and evaluate the potential training effectiveness of the system. The use of golden-arm pilots or subject matter experts as test subjects, however, may result in difficulties generalizing the results to the ultimate user population. For example, even highly experienced instructors may not detect a significant design flaw until they actually use the system to teach real students.

When test participants are selected on the basis of their expertise (either on the basis of their extensive experience or because of their lack of it), it may be difficult to obtain subjects in sufficient numbers. Although not impossible, small sample sizes make the evaluation of test data exceedingly difficult. How is a tester to interpret the finding that 1 of the 5 subjects performed poorly or rated the equipment operability as poor? It is impossible to conclude whether this represents a potential design difficulty for 20% of the intended population or whether it is an anomaly specific to that subject. Here again, generalization of the results of the test to actual operation of the product or system is compromised.

Finally, when actual system users are selected as representative subjects and are temporarily reassigned, or their operational duties are interfered with for purposes of a test effort, their motivation to participate may be quite low and, once again, may adversely affect the generalizability of the test results thus obtained. Much of these negative motivations can be neutralized through careful explanation of the purposes of the test in terms of its ultimate impact on the design of the system or product they may one day use and by designing the test to disrupt their jobs as little as possible. No single solution to the test participant issues described above exists. The best the human factors tester can hope for is that adequate numbers of representative subjects (or subject matter experts, if need be) will be available and cooperative.

In this chapter, we have already alluded to the issue of maintaining operational fidelity. Briefly, equipment configurations that differ substantially from the eventual production item, insufficient or absent operating procedures, frequent equipment malfunctions, intrusive test equipment and procedures, and test environments or scenarios that do not reflect actual field use all affect the thoroughness of the test and the generalizability of the results. It should be apparent to most testers that the degree of fidelity available, and necessary, is dependent on the phase of system development and the type of test. Representative equipment may be unavailable during early DT&E or an operational assessment, whereas it is more likely to be available, and much more important, for OT&E.

Crowding resulting in increased noise levels in data collection areas can compromise the operational fidelity of the test. Similarly, requiring test participants to complete lengthy real-time questionnaires or communications logs may adversely increase their workload or actually change the way they perform their duties, which also compromises operational fidelity. Data collection activities should be nonintrusive such that they do not affect the operational performance of the system under test and thus reduce the representativeness of the data collected.

Fourth, and finally, we come to the problems associated with test design. Problems in this area can result from a wide range of issues, including vague or incompletely prepared test plans; insufficient numbers of test trials; insufficient pretesting of materials, instrumentation, and data collection procedures; insufficient numbers of trained data collectors; and unsafe testing environments. Many of these problems result from inexperienced testers or attempts to test on the fly without sufficient time and resources dedicated to test planning.

The key to a successful test plan is to begin planning early in the development process, which we discuss in more detail later in the book. A number of test issues need attention as early as concept demonstration and validation, including issues concerning readiness for testing, such as schedule adequacy and the availability of resources; the status of documentation, with emphasis on user requirements documents, including completeness, clarity, sufficiency, priority, rationale, or other factors that could affect testability; and identification of system development and system maturity trends that would impact the ability to conduct the test or that could impact the ability of the system to meet user requirements (i.e., identification of programmatic voids).

In order to adequately plan for a test effort, one must also consider the limitations and constraints of the manpower available for the data collection effort. A test planner should not select human factors methodologies that require a data collection effort beyond available manpower resources. For example, observational data collection is very labor intensive; an individual data collector cannot be expected to collect large amounts of diverse data without degrading their observational reliability and accuracy. As an alternative, the analyst may wish to explore the potential for using automatic system recording devices, either through the use of system software or video or audio tapes of system operations. The allocation of data collection duties must take into account both peak workload and the duration of data collection activities. The test planner also needs to take into account the possibility (inevitability) of data loss as a result of bad weather, personnel turnover, and equipment failure. It is not uncommon for testers to plan for as much as 25 to 50% data loss, depending on the type of system being tested. Finally, the structure of the human factors test must be compatible with the test structure developed for the rest of the system and lend itself to aggregation of the data to answer the evaluation issues.

Evaluation Issues

The formal development process described earlier, with all of its phases, reviews, and milestones, was designed to reduce risk in producing and fielding a product or system. In the analysis of test results, the quantification of the decision risk associated with the evaluation often becomes an issue. There are three sorts of evaluation issues that arise in this regard: issues associated with the test criteria, issues of statistical inference, and issues of logical inference. Criteria issues arise because many of the human factors measures used in test and evaluation, particularly those involving cognitive limitations, such as workload, situation awareness, and decision making, do not possess definite criterion values or red-line limits (Charlton, 1988; Mackie, 1984; Meister, 1986).

Two ways of managing problems of test criteria exist. The first is simply to not use criteria and report the human factors results in a narrative fashion, without making an explicit pass-fail judgment on the human factors measure in question. The results are described in terms of the nature of the problem identified, its effect on the operators, and its impact on system effectiveness and suitability. This linkage to operator and system performance can be established either logically or through quantitative statistical methods, such as multiple regression analysis (Charlton, 1992). The alternative is to define a system-specific or test-specific criterion, based on research results or tests of similar systems and agreed on by all of the agencies involved.

The second criterion issue concerns the statistical methods used to determine if the test criterion was in fact met. For example, if the minimum acceptable operational requirement for some system function, such as the time to compose and send a message were set at 5 minutes, the system under test must actually perform better than that to demonstrate with some statistical certainty that the system could meet this requirement when fielded. The performance level required for the test sample will be a function of the degree of statistical certainty required (Type I error rate or alpha level) and the sample size used for the test. This situation assumes that the null hypothesis for the analysis is that the system doesn't work, and the purpose of the test is to prove that it does meet the operational requirements, an entirely reasonable position for a testing agency to take (see Fig. 2.2).

A quite different assumption often brought to the test by the developing agency is that the burden of proof is on the tester to prove that the system doesn't work. With large sample sizes and results that are clearly more or less than the requirement, these two positions are not in contention. With small samples and results that are slightly above or below the criterion, however, the tester and developer can come into conflict over the conclusions to be drawn from the test data. In the case of an average result slightly

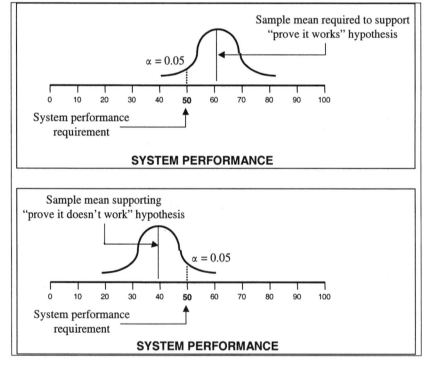

FIG. 2.2. The opposing interpretations of test results represented by the "prove it works" and "prove it doesn't work" hypotheses (note that test results ranging from 39 to 61 would be undetermined and would require additional testing to resolve).

below the criterion, assume 4 minutes and 45 seconds in our example, the tester may conclude that the system will not meet the operational requirement. The developer may argue correctly that in order to draw that conclusion with any degree of statistical certainty, the system performance would have to have been 4 minutes and 20 seconds or worse (based on a null hypothesis that the system works, a given sample size, and some desired alpha level). A similar, but opposite, disagreement over interpretation can result if the test average were to lie slightly above the criterion.

To reduce the potential for this sort of conflict, the tester should plan for a sample size large enough to make the range of results that will be in contention as small as possible. Further, an evaluation plan that clearly states how such equivocal findings will be resolved (in terms of simple point estimates or performance thresholds) should be agreed on by the tester and developer prior to the start of the test.

The third disagreement over logical interpretation of test data can arise between the human factors tester and those responsible for another portion of the test. For example, it is a common finding that operator error rates are often higher than system hardware and software error rates. This is, in no small part, caused by operator-initiated workarounds necessary to keep the developmental system functioning effectively. The consequences of these workarounds are a corresponding increase in operator workload and a masking of poor system hardware and software reliability. In these cases, the human is frequently—albeit mistakenly—portrayed as the bottleneck or limiting factor in system throughput and effectiveness. In these cases, however, the human is the limiting factor, not because of poor operator performance but because of poorly designed hardware or software. The hardware and software in a well-designed system should be transparent and serve only to enable the operator to do his or her job. The disagreement that arises is whether to rate the system design as not meeting requirements or simply to blame poor performance on uncharacteristically poor operator performance, which requires correcting the problems with better training, better operator manuals or both). Even the best-laid human factors test plans may not prevent these sorts of disagreements, but a comprehensive test design can provide the tester with a compelling case with which to rebut these arguments.

SUMMARY

Though the details of the system development process itself may vary considerably depending on the nature of system requirements, the basic process remains constant: a user decides there is a need; the need is addressed by developing a system concept; this concept is further developed into models and prototypes, which are tested against preliminary technical and operational specifications; the decision to proceed to full scale engineering is made; advanced engineering leads to a final design; manufacturing specifications are generated; the product is manufactured and delivered to the user; and finally, postproduction testing considers product improvements and ideas for next-generation products.

The message for human factors testers should be clear: Timing is everything. Early involvement is necessary so that we can identify problems or a need for improvement and make these improvements before the system is well into production. The more complex the system, the earlier the testers must get involved. Testers must sometimes plan years in advance for test assets, ranges, instrumentation, and so on. This early involvement will reveal any problems in a system's schedule that might prevent us from getting the data required to support a major decision or program milestone. By becoming involved early, we can also ensure that requirements are definable and testable. The process of testing involves more than just showing up to collect data. To the degree possible, we need to provide early human factor inputs by reviewing system designs and looking for potential difficulties involved in generation and fielding of products and systems.

REFERENCES

Andre, T. S., & Charlton, S. G. (1994). Strategy to task: Human factors operational test and evaluation at the task level. *Proceedings of the Human Factors and Ergonomics Society 38th Annual Meeting,* 1085–1089.

Bittner, A. C. (1992). Robust testing and evaluation of systems: Framework, approaches, and illustrative tools. *Human Factors, 34,* 477–484.

Charlton, S. G. (1988). An epidemiological approach to the criteria gap in human factors engineering. *Human Factors Society Bulletin, 31,* 1–3.

Charlton, S. G. (1992). Establishing human factors criteria for space control systems. *Human Factors, 34,* 485–501.

Charlton, S. G. (1996). SITE: An integrated approach to human factors testing. In T. G. O'Brien & S. G. Charlton (Eds.), *Handbook of human factors testing and evaluation* (pp. 27–40). Mahwah, NJ: Lawrence Erlbaum Associates.

DOD Instruction 5000.2. (1991). *Defense acquisition management policies and procedures.* Author.

Kantowitz, B. H. (1992). Selecting measures for human factors research. *Human Factors, 34,* 387–398.

Mackie, R. R. (1984). Research relevance and the information glut. In F. A. Muckler (Ed.), *Human factors review 1984* (pp. 1–12). Santa Monica, CA: Human Factors Society.

Meister, D. (1986). *Human factors testing and evaluation.* Amsterdam: Elsevier.

Chapter 3

Selecting Measures for Human Factors Tests

Samuel G. Charlton
Waikato University
and Transport Engineering Research New Zealand Ltd.

It has been said that it is the little things in life that matter most. The same can be said for the field of test and evaluation; little things often take on a significance that one could not have predicted. So it was with the approach described in this chapter. The origin for what became known as the SITE (Situation, Individual, Task, and Effect) approach came from a practical need to explain human factors test results to people unfamiliar with human factors. Along the way, it also proved useful in other contexts: as a simplified structure for teaching test planning, as a post hoc or anecdotal data collection framework, and as a method of organizing human factors test findings. The approach was not intended as a comprehensive or integrated approach to HFTE; rather, it just seemed to evolve over a number of years and insinuate itself into any number of different test procedures, methods, and tools. SITE is presented here in the hopes that other members of the human factors test and evaluation community may find it useful as well.

Even a cursory introduction to the field of human factors test and evaluation reveals a dizzying array of test issues, measurement methodologies, and analysis paradigms. As a result, human factors analysts planning their first test often find themselves in a quandary about what data to actually collect. As indicated in the preceding chapter, the selection of test issues and measures has historically been a matter of the expertise and experience of individual testers. Guidelines for selecting human factors methods and measures for test and evaluation have been proposed over the years and have been organized around test considerations (Bittner, 1992), system considerations (Charlton, 1988), and theoretical frameworks (Kantowitz, 1992), or a combination of all three (Meister, 1986).

Although each of these approaches attempts to provide guidance for selection of human factors test and evaluation measures, it still can be a daunting task for a new tester to move from these general guidelines to the specifics of a given test. It is often useful in these circumstances to have the new tester consider the kinds of statements they want to make at the end of the test and, working backwards from there, select the test measures necessary to support those statements. For example, common complaints with regard to human factors testing are that it fails to provide answers to why certain human factors design features are important, what impact they will have on the users of the system, how the design will affect the way people use the system, and what the implications of a particular design or test result are in terms of the product's bottom line: user satisfaction, effectiveness, or both. The new human factors tester is typically ill prepared to report these human factors test results in a way that makes sense to decision makers outside the human factors profession.

Motivated by the fear of missing something important, combined with an inability to foresee what will be important, new HFTE analysts often design a test that includes as many different measures as they can think of. Confronted with the realities of limited test resources, their wish list of measures is trimmed to a more manageable number of measures, typically resulting in scattergun test design. With luck, at least some of these measures will provide useful data and inevitably form the foundations of personal preferences in test design that will guide the tester throughout the rest of his or her career.

Recognizing the importance of these early testing experiences, and not wishing to leave the design of those early tests to schoolbook theory, personal preference, or luck, we tried to provide new testers with a modicum of structure by dividing the universe of possible measures into four broad categories, each of which should be represented to some degree to make sense of the results and convey the findings of a human factors test to a wider audience (Charlton, 1991). The resulting approach, called SITE, attempts to support the design of human factors tests and the interpretation of test results by presenting human factors test issues within a context of situation, individual, task, and effect attributes. An overview of the SITE structure is shown in Fig. 3.1.

The situation category is composed of human factors issues associated with the environment in which the human operator or user of the system is placed. This includes attributes of the operational setting, such as software, hardware, tools, training, environmental conditions, and number and type of operators. It will be readily acknowledged by most readers of

Situation	Individual	Task	Effect
What are the relevant elements in the environment, stimuli, setting events, system functions, or goals?	*Who is using the equipment of operating the system? (including their experience, skills, and momentary cognitive states)*	*How is the equipment being used and what behaviors are occasioned? (how hard, how fast, how much)*	*Success or failure?* *Satisfaction or disappointment?*

FIG. 3.1.   The SITE structure.

this chapter that the controls, displays, and software interface will, to some degree, determine the extent to which typical users will be able to operate the system effectively. Similarly, the anthropometric design, habitability (temperature, lighting, and noise levels), publications (manuals, written procedures, and decision aids), manning levels, organizational structure, allocation of functions, task design, and training programs will have significant consequences for effective use of a system or tool. For many human factors and ergonomics practitioners, this is the point at which the world of potential issues and measures ends. Without explicitly considering how these tangible features of the environment translate into performance via their effect on human operators is tantamount to viewing human factors design as an end in and of itself. Although most, if not all, human factors and ergonomics professionals will recognize the obviousness of this statement, far too many will base their test and evaluation procedures solely on measurement of these environmental states and objects, relying on the comparison of the measures to acknowledged standards for the correct height, weight, color, and other characteristics of the tools and objects in the system or product to be tested. The SITE perspective urges the tester to consider the further implications of these tangible aspects of the situation. Compliance with ergonomic standards for workstation design in and of itself means little until we turn our attention to what the users bring to the workstation and the types of tasks the workstation enables the users to perform.

The individual category of the SITE approach involves measurement of the attributes of the individual users of the system. Cognitive workload and physical fatigue, for example, are two of these characteristics or states of the operator that, once again, most readers will readily agree are important human factors measures. The proficiency or skills of the operators also merit examination for most tests, particularly where the user population is diverse and when there are explicit training, help, or diagnostic functions of the system or product. Yet these measures cannot stand on their own any more than the situation measures described above. Although some human factors testers (especially those with a cognitive psychology background) do indeed collect these types of data, the linkage of workload, fatigue, and other states to the user performance that results and, more important, to the environment and tool that gave rise to those operator states is all too often left as an implicit assumption rather than explicitly explored in their test measures. SITE reminds the tester to examine these relationships specifically to aid in clearly and convincingly interpreting and conveying the meaning of the test results.

The task category includes elemental measures of simple or choice reaction times, complex movement times, and the sequence or accuracy of the users' control actions. Similarly, the larger components of task performance may be of interest; longer goal-directed series of responses can be measured in terms of their speed, accuracy, timing, or quality of performance. Some testers might also include cognitive heuristics or mental transformations involved in problem solving or decision making and thus include measures to capture outward manifestations of these behaviors by specialized techniques, such as protocol analysis or thinking out loud (Ericsson & Simon, 1984; Ohnemus & Biers, 1993). Regardless of the tasks involved in the system or product under test, they cannot be viewed independently of the situational conditions and operator characteristics that afford their expression. By the same token, measurement of user performance is essential to evaluating the design of tools, technologies, and procedures.

The final category is the effect, the product or consequence of the interaction of the situation, individual users, and their actions in terms of some outcome measure of human-system performance or user satisfaction. Effect data are those outcomes that matter most to the producers and ultimate users of the system or product. Measures such as total time required to run a load of laundry, write a software subroutine, or contact a satellite or long-term measures, such as injury rates, staff turnover, or even customer loyalty, are what matter to the users and producers. The goal of measuring effect data is the way in which human factors design becomes meaningful for the typical consumer of human factors test and evaluation information.

We are not proposing that most human factors and ergonomics professionals are unaware of these linkages between situation, individual, task, and effect. In fact, it may be because these linkages are so apparent or obvious for human factors professionals that, in far too many cases, they may have been left out of a test design. These linkages are not, however, immediately apparent to people outside the human factors and ergonomics disciplines. Human factors engineering is not in and of itself a principal goal of most system designs, and as such, human factors testing is all too often omitted from design activities unless the linkage between those designs and the "things that matter" are made evident to system developers. In many years of teaching and training human factors testing and analysis, one of the hardest messages to impart is the importance of building a complete human factors analysis, that is, grounding human factors data to system performance, user satisfaction, and other measures by which the entire system will be judged a success or failure. The SITE perspective tries to reinforce the logic that human factors design is important only inasmuch as it serves to contribute positively to system performance and that the linkage from design to outcome must be made explicit each time a system or product is tested.

Thus, the goal is to encourage a comprehensive human factors test by providing new analysts with a structure for selecting and employing a complete suite of test and evaluation measures and, most important, an interpretive context for understanding human factors data and explaining their significance to others. The remainder of this chapter will illustrate this approach across a variety of human factors test planning, data collection, analysis, and reporting examples.

TEST PLANNING

It should be no surprise at this point that we feel that the most important determinant of a successful human factors test is developing a test plan that includes a complete suite of human factors measures. Unless you plan for collection of a measure, it will not be collected, cannot be inferred post hoc, and thus will not figure into the evaluation. Finding out at the end of a test that a key piece of datum was not collected presents the tester with the expense of repeating all or part of a test (if, in fact, the test can be repeated or that the tester even realizes what he or she is missing). To aid in the selection of test planning issues, the SITE mnemonic provides a template of the four measurement categories described above (and illustrated in Fig. 3.2): Situational conditions, individual characteristics, tasks (operator or user behaviors), and the effect or outcome of the combination of the preceding three conditions.

Although it is clearly unreasonable to collect all of the data shown in the figure, a complete human factors test plan should include measures addressing at least some of the issues from each of the four SITE categories.

Situation	**Individual**	**Task**	**Effect**
Controls & displays	Fatigue	Reaction time	System output
Shift length & hours of rest	Workload	Completion time	System accuracy
System modes & software	Skill level	Accuracy	User satisfaction
Habitability & anthropometry	Experience	Sequence	Cost effectiveness
Manning levels	Situation awareness	Reliability	
Documentation		Repetitions	
		Forces applied	

FIG. 3.2.   SITE test planning issues.

To focus the selection process and narrow the resulting set of measures, we usually recommend that a planner work from right to left, from the effect measures to the situation measures. The planner should first identify the most important functions, outcomes, or performance criteria for the system or product and then select human factors measures that have relevance to, and a potential impact on, those functions. For each possible effect measure, the planner should ask, "Is this particular system performance measure important to the overall function and purpose of the system?" and "Does this system aspect or function have a significant human component?" If the answer to both questions is yes, then the measure is a good candidate for inclusion in the effect category of the test and will be used as the ultimate grounds for judging the impact of a human factors design. Potential effect measures can often be found in early system design documents, such as requirements analyses or product need statements.

Selection of user task measures should be based on identification of the most important operator actions associated with each system function selected. For each candidate task, the analyst should ask, "Is this user activity truly relevant and important to the system function (i.e., nontrivial)?" and "Is this activity representative of how users will ultimately operate or interact with the system?" To help answer these questions a planner may have to seek the opinions of subject matter experts, whether they are professional word processors, air traffic controllers, or machinists. As described earlier, the potential user-task measures test may range from discrete measures of simple reaction times to a complex series of responses and to higher-order cognitive problem solving and decision making. Planners should be reminded that it is not necessary or desirable to measure all possible user activity; if the tasks are truly relevant and representative, their number can be kept to a small and manageable amount.

After selecting the relevant operator tasks, the next step is to select the individual user characteristics to measure. In general, the planner should include the characteristics that they think will most directly affect the user's performance on the tasks that have been selected. For example, high workload or fatigue may contribute to slower reaction times and variable task performance for systems with extended periods of operation. Measures of situation awareness are often of interest when vehicular control or supervisory control of objects moving in three-dimensional space is involved. Task and skill analyses may be useful in determining levels of training or practice required to operate the system or when there may be a potential mismatch between the target user population and the skills that will be required of them to use the product.

Finally, we come to the selection of situation measures or design characteristics. One can use two routes to identify candidate issues from this category. A planner can either select measures on the basis of the hardware and software features that will be used most frequently or are associated with important functions (and are thus design aspects that will most likely have a significant impact on users), or a planner can focus on situation measures associated with the greatest design risk (i.e., new, unproven technology or design applications that are novel or unique for the system under test). Ergonomic issues, such as habitability, workstation layout, and tool design, are particularly important in the design of working environments, frequently used tools, and workplace equipment that is used every day. Similarly, system safety features, maintenance equipment, emergency procedures, or decision support for nonroutine operations will be important measures for gauging how well the system design can minimize error or faults.

The goal of the selection process is to develop a test plan that includes relevant human factors issues that may have an impact on the ultimate success or failure of the system, that is, the effect measures. To ensure a comprehensive human factors test, measures from each of the four categories should be considered. Thus, for the purposes of test planning, SITE can be thought of as a convenient organizational tool for selecting the human factors issues appropriate for any given test. The approach has proved useful in teaching new human factors analysts the principles of test design, as well as providing a convenient reference point for experienced human factors testers.

For example, consider a recent test conducted by three members of a graduate class in human factors. The project the students selected concerned a small but high-output dairy product factory. The factory possessed a high degree of computer-controlled automation and was run with only 15 full-time staff. Interviews with the production manager ascertained that the employee injury rate, as reflected in time lost because of injury, was the principle outcome measure of interest to the factory management. All human factors recommendations would be considered, but any capital expenditure required would be judged against the capital return from improved productivity resulting from lower injury rates. A test plan was then developed by identifying the high-frequency injuries (burns, mechanical injuries to fingers, and back strains), the tasks associated with injury (cleaning, online maintenance, and product loading and unloading), relevant employee characteristics (fatigue, stress, physical workload, training, etc.), and the environmental situation (lighting, flooring, noise levels, etc.). A relatively brief period of data collection followed and consisted of observation, questionnaires, and archival analysis of injury reports. One of the first things noticed was a seasonal pattern that showed peaks near the ends of summer and winter for some types of injury (burns and finger injury). Back strains appeared to be relatively constant throughout the year. Because there was no seasonal difference in the factory products or output, it was not immediately clear why these types of injuries would be seasonal. Looking at the tasks associated with injury, the relationship between back strain and lifting and stacking was relatively straightforward (stackers lift and place 5 boxes/minute throughout the day, for a total of 37 tons of whipped cream, liquid cheese, and ice cream mix per day). Tasks associated with burns and finger injuries appeared to be the result of intentional errors or short cuts, such as placing a hand on a pipe to see if it was hot and active or clearing a jammed piece of equipment without first shutting the equipment down. Why should these injuries be seasonal?

To make a fairly involved story shorter, the students next examined the employee questionnaires and noted high levels of stress and fatigue, which in turn were associated with work in very noisy areas of the factory that were extremely hot in the summer and cold in the winter. The scenario, developed through analysis of these measures, was one where heat and noise contributed to worker fatigue and stress. This in turn led to employees taking unnecessary and unsafe shortcuts, particularly in areas where communication was difficult or where a long trip to the control room would be required to shut down equipment or check equipment status. Recommendations included improving air conditioning and hearing protection to reduce fatigue and stress, placing temperature gauges for important system functions on the factory floor, and processing interrupt switches to temporarily disable jammed portions of the processing equipment. All recommendations were designed to reduce the incidence of employee shortcuts and, thus, time lost because of injury. Only by collecting information from each of the SITE categories were the students able to identify the linkage between environmental conditions and seasonal injury rates associated with nonseasonal tasks.

DATA COLLECTION

In addition to its use as a training and test planning aid, the SITE structure can be used to elicit information from users during data collection. One of the initial applications of SITE to human factors test and evaluation was as a data collection tool (Charlton, 1985). The specific application was intended to better understand and document procedural changes introduced by medical staff on the job. It is a common observation that users adapt and alter prescribed procedures to suit their own way of doing things. These work arounds, or instances of intentional errors, can have significant impact on the overall effectiveness of the operator-machine system, but they are rarely analyzed or documented during tests. As a data collection method, SITE was used to capture the behavioral changes as they occurred; their relationship to operator workload, stress, and fatigue; and the surrounding stimulus situations and environmental events that occurred contingent on or contiguously with the behavior changes.

The test involved military personnel operating a field medical facility, specifically their levels of fatigue and stress and, most important, their use of work arounds and nonstandard procedures (Charlton, 1985). The outcome, or effect, measure of interest for the test concerned the rate of processing of injured personnel into the medical facility. In general terms, the tasks included removing dirty clothing and debris from patients and preparing them for emergency care. As already mentioned, fatigue, stress, and workload were the individual or crew measures of interest. A wide range of situation variables were involved in the test, including time of day (or night), temperature, noise, patient processing tools, and manning levels. However, there were also some significant constraints on data collection, making instrumentation and observation unavailable during test conduct. The procedural change data were, by decree, to be limited to what could be collected via a questionnaire administered at the end of each shift.

After some deliberation and preliminary testing, a questionnaire was developed that asked the test participants to describe three patient-processing activities in their last duty shift that presented some difficulty. For each activity the subjects were asked to describe what procedure or task was being performed, what was normally supposed to happen in that procedure, what actually happened, what preceded and followed the procedure, and how they were feeling immediately before the problem (with examples from workload, fatigue, and stress rating scale anchors).

The questionnaire was surprisingly successful in eliciting a range of staff work arounds and the situations leading up to them. Some of the more frequently reported cases of behavior change and high workload resulted from patient backlogs (i.e., patients waiting to be processed into the facility). For example, large patient backlogs resulted in some members of the medical staff moving litter patients through facilities reserved for ambulatory patients. This resulted in a reduction in the number of times the crew had to lift the litter patients as well as some increase in speed of patient processing, reducing the patient backlog. Once begun, this work around was repeated throughout the test. The effectiveness of this procedural change with regard to patient care is unknown. Another case of behavioral change associated with patient backlogs concerned passing medical instruments that had been dropped on the ground to the next duty area without first cleaning them properly. In both cases, medical staff facing patient backlogs found short cuts in patient processing.

Other behavior changes associated with high workload and stress involved communications procedures. Heavy radio traffic frequently prompted the medical staff to send a runner to a different duty area to convey a message. This lack of effective communication between various areas in the facility resulted in a loss of manpower because staff had to leave their duty areas to communicate with other medical personnel. The time pressure and frustration produced by ineffective communications also led staff in some areas to ignore their communications units, physically muffle them, or both. This alleviated what was perceived as an auditory nuisance, but it created a potentially hazardous situation when important messages to these personnel went unheeded.

The final test report described these findings and how the medical staff's patient processing capacity (measured in terms of patients processed per hour) was compromised by work arounds resulting from noise, stress, and high workload. These work arounds, and their contributing situations, would have otherwise gone unreported.

Subsequent applications of this data collection technique have been used to capture data on a number of operator activities that may be of interest during a test. For example, one testing application resulted in the development of a standardized reporting tool called a critical incident form (CIF). The CIF was developed to capture significant problems in the command and control of the Global Positioning System constellation of satellites, either by means of direct observation or operator reports, as they occurred. Over the years, the CIF form has been refined for ease of use while ensuring that the situational, personnel, behavioral, and outcome conditions associated with an operational problem are captured for subsequent analysis. An example of the CIF is shown in Fig. 3.3. This application of the SITE method is essentially a form of the critical incident

---

CRITICAL INCIDENT FORM

Date _____   Time _____   Observer _____

Task description _____

Operator position _____   Workload score (if any) ____   Fatigue score (if any) ___

"This form is intended to document any difficulty in operating the system under high workload and fatigue. The questions are used to evaluate the system, not you, so please answer the questions as accurately as possible. Your name will not be used at any time."

Situation: *"Briefly describe the situation leading up to this task."*
　　　(Subject's description should include system status, preceding task, display/console
　　　configuration, and any relevant environmental conditions)
_____
_____

Subject condition: *"Describe how you were feeling during the task."*
　　　(Subject's condition should include fatigue, stress, sleepy, unprepared (in terms of
　　　training or practice).
_____
_____

Task incident: *"Briefly describe what happened, to your best recollection."*
　　　(Subject's description should include response description and dynamic information
　　　how hard, how fast, how long, etc.)
_____
_____

Consequences: *"Briefly describe the events that followed the task."*
　　　(Subject's description should include system status, preceding events, and
　　　any relevant environmental conditions)
_____
_____

Notes: _____
_____

FIG. 3.3.   SITE data collection form.

data collection techniques first described by Flanagan (1954) and more recently summarized by Shattuck and Woods (1994).

In addition to the CIF form and the self-report questionnaire described previously, other applications of the SITE data collection method have included structured interviews conducted at command and control facilities and an observation checklist developed for command and control facilities. For a complex system, where every possible operational situation can never be correctly anticipated or completely tested, the CIF is a particularly effective method of documenting work arounds with a structure that provides all of the information necessary for subsequent analysis. The results of the analysis are evaluated in terms of the impact that behavioral changes and high workload have on system effectiveness. In all of these cases, the key feature of the SITE data collection structure is the elicitation of information on situations and user activities that, because of their spontaneous nature, cannot adequately be captured by instrumentation or other preplanned data collection techniques. The data collection technique focuses on capturing these novel user behaviors—along with their antecedent situations—the individual characteristics of the users, and the consequences or results of the actions.

DATA ANALYSIS

The stimulus for applying the SITE structure to data analysis came from two problems inherent in evaluating the human factors tests of large satellite control systems. Owing to their size, complexity, and cost, these systems are typically one-of-a-kind, evolutionary systems. Although the human factors design considerations for smaller products or systems can often be researched and tested in advance, these large systems are unique and are frequently reconfigured. Testing of these systems is iterative; human factors solutions are developed as the problems are identified, and operational procedures are changed as task loads and missions are defined and redefined. Analysis of human factors issues is hampered by the frequency of these changes and by test restrictions that do not allow tests to interfere with ongoing operations.

A second analysis problem was the lack of human factors standards or criteria on which to base the evaluation of system designs. The lack of human factors test criteria is a problem pervasive throughout human factors test and evaluation (Mackie, 1984; Meister, 1986). Although design standards do exist for the optimal physical layout and anthropometry of an operator's workstation, similar standards do not exist for presenting information and designing tasks so that operators can correctly assimilate and respond to system information. Because of the problems associated with frequently changing system configurations and lack of appropriate human factors criteria, the SITE perspective was applied to the development of a statistical approach to identifying human factors design risks in the absence of criteria (Charlton, 1992). This approach establishes a circumstantial case for the identification of situations and operator tasks that are at risk or highly correlated with human error and poor system performance. This approach is analogous to the medical epidemiological approach specifying risk factors associated with a disease in the absence of causal data specifying the physiological mechanism of the disease (Charlton, 1988).

As shown in Fig. 3.4, the goal of SITE data analysis is to identify candidate measures of operator performance, individual operator characteristics, and situation measures that may have a significant impact on system performance. Candidate operator task measures might include operator response times, decision times, response series times, and error frequen-

Effect	**Task**	**Individual**	**Situation**
Did the system meet the performance requirements? *Were the users satisfied with system performance?*	*Which operator tasks had the greatest impact on system performance (or user satisfaction)?*	*Which user characteristics affected task performance?*	*Which design considerations & environmental conditions affected the users & task performance?*

FIG. 3.4. SITE analysis structure.

cies representing several levels of task complexity. (Task measures are drawn from frequent user activities and human-intensive system functions so that they are both representative of operator activities and relevant to system effectiveness.) The situation and individual measures may include the results of human factors questionnaires or checklists completed by testers observing system operations. These situation features include ratings or measurement of display formats, data entry procedures, system documentation, training, ambient noise, illumination, and any other issues relevant to operator performance. Individual operator characteristics include measures of workload, fatigue, situational awareness, and skill levels. Finally, the measures of system performance include system throughput times, failure rates, and other key indicators related to the effectiveness of the system undergoing evaluation. Unlike the human performance and human factors design measures, these system measures typically do possess objective criteria of success or effectiveness.

Once test data have been collected, the data analysis is directed at establishing the degree of association between the human factors parameters and system effectiveness via multivariate statistics. The objective of the analysis is to link the criterialess human factors measures to the objective criteria for system performance. An example of this linkage is shown in Fig. 3.5. In the example, two operator tasks were found to have a significant predictive relationship with the system function of completing the portion of a satellite contact known as pre-pass. The two tasks were identified by entering time and error rates for nine operator tasks associated with pre-pass into a stepwise regression analysis to identify the optimal linear model (i.e., the smallest number of predictors resulting in the greatest predictive statistic, $R2$). These two tasks were then used as the focus of the analysis of the remaining human factors issues. For the first task measure

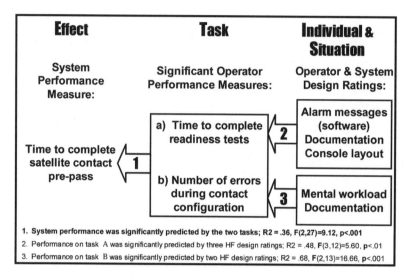

FIG. 3.5.   SITE analysis for satellite control example.

identified, the time to complete readiness tests, the situation issues with the best predictive power were the design of software alarm messages, accuracy and completeness of documentation, and the layout of the operator consoles, as measured by operator questionnaires. The second task measure, error rates during configuration, was found to be most closely associated with the operator characteristic of high workload and the situation issue of accuracy and completeness of documentation.

The results of the analysis were validated through both interviews with subject matter experts and replications of the test (Charlton, 1992). The measures of individual characteristics were combined with the situation measures because there were too few of them (i.e., only workload and fatigue) to be analyzed separately. The findings from the analysis were incorporated into the system design through numerous upgrades to the system software, hardware, and procedures. Further, as the data collection and analysis were repeated over time, the results were sensitive to the changes in design and operator performance. As the human factors problems were corrected, or procedures were changed, the data analyses correctly reflected the changes and identified new problems, where they existed. Thus, the SITE analysis method identifies the operator tasks having the greatest relationship to system performance deficiencies and the human factors engineering and personnel issues with the greatest impact on performing those tasks. The test criteria for human factors engineering in these systems can be operationally defined as a positive contribution to overall system effectiveness.

It should be remembered that this data is essentially correlational in nature. Much of human factors testing involved with these complex systems is conducted without true experimental protocols, such as random sampling and factorial manipulations that would be required to draw firm conclusions about causality. However, as the development effort progresses, or as the operational system matures, the human factors tester can attempt to derive explanatory or logical connections between system design, operator performance, and system operations through repeated testing. These explanatory connections can then be used to establish human factors design criteria (presumably specific to the type of system and type of operators involved) to be used in the design of future systems or upgrades to the current system.

The SITE approach to data analysis has enjoyed considerable success where it has been employed. The lack of human factors design criteria is remedied by using overall system effectiveness as the measure of performance by which human factors design issues are judged. In other words, areas of human factors design are identified as being problems only inasmuch as they are statistically associated with poor system performance. The approach also has the advantage of considering a variety of human factors issues simultaneously, allowing the identification and evaluation of interactions among various human capabilities and system design features. Further, the analyses can provide periodic snapshots of dynamic human factors issues as they evolve in the frequently changing environment of a large control system. Finally, the approach can serve as a starting point for the development of a taxonomy of human factors issues and design criteria. Through the systematic use of this and similar evaluation techniques, an empirical database of human factors standards can be collected and generalized for application to the design of new systems. Faced with a need to field effective operational systems, and in the absence of definitive human factors criteria, SITE analyses have shown much promise in the human factors test and evaluation of complex systems.

REPORTING A SITE STORY

The application of SITE to test reporting brings our discussion full circle and reemphasizes the underlying message presented in this chapter. The hallmark of the SITE approach can be thought of as the development of a compelling human factors narrative or story (Charlton, 1991). As any experienced storyteller knows, the ingredients to a good story are fourfold: Every story has a beginning, a middle, an end, and a moral. In the case of a human factors test report, particularly if the intended audience does not have a background in human factors or ergonomics, the story must include the situation, the people involved, what they did, and what happened as a result.

The beginning of the story, the description of the situation in terms of the hardware, software, and procedures, can be thought of as the narrative hook. This is the part of the story that must set the stage for the audience and pique the interest level for the system aspect and human factors issues being described. The individual characteristics describe the protagonists of the story, and the task measures describe in detail what they did. The final element of the story, the consequence or effect measures, places the findings in perspective by describing what happened as a result, not unlike the conclusion or moral of a story. This final component defines the importance of the human factors test results within the larger context of system performance.

If any of the four SITE elements are missing, the story begins to break down. For example, without knowing the details of the situation, it will be difficult to know how to formulate recommendations for fixing the problem or to predict when the problem is likely to occur in the future. Information on the people involved is important so that conclusions can be made about who is likely to experience the problem or how widespread the problem is. Detailing the response, task, or activity will document for the record how the operators are using the system or equipment and may include tasks or response sequences that the design engineers never anticipated. Finally, and most important, describing the consequences or effect of the problem will enable the human factors tester to answer the inevitable "so what" questions: "So what if the on-screen menu function appears on top of the error message window? What's the mission impact of that?"

In summary, the SITE approach has been used to accomplish several test and evaluation goals. It has been used to help focus test planners' thoughts when beginning a test planning effort by providing a template of potential human factors test issues. During data collection, SITE CIFs have been used as a supplementary data source to capture information from users about unanticipated human factors problems or operator work arounds. In dynamic, one-of-a-kind systems, the SITE perspective has been used as an analysis structure to address the problem of criteria by linking human factors to mission effectiveness through a four-part narrative. The approach relies on combining the essential ingredients of a good story. Only with all four SITE elements can you ensure a complete story—and a complete human factors test and evaluation.

REFERENCES

Bittner, A. C. (1992). Robust testing and evaluation of systems: Framework, approaches, and illustrative tools. *Human Factors, 34,* 477–484.

Charlton, S. G. (1985). Behavior analysis: A tool for test and evaluation. *Proceedings of the Human Factors Society 29th Annual Meeting*, 188–192.

Charlton, S. G. (1988). An epidemiological approach to the criteria gap in human factors engineering. *Human Factors Society Bulletin, 31*, 1–3.

Charlton, S. G. (1991). *Aeromedical human factors OT&E handbook* (2nd ed., AFOTEC Technical Paper 4.1). Kirtland AFB, NM: Air Force Operational Test and Evaluation Center.

Charlton, S. G. (1992). Establishing human factors criteria for space control systems. *Human Factors, 34*, 485–501.

Ericsson, K. A., & Simon, H. A. (1984). *Protocol analysis: Verbal reports as data*. Cambridge, MA: MIT Press.

Flanagan, J. C. (1954). The critical incident technique. *Psychological Bulletin, 51*, 327–358.

Kantowitz, B. H. (1992). Selecting measures for human factors research. *Human Factors, 34*, 387–398.

Mackie, R. R. (1984). Research relevance and the information glut. In F. A. Muckler (Ed.), *Human factors review 1984* (pp. 1–12). Santa Monica, CA: Human Factors Society.

Meister, D. (1986). *Human factors testing and evaluation*. Amsterdam: Elsevier.

Ohnemus K. R., & Biers, D. W. (1993). Retrospective versus concurrent thinking-out-loud in usability testing. *Proceedings of the Human Factors and Ergonomics Society 37th Annual Meeting*, 1127–1131.

Shattuck, L. G., & Woods, D. D. (1994). The critical incident technique: 40 years later. *Proceedings of the Human Factors and Ergonomics Society 38th Annual Meeting*, 1080–1084.

Chapter 4

Test Documentation: Standards, Plans, and Reports

Thomas G. O'Brien
O'Brien & Associates

Thomas B. Malone
Carlow International

Building a house would be nearly impossible without first preparing the blueprints. Like a blueprint, the human factors test plan establishes the basic shape and structure of the test to give a clear understanding of what, why, and how testing is to proceed. A properly planned test will facilitate the process of conducting a logical and organized investigation of the human-system interface. A test plan may range from single subject performing a simple task and noting his or her gross performance to volumes of documentation addressing issues in data collection, reduction, and analytic methods and procedures. Although test planning may involve any level of complexity, all test plans should have three objectives: to address why you want to conduct the test, what is to be tested, and how to carry out the test. Also, test plans are developed to provide management with a description of the test you wish to conduct (Juran & Gryna, 1980). This is especially true in large research and development programs in which human factors is one among many elements that must be considered before a decision can be made to proceed from one phase of development to the next.

This chapter presents three parts: Part one discusses the test support documentation used to support the plan and provides some examples of the kinds of documentation available to the tester. Part two describes the structure and contents of a human factors test plan. Part three examines methods for preparing test reports.

TEST SUPPORT DOCUMENTATION

Test support documentation provides essential information needed to support the planning, conducting, and reporting of HFTE. It may include standards, criteria, or guidance used in assessing the implications of human factors evaluations and the processes or procedures for ensuring that HFTE programs will produce reliable and valid data. From the early 1950s to the early 1990s, a substantial part of formal HFTE documentation available in the United States focused on testing and assessing military systems and equipment. An extensive ensemble of policy, standard, and guideline documentation emerged to define HFTE criteria, tasking, and methodology. As the Cold War drew to a close, and with the introduction of acquisition reform in the Department of Defense (DOD), development and use of HFTE and other documentation diminished. At the same time, HFTE imperatives beyond military needs arose and included applications in medical equipment, nuclear power, transportation, information systems, and other commercial products. Indeed, the 1980s saw the development of human engineering design criteria as yardsticks for acceptance of human factors design in various product lines.

In many cases, military human engineering and HFTE standards and guidelines served as seminal documents for development of the commercial counterparts that have become prominent in the world of commerce and information technology. Not surprisingly, as more of today's products are controlled by computer processes, the focus on HFTE interests and documentation is transitioning from hardware to software and from manual operations to automated systems. HFTE documentation has also become more accessible in electronic form through media such as the Internet.

Classes of HFTE Documentation

HFTE documentation can be classified by its purpose. Documentation that provides design criteria against which the design of a system, device, or human-machine interface can be compared is termed an HFTE *design standard* or *convention*. (Hereinafter we use the term *standard* to include *convention*). Design standards are used to provide metrics and criteria against which the physical attributes of a design concept are compared. Design standards are concerned with evaluating or assessing physical attributes of a design representation, such as equipment dimensions and anthropometry; levels of force and torque; elements of the mechanical, thermal, and light environments; arrangements and layouts; colors, markings, labels; and auditory signals and messages.

An implicit assumption is that, at some time in the dim and distant past, empirical tests were conducted that produced human factors design standards that enhanced human performance and safety. A second, related assumption concerning the need for design standards is that human performance is enhanced when the human is presented with a design configuration based on accepted, and possibly empirically based, design conventions. Adherence to common and standardized design approaches for human-machine interfaces is a widely accepted practice for reducing the incidence of human error.

Documentation that addresses the performance of humans on specified systems tasks and task sequences constitutes a performance standard. Human factors performance standards usually are concerned with how well an individual or team performs as evaluated against system-specific performance specifications. Human performance standards may include time to select an action; time and accuracy in making decisions; accuracy of interpreting information, problem solving, diagnosis, action selection, team performance, communication, and control activities; time to respond; and work loads associated with a control sequence.

Although most of these performance standards are system specific, one class of performance standard is somewhat more general. The usability of a computer interface (i.e., facilitating human–computer interaction by applying graphical- or object-oriented user interface conventions) is an increasingly important human performance issue, and several standards have been developed to address designs that fit users' cognitive, perceptual, and memory capabilities; displays that are standardized, easily read, and interpreted; procedures that are logically consistent; documentation that is clear, easily accessed, and readable; and online help that is available and responsive.

The third class of HFTE standards is process standards. These standards define the requirements for HFTE processes and procedures for two different situations: where human factors issues must be addressed in the context of a system-level test and evaluation exercise and where human factors processes and procedures must be considered in the context of a human factors test or evaluation. Examples of design, performance, and process standards are presented in the sections that follow.

Human Factors Design Standards

Suffixes to the references standards and guides are not included because they are periodically updated, some more frequently than others. There have been a number of government and industry standards published over the years that deal with human factors; some have been incorporated into others while others have been outright cancelled. For example, when the second edition of Handbook of Human Factors Testing and Evaluation was first begun, MIL-STD-1472 had been in its *D* version. Now, several years later, it is in its *F* version. Thus, when searching for a human factors standard or guide it is best to search for the source leaving off the suffix. Indeed some internet search engines are very particular and may respond negatively if the suffix were included for a publication that has been superceded by a newer version.

MIL-STD-1472.   The fundamental and most comprehensive set of human factors design standards is contained in MIL-STD-1472, *Human Engineering Design Criteria for Military Systems, Equipment and Facilities*. This document presents human engineering principles, design criteria, and practices to integrate humans into systems and facilities. This integration is required to ensure effectiveness, efficiency, reliability, and safety of system operation, training, and maintenance. This document contains standards for the design of human-machine interfaces and includes data and illustrations of visual fields, controls and displays, physical dimensions and capabilities of humans, work-space design requirements, environmental conditions, design for maintainability, design for remote handling, and safety considerations.

DOD-HDBK-743.   This DOD handbook, *Anthropometry of U.S. Military Personnel*, provides anthropometric data on U.S. military personnel. The body size and proportion information is of great importance in the design of clothing and personal equipment as well as military vehicles, such as tanks, submarines, and aircraft. Included with the data are the data sources, diagrams of body measurements, and definitions of related terminology.

MIL-HDBK-759B.   This military handbook entitled *Human Factors Engineering Design for Army Materiel* provides nonbinding human factors guidelines, preferred practices, and reference data for design of Army materiel. This handbook also provides expanded, supplementary, and Army-relevant human factors engineering information that may be too detailed, lengthy, or service oriented for inclusion in military standards, such as MIL-STD-1472.

ASTM 1166.   This standard, entitled *Standard Practice for Human Factors Engineering Design for Marine Systems, Equipment, and Facilities* is produced by the American Society for Testing and Materials and provides commercial human factors design standards directed at ships and marine systems.

NUREG 0700.   This standard, *Human-System Interface Design Review Guideline*, published by the Nuclear Regulatory Commission (NRC), and represents a tailoring of the design standards from MIL-STD-1472 to the design and evaluation of nuclear power plant control consoles and human-machine interfaces.

NUREG/CR-5908. This standard, entitled *Advanced Human System Interface Design Review Guideline* and published by Brookhaven National Laboratories represents an attempt to generate a design standard for advanced, computer-based systems in nuclear power plants. Advanced control rooms use advanced human-system interfaces that have significant implications for plant safety in that they affect the operator's overall role in the system, the method of information presentation, and the ways in which operators interact with the system. The purpose of this document is to develop a general approach to advanced human-system interface review and the human factors guidelines to support NRC safety reviews of advanced systems. Volume 1 describes the development of the Advanced Human System Interface Design Review Guideline. Volume 2 provides the guidelines to be used for advanced human-system interface review and the procedures for their use.

NASA Standard 3000. This standard, the *Man-Systems Integration Standards*, provides a good synopsis of the rationale behind design requirements. It provides guidelines, recommendations, and other nonbinding provisions (the "shoulds"), as well as contractually binding standards, requirements, and criteria (the "shalls"). It also has sections titled Example Design Solutions, which illustrate and describe typical examples of how requirements have been implemented in manned spacecraft.

DOT/FAA/CT-96/1. The FAA's *Human Factors Design Guide for Acquisition of Commercial-off-the-Shelf (COTS) Subsystems, Non-Developmental Items (NDI), and Developmental Systems* provides reference information to assist in the selection, analysis, design, development, test, and evaluation of new and modified FAA systems and equipment. The standards are primarily focused on ground systems and equipment, such as those managed and maintained at Airways Facilities. It is nearly identical to MIL-STD-1472 but has been modified to reflect air- and aerospace-related system applications.

Human Factors Performance Standards

ISO 9241 Series. *Ergonomics of VDT workstations, Part II—Guidance on Usability* covers how to specify and measure usability. Usability is defined in terms of components of effectiveness, efficiency, and satisfaction for the user. Goals and the context of use must be described in order to measure usability. Guidance is given on how to choose and interpret usability measures, how to specify and evaluate usability in the process of design, and how to specify and measure usability for an existing product or system.

Human Factors Process Standards

MIL-HDBK-46855. *Human Engineering Requirements for Military Systems Equipment and Facilities* details requirements and tasks during development and acquisition programs. Topics covered are analysis of human performance, equipment capabilities, task environments, test and evaluation, work load analysis, dynamic simulation, and data requirements. An application matrix is provided detailing program phases when each task is appropriate.

Test Operating Procedure 1-2-610 (TOP). The U.S. Army Test and Evaluation Command published the TOP in two volumes. Volume I is the test operating procedure itself, consisting of guidance in planning a human factors test or evaluation. The test-planning guidance is presented as a process in a series of steps. Sections of the TOP are preparation for test, data requirements and analysis, and specific test procedures for such evaluation activities as assessment of lighting, noise, temperature, humidity, ventilation, visibility, speech intelligibility, workspace and anthropometrics, force and torque, panel commonality, maintainability, individual performance, error likelihood, crew performance, information systems, and training. Volume II contains the Army's Human Factors Engineering Data Guide for Evaluation (HEDGE). This document contains evaluation criteria annotated from MIL-STD-1472 to support the identification of criteria to be used in HFTE.

PREPARING A HUMAN FACTORS TEST PLAN

Purpose of the Test

This first section of a test plan states why you are conducting the test. For example, the decision to proceed to production with a new product, to declare an initial operational capability for a new system, or to transfer responsibility from a developing agency to an operating agency are all supported by test results. The nature of the decision or milestone being supported by the test is described at the outset. For human factors-specific tests there may also be human factors-specific goals or objectives. Consider, for example, the following statement of test purpose: "When considering the operator-system interface the fundamental requirement is that this interface result in safe, efficient, and effective system performance.

Consequently, testing against this requirement involves a thorough examination of the hardware, software, and procedures associated with the system's operation and maintenance to determine whether human factors problems uncovered during previous development (assuming the system under test was designed as an improvement over a predecessor system) have been addressed. It also confirms that the system can now meet all of its operational requirements in all mission profiles using representative samples of the user population" (O'Brien, 1982). Some of the more common reasons for conducting a human factors test include the following:

1. To provide technical or operational human factors data (as a stand-alone human factors test or as part of a larger systems test) to determine a system's readiness for transition into the next phase of development or operation.
2. To determine whether any design or operational problems exist with the system or product.
3. To support the evaluation of the degree to which system specifications and user requirements are met or exceeded, whether or not system shortcomings and deficiencies identified previously have been remedied, and whether any operator-system interface strengths found during earlier development have been retained.

System Description

This section describes what is to be tested. It should contain a detailed description of the system, or at least enough detail to give the reader a good idea of how the human is to interact with the system. Often, when systems are in the earliest phase of development (the concept-development phase), a concept of operation may be the only source of information available to the test specialist. In other words, the system is still an idea in the designer's mind. For example, the concept for a new vehicle may include a description of a capability to provide the driver with navigation aides, such as automated map displays, or with a steering device that might replace the traditional steering wheel. Consequently, the system description may contain only a sketch of how the operator and maintainer might interact with hardware and software.

For systems in more advanced stages of development, a more detailed system description would be appropriate. Here, the test specialist might wish to include a more sophisticated view of not only the system but also its components, particularly components that require human control. For example, software-driven display components may require their own description, which might include the type of information to be displayed.

Human Factors Test Issues

A test issue is simply an unanswered question about the technical or operational performance of components, subsystems, and systems. Test issues are typically technically oriented for developmental testing and operationally oriented for operational testing. They may include both technical and operational perspectives for combined developmental and operational tests. Test issues usually begin as general statements, and, as the system progresses from its early, concept-exploration phase of development to full-scale engineering, more detailed issues are identified and used as a basis for further testing. For example, at the program outset, the issue may broadly state the following: To what extent do the design and operation of hardware, software, training, and operational/maintenance procedures associated with the system affect the human-system interface relative to safe, efficient, and effective mission performance, neither causing or encouraging errors nor introducing delays in human performance? Certainly, this would cover most any human factors issue. It is only later, when the system's design becomes firm, that detailed issues are developed. An example of this would be the following: To what extent does the system's human-software interface design affect operator response time in a realistic environment when compared with the predecessor design? In our example, the test issue would eventually lead to a comparative test of information presentation modalities to test whether the new design was indeed an improvement over the old.

A test issue also may also be simply the following: Is the customer happy with a product? Or, it may focus on a functional aspect of a system's human interface, such as the following: Can the emergency indicator light be seen from 20 meters in a smoke-obscured environment? Thus, the specific issue, or issues, depends on the nature and complexity of the system under test and the nature and complexity of the human-system interaction.

It is important to avoid test issues that focus solely on whether the system satisfies specifications, designs, or standards. Suppose, for example, the test specialist finds that engineering design criteria are met but only by the barest margins. Would this mean that the system passed the human factors test and that the system's human engineering design is safe, efficient, and effective? Not necessarily, because even if hardware design criteria are met, testing has not evaluated the workload demands, skills and training requirements, and other human dimensions. It is better to express test issues in terms of the extent to which the system meets or exceeds functional requirements. This way, even if certain design criteria are met, and if there is some doubt about human performance for system parameters that have not yet been tested, the evaluator can caveat the conclusions as such or suggest that more tests be conducted before drawing definitive conclusions.

A comprehensive human factors evaluation should rely on more than one set of test dimensions, if possible. In developmental testing, compliance with standards should certainly be examined, but compliance with standards alone is not enough. User satisfaction and human performance data on mission critical tasks should be considered in combination with standards compliance when preparing test issues.

Data Requirements

This section of the test plan translates the broadly stated test issues into specific data elements that must be gathered to answer the test issues. Depending on whether a test is developmental, operational, or a combined developmental-operational test, human factors data may include any one or a combination of the following: human engineering measurements, ratings of user acceptance and opinions, and human performance measures. Human engineering measurements are almost exclusively associated with developmental testing. Typically, they are used to determine compliance with standardized human engineering criteria, such as MIL-STD-1472, or with Occupational Safety and Health Administration (OSHA) regulations. User acceptance and opinions are usually used as an indicator of user satisfaction, or they may point to other problems with system design and operator/maintainer performance. Human performance data will typically consist of time and error measurements. However, other measurements may be appropriate, including the following: potential error impact, concurrent tasks, associated controls and displays, and workload difficulties associated with critical operational and maintenance tasks.

Specific terms are used by various testing agencies to refer to data requirements. For example, some DOD agencies refer to measures of effectiveness (MOE) as a measure of a system's degree of accomplishment of a specific operational task (AFOTECI 99-101, 1993). To support the MOE, associated measures of performance (MOPs) are developed. A MOP is a qualitative or quantitative measure of a system's capabilities or characteristics. It indicates the degree to which that capability or characteristic performs or meets the requirement under specified conditions. MOEs and MOPs may or may not include evaluation criteria. The following is an example of a MOP:

MOP-1. Human performance on mission-critical task sequences. Evaluation criteria: The sum of the individual operator performance task times (averages) on mission-critical task sequences must be less than or equal to the total system throughput time requirement. For the communication system, the sum of the average operator completion times for sending and receiving 1,500 messages cannot exceed 24 hours.

Here, the human factors MOP describes a characteristic of the system that is evaluated against a system functional requirement. In other words, the system's design should allow the operator to perform certain functions within a given period; otherwise, the system is judged incapable of meeting this particular functional requirement. MOEs or MOPs like the previous one may be supported by other data elements. In our example, MOP-1 may be supported by the following:

> *MOP 1-1.* Average time for a system operator to complete a communications health check.
>
> *MOP 1-2.* Average time for a system operator to prepare a Level-1A Tactical Air Support Message.
>
> *MOP 1-3.* Average time for a system operator to initiate a communications sequence.

Notably, when developing measures of performance, it is important to be able to trace each measurement directly to the MOE, which in turn can be directly associated with the test issue.

Test Conditions and Test Scenarios

Test conditions are typically identified with developmental testing. They describe certain prerequisites under which the test is to be performed. Test conditions should be identified for all kinds of human factors testing, regardless of the type of data to be collected (i.e., human engineering measurements, user satisfaction and opinions, or human performance). For each test condition, a reason for having identified that condition should be provided (usually, this is tied to system specifications or user requirements). For example, some systems may require operation in the pitch black of night or in the extreme cold, such as the arctic. Testing might include measuring crew performance while operating in special clothing and equipment, such as cold-weather clothing, nuclear, biological, chemical (NBC) protective gear, and night vision devices.

The term *test scenario* is associated more typically with operational testing than it is with developmental testing. A test scenario can be best described as a short story that specifies the conditions of realism under which the operational test is conducted. Test scenarios should be selected to represent not only how the system or product will be used 90% of the time but also the 10% extreme or emergency scenarios. Scenario selection should also consider cost and the return on investment with regard to the type and amount of information gained for the money spent. In other words, the cost of testing generally rises as operational fidelity rises.

Sampling Methods

Related to the subject of test conditions and test scenarios, the method of determining the number of subjects required to satisfy a statistical treatment of the data should be stated. This section should also indicate the pool from which test subjects will be drawn and the methods of selection.

Test Articles

The number of test articles required to conduct the human factors test should also be given. For example, a test to address whether the design of a new type of glove is adequate for gripping metal tools should specify that n number of gloves is required.

Test Constraints and Restrictions

A test constraint is any factor that might unduly influence the outcome of a test. Constraints tend to reduce test validity or generalizability because of decreased operational fidelity, a lack of qualified subjects, insufficient trials, or confounding influences. Meister (1986) identifies a number of test constraints that affect either the control of the test or operational fidelity (one is typically traded for the other). Generally, the more fidelity, the less control, or the more fidelity, the greater cost of achieving a high degree of control (as mentioned earlier). Naturally, many of the constraints posed in testing are unpredictable, such as uncooperative weather. Many of these constraints were described in an earlier chapter on the role of human factors in the systems development process. Identifying predictable test constraints in the test plan is important; it puts management on notice that certain unavoidable problems may arise which might influence the test validity or generalizability.

By contrast, test restrictions are boundary conditions under which the tester is willing to conduct the test. For example, test subjects should possess normal intelligence or demonstrate some minimal level of training before being allowed to serve as test subjects, test articles should be operational, ambient temperatures and humidity should be within a certain range, and so on.

Methods of Analysis

This part of the test plan provides a general description of how the human factors data will be treated after it is collected. The test specialist should identify specific methods for comparing data with test criteria. If criteria do not exist, the methodology section should explain how the data are to be summarized and presented in answering the test issues.

For developmental tests, the human engineering design measurements should be compared with criteria in appropriate military or industrial standards and specifications or with those in system specifications or other design and user requirements documents. The plan should state how the results will be characterized for measures that do not meet criteria. For both developmental and operational testing, data reflecting user acceptance and opinion, situation awareness, training effectiveness, and other results should be used to learn, from users and other test participants, of those operating and maintenance characteristics that seem to cause problems or limit system performance.

Other Considerations

The tester may find it necessary to include other test requirements to fully and accurately assess or evaluate the system. Additional requirements will depend on the system's phase of development and whether it is a developmental, operational, or combined developmental-operational test.

Anthropometric Requirements. Applicable mostly in developmental testing, anthropometric requirements of test subjects should be based on the system's target population. Many of today's automobiles are designed to accommodate the fullest range of statures, roughly 5th percentile female to 95th percentile male (Peacock & Karwowski, 1993). Also, the U.S. military continues to require systems to be designed to accept 5th percentile female to 95th percentile males. Thus, prior to any test planning, the specialist should research the target population.

Task Analysis. Task analysis is an important activity for both developmental and operational testing. A task analysis is often conducted in two phases: gross task analysis and detailed task analysis. The gross task analysis results from a general examination of operator-system interfaces. It includes gross operator responses to system stimuli and provides the test specialist with a general idea of tasks involved. The detailed task analysis is a refinement of the gross task analysis. It provides a minute description of operator activities, which can be reliably observed among all operators; that is, the actions observed from one operator will be essentially the same as those observed for any operator for that specific interface. Task analyses should include task difficulty, its importance relative to other tasks, its impact on mission accomplishment and training, and similar information. Military Standard-1478 (1991) provides robust procedures for conducting task analysis.

Criteria and Standards. A criterion is some measurable level of human performance or design attribute required of a system. A standard is an established criterion that may be system independent. Viewed from another perspective, a criterion may or may not be a standard (i.e., accepted by everyone as a minimal level of performance). In more practical terms, a standard may be an agreement among developers of some minimal level of performance, some level of performance established by a comparison system, or a threshold established by the engineering community, the value of which is based on research findings and accepted by users. Thus, the terms *standard* and *criterion* differ in that a standard might be considered permanent until new research shows that it should change or that agreement among professionals deems that it should, whereas criteria may be unique to a system or a stage of development.

Military and industrial standards (described earlier in this chapter) provide certain criteria for the physical layout and design of systems for which the human plays a direct role: workstations, facilities, and individual clothing and equipment, to name some. Yet, performance standards are conspicuously lacking. This is probably because as human-system interface designs change, so do performance expectations. Another way of expressing this is as technology changes, so do the rules governing what is a safe, efficient, and effective human interface. The criteria upon which standards are founded are elusive and ever shifting. Even so, most practitioners will agree that standards are a necessary part of the evaluation process.

Selection of Test Measures. Typically, the test specialist will determine measures based on system requirements and test issues, which are used to evaluate the performance of the system as a whole. Multiple measures may be associated with a single system requirement. Take, for example, the time requirement for an operator to successfully diagnose an engine failure within 1 minute from the time the error code was displayed on an aircraft console. The tester may decide to record time lines and errors across some number of trials. The choice of measures depends on two things: what is possible to measure given technology and operational constraints, and what is the best or closest measure representative of the performance you wish to evaluate. For example, heart rate probably is not as good a measure of effectiveness for exiting an airliner's emergency hatch as the time to open the hatch is, although the two measures are probably related. However, if we want to measure anxiety as an index of crew effectiveness—and we have shown that anxiety is directly related to effective or ineffective behavior—heart rate would probably be a better measure of performance.

Data Collection Methods. A well-planned test will include a clear and concise data collection section. It should include as a minimum the following:

1. Standard forms. Data sheets should be prepared that allow a tally of errors, time lines, and space for comments by the data collector. Interview forms should meet the criteria for developing questions found in the previous chapter on questionnaire-development methods. Forms are referred to as standard only because the same form should be administered from one operator or maintainer to the next. Data collection forms should also accommodate lists for scheduling subjects, test equipment and facilities, and test instrumentation.

2. Instrumentation list. A list of instrumentation should be provided. This may include anything from pencils, pens, and notepads, to sound- and light-level meters, and other instrumentation. For each instrument, where it is appropriate, the model number and last date of calibration should be included.

3. Test personnel list. A list of data collectors and test subjects should be developed. The list should include a brief description of qualifications required of each data collector, including a minimum understanding of the system to be tested and of tasks for which subjects will be measured, as well as experience in recording events and personnel interviews. The list of test subjects should include a description of the minimum acceptable level of education, the training of the test subjects, or both.

Step-by-Step Data Collection Procedures. This section, sometimes referred to as detailed test procedures, describes how to conduct the test. Often, this section is prepared separately from the higher-level test plans, usually as an addendum or appendix to the test plan. Here, the human factors test specialist must devise for each trial a detailed scenario that describes exactly what the data collector is expected to do. It includes test setup and preparations prior to any data collection, use of instrumentation if instrumentation is part of the data collection process, instructions on how to record data, and procedures for concluding each test scenario. It may also include procedures for handling emergencies, such as a fire or severe weather. The following is an example of step-by-step procedures:

1. Check to make sure that the system is fully operative.
2. Just prior to the start of the test, introduce yourself to the subjects and explain the purpose of the test.
3. Set up cameras for recording trials and check to make sure the cameras are operational.
4. Start cameras. Begin test.
5. Signal to the test subjects to initiate actions.
6. Record data as appropriate.
7. Repeat procedures as necessary.

8. At the end of trials, debrief subjects. Issue prepared questionnaires. Gather instrumentation and secure it properly.

Compilation and Storage of Data.   Steps for compiling data should describe what each data collector should do with data collection sheets after each trial. It should also describe where data sheets should be stored safely for later reduction and analysis. This is especially important if multiple data collectors are involved. If electronic data (i.e., computer-based data collection methods) are employed, it is important that each data collector have the same, or at least compatible, data collection software.

Data Reduction.   This section should explain how the compiled data is to be reduced for statistical treatment. For example, time lines in minutes and seconds across critical tasks might be retrieved from automated recording devices (i.e., time-tagged computer files). The data may further be summarized as measures of central tendency in preparation for statistical treatment. Interview data in the form of questionnaire ratings may be summarized as frequency distributions based on measures of central tendency. Once the data is reduced (i.e., extracted from data sheets or automated media) it is ready for analysis.

Data Analysis.   This part explains how data will be treated. Analysis methods will depend on the test. For example, a compilation of median scores on a survey may require a simple table of percentile distributions or perhaps a graph with accompanying statements that describe the sampling error and confidence interval. Experimental designs may require a description of the treatment effects and interactions among variables, with accompanying F statistics and degrees of freedom. Where human engineering and ergonomics checklists were used to report compliance with standard criteria (e.g., measurement of workstation layout and design) the analysis may involve a simple comparison of those measures to published criteria.

References.   Here the tester might include appropriate references, such as military and industrial standards, design, and user requirements documents for human engineering and human performance.

Requirements Test Matrix.   A requirements test matrix (RTM) is useful both as a management and as an actuarial tool. Earlier we stated that testing is (or should be) done with a very specific purpose in mind—to address test issues, such as work load and human engineering design adequacy. Human engineering and ergonomic criteria describe those aspects of human performance or design for which there are personnel and system requirements. From these, we derive measures that can be compared with standards. Although standards sometimes exist, they often must be established. A test matrix should capture all of this by tracing human factors test issues in the evaluation plan all the way to data collection procedures in the detailed test plan. One very important consequence of building an RTM is that it prevents unnecessary and redundant testing, and it shows that all human factors test issues have been accounted for.

Management and Resource Allocation Addendum

In addition to the top-level test plan, there are frequently companion test documents. One of these is called a management and resource allocation addendum, or test resource plan. Often the addendum is published separately, with copies going to management, to a company's financial officer, or to other professionals for special consideration. These documents typically contain the following sections:

Management.   Not all test plans require management documentation; it would be simple enough to identify one person as the human factors test specialist and leave it at that. However, in many programs in which human factors is but one part, project managers might want to know what role human factors testing and evaluation plays in the greater scheme of things. A block diagram showing whom the human factors specialist reports to, and who reports to the specialist, is usually sufficient.

Test Subjects.   Identify very briefly the number of test subjects required. Identify any special attributes, such as anthropometry, age, physical and mental abilities, and training.

Funding.   Funding requirements are perhaps the most difficult to determine. A checklist often helps. It should include labor hours for both test personnel and paid subjects, lease items, and expendables. It might also include travel costs, such as air and ground transportation, motel, and per diem expenses.

Test Articles.   Identify the required number of components, subsystems, or systems required for the human factors test. A complete description of the article should be provided (e.g., "a fully equipped automobile" or "a helicopter with rotors").

Test Facilities.   Identify any necessary test facilities. For example, a human factors test may require the collection of performance data in specially constructed climatic chambers.

Ancillary Support Equipment. Ancillary equipment includes such things as power generators, wiring harnesses, clipboards, pens and pencils, and most anything that falls outside of the other test support categories.

In summary, the human factors test plan can be thought of as a contract between management, the developer, and the tester as to the scope of testing and the information about the system's human-system interface that the tests can be expected to yield. Once the plan is approved, the test specialist may then conduct tests.

PREPARING HUMAN FACTORS TEST REPORTS

Generally, the test report will echo the structure and content of the test plan. In a sense, the former is the past tense of the latter: "We tested for these things, and this is what we found." The importance of a clear, concise test report is emphasized. It is, after all, the only viable means the test specialist has to communicate results, and it becomes a permanent record for those who follow. When planning a test, the tester should keep in mind that at some point the plan will serve as a basis for preparing the test report and that an incomplete or muddled plan will only contribute to an incomplete and muddled report. Test results can be reported in a number of ways, depending on the size and scope of the test, the phase of development in which the test was conducted, and the urgency of the report.

Preliminary Reports

The purpose of a preliminary report is to convey summary information immediately after testing and prior to the issuance of a formal test report, which might take months in preparation. Preliminary reports are often used to report safety-related problems or human-systems integration anomalies that might have a severe impact on cost and schedule. For whatever reason, this form of report is issued when the test specialist feels the information is so urgent that it cannot wait for the final report. Preliminary reports may be issued as early as the concept-development phase of systems development to include problem forecasts based on computer modeling of the planned system or think-tank sessions with subject matter experts (Parente, Anderson, Myers, & O'Brien, 1984). Such reports might also reveal unforeseen advantages of a proposed design.

A preliminary report may also be generated at the beginning of each successive stage of development. For example, a postproduction market survey might sample customer satisfaction concerning operability of a certain product. Regardless of the developmental phase, preliminary reports are most beneficial in assisting others of the engineering and science teams in solving complex systems problems while it is still cost effective to do so.

Preliminary reports are not without their own inherent liabilities. One drawback is that they are preliminary; some readers may simply put little weight in test results that seem premature. Another potential problem is that the data from which analyses are drawn may be weak in terms of the quantity and quality of information available about the human-system interface so early in development. The evaluator is cautioned to take special care in not stretching conclusions beyond reason. General conclusions about human performance based on feedback from subject matter experts, for example, are possible but should be caveated with the appropriate statements about the limitations those inferences subsume. Typically, a subject matter expert is sharing a best guess based on the same information the test specialist has—with the benefit of experience on prior or similar systems. Other data, such as measures of operator response over some set of tasks, would provide a stronger, more believable base for which to draw early conclusions about a system's human-system interface, even if those tasks are only an approximation of those of the end system.

The release of a preliminary report assumes that a minimum of testing has occurred. As it was stated earlier, a test report generally includes the elements of the test plan. Such a report should contain as a minimum the following:

1. Purpose of the test. States briefly why the test is being conducted.
2. System description. Identifies what is being evaluated and, more important, what is not being evaluated (e.g., training devices, nonproduction maintenance equipment).
3. Test constraints and limitations. Identifies test constraints relative to the system being developed. These might include such things as partial system completeness, mock-ups and simulators that provide only an approximation of the final system, nonoperational systems, and others. Many preliminary tests are limited to subsystems, or certain aspects of human performance, and should so be stated.
4. Test methodology: sampling, data collection and analysis. Summarizes test procedures.
5. Results and analysis. Includes summary data, and may include further statistical analyses of data.
6. Evaluation and recommendations. States the evaluator's conclusions about the preliminary test data and infers what those data mean with respect to the system under development. Includes a statement of deficiencies and recommends appropriate corrective actions, if appropriate. For human engineering design and human performance

issues that may have no obvious solutions, further testing may be required to investigate methods of correcting those problems.

In operational testing, preliminary test reports also serve to assist the evaluator in narrowing down certain aspects about the system that will require investigation as part of formal testing. For example, combat ergonomics assessments consist of a two-part methodology. The first part entails a preliminary examination of the system to identify specific human-system integration features from which formal test plans are developed. Human factors engineering features (i.e., workstation design, software usability, control and display layout and design, and others) are noted and measurements are made. The evaluator also will discuss with crews system operation and maintenance features from which crew surveys may be developed and administered as part of the formal testing activities. Also, human performance requirements are examined so that human time and error parameters can be included as part of formal testing procedures. Often, a preliminary combat ergonomics assessment, published far enough ahead of formal testing, will identify problems related to safety or perhaps features that cause or contribute to an unacceptable level of human performance degradation, so by the time formal testing occurs, the developer will have had the opportunity to address these problems.

Intermediate Reports: Interim, Summary, and Quick Look Reports

The distinction between intermediate reports and preliminary reports is that although the latter is typically issued in the early part of a developmental phase, the former may be issued at any time during the course of testing. The primary purpose of intermediate reports is to bring problems that may surface during the normal course of testing to the attention of managers, key engineers, and scientists involved in the development and operation of the system. In one sense, these reports are considered informal, a snapshot of the progress of the test. Test results may be couched in tentative terms, for example, "What appears to be going on at this particular point in testing is a work overload experienced by the operator when performing multiple, concurrent tasks. However, we should wait for further results before drawing firm and lasting conclusions, or before recommending design modifications."

Despite their informal tone, intermediate developmental and operational test reports are important tools, serving as feedback to the development and using communities. In developmental testing, they are one of the most important tools the human factors test specialist has to effect human-system interface improvements prior to any firm commitment to a system's final design configuration. These provisional reports have another valuable role that is unique to operational testing: to provide the user with a qualitative and quantitative indication of the system's capacity for meeting user requirements long before the system is fully developed and released for field use. It is not unusual to see a design department or a group of users holding their breath in anticipation of an intermediate test report and, once it is issued, to see them discuss with great animation the system anomalies discovered by the test specialist, or even unexpected enhancements in human performance.

Key points to remember about interim, summary, and quick look reports are that they should be concise and exact about what was tested and about any problems discovered, and they should contain the proper disclaimers regarding the limitations of the testing itself. An intermediate report should include as a minimum:

1. Person responsible for the report. This should be the same person who conducted the test and should include an office address, phone number, e-mail address, or website.

2. System or subsystem involved in the measurement. Describe briefly the system or subsystem involved in the test. This section should also state whether this was a dedicated human factors test or whether the measurement was conducted as part of another test (e.g., system safety test).

3. System anomalies and deficiencies, if any, and date and time of occurrence(s). State any out-of-tolerance events or unexpected performance by the operator maintainer or control loop feedback from the system. State any deficiencies, such as aspects of the human-system interface design or operating procedures that could be considered missing, out of place, and in error. An example would be a system's failure to provide an operator with the recommended amount of outside visibility in a newly developed automobile.

4. Raw data and analyses. Raw data may not always be an appropriate inclusion to an intermediate test report. However, the test specialist may feel that it is appropriate to provide reviewers with that data to double-check for themselves. Analysis of data should accompany the raw data and be identifiable as a separate section in the report. These data may include test logs.

5. Evaluator's conclusions. Here, the evaluator should state precisely only those conclusions supported by the measurements and available performance criteria. Where criteria and standards are not available, an intelligent statement of conclusions may be made based on the specialist's data, together with personal impressions about what really took place. In the latter case, however, use caution to avoid inferring beyond reasonable bounds.

6. Recommendations. If anomalies have been identified during human factors testing, and if enough information about the problem is available to correct it, a recommendation as to corrective actions is appropriate. If information about how to fix the problem is not available, or if the test specialist is not sure of the proper solution, the recommendation should be that further investigation is needed to effect the proper solution.

Final Reports

In some respects, a final report is the anticlimax to many weeks, months, or even years of testing. If the test specialist has done a proper job, any problems would have been reported and corrected long before the issuance of the final report—that is, preliminary and intermediate reports. So then, why a final report?

Two very good reasons for issuing a final report are, first, to pull together in summary form the entire human factors test and evaluation effort, documenting raw and treated data and the evaluator's conclusions and recommendations. What were the original test issues and were they satisfactorily addressed? Were there any deviations from the original plan? What were they? What are the results and evaluator's conclusions, in a nutshell? The second reason is to document the test for future research and development programs and for the benefit of others who may be faced with similar test problems. This latter point might also include the cost of the human factors test and evaluation effort for future reference.

In many cases, final reports are viewed as more than an historical account of events. In fact, many developmental programs wait to make important decisions, for example about advancing to the next phase of development, until the final test reports are in. For this reason alone timeliness of the final report is important.

Format. The final report should mirror the subheadings developed as part of the test plan—that is, it should include a description of the system that was tested, human factors test methods applied, results achieved, conclusions and recommendations drawn, and so on. The only addition might be a section explaining any divergence from the original test plan.

Results. Depending on the scope and complexity of the test, results should be summarized and included in the results section of the final report. Generally, results are organized around the original test issues, measures, and criteria. At a minimum, we recommend that the following be included as test results:

1. Summary human performance data (data that describes the performance of subjects tested). Raw data should be appended.

2. Human reliability predictions based on performance measures.

3. Table of human engineering design standards with corresponding measurements taken of the system under test. A discrepancies column of the table should show where criteria were *met, not met,* and *partially met.* Recommendations for resolution of discrepancies may be provided as part of the conclusions section that follows.

4. Table of test subjects' responses on surveys and questionnaires.

5. System aspects that were most significant in influencing human performance.

6. The contribution of human performance to overall system performance, in quantitative form, if possible.

In the first part of this chapter, we described how three types of human factors data are analyzed. Human engineering design measurements should be presented in tabular form, along with criteria in appropriate military or industrial standards and specifications or those in systems specifications or other design and user requirements documents. User acceptance and opinion data should be reported in summary fashion with appropriate ratings of user acceptance, confidence, and other attributes as measured. Finally, the human performance data can be reported in appropriate summary form. The report might state the evaluator's conclusions based on a comparison of measured human performance against criteria specified by the developer or, more specifically, in the design and user requirements documents.

Conclusions. The most important thing about a final report is the evaluation itself. It is answering "so what?" regarding the original test issues. As was explained in an earlier chapter, an evaluation draws on the test results to judge whether original test requirements were satisfactorily addressed. It would be simple enough to construct tables showing certain human engineering standards that were met, not met, or partially met; summary statistics charts showing how users were happy with how something operated; or statistics that operators performed with enough speed and accuracy to achieve minimal mission requirements. But what if the system barely passed human engineering standards? So what if users opined their happiness or unhappiness with the system? Or that test subjects performed satisfactorily over tasks? Does this mean that the system passed?

To address the test issue typically requires more than verifying whether or not certain hardware standards were met, as discussed earlier, or whether test subjects who may or may not have been representative of the user population were happy, or even if they performed satisfactorily. An example comes to mind where a system marginally met all of the appropriate standard human engineering design criteria; when measured over some set of performance criteria, subjects performed adequately. Yet they responded on postevent questionnaires that they absolutely hated the new system and had a very low confidence in it. How would you evaluate these seemingly contradictory results?

Satisfaction of established human engineering design criteria, human performance, and the opinions of test subjects should be considered within a total systems framework, never individually, unless they are the only data available. In our example, the evaluator gave a positive evaluation because the test issue weighted user performance over the other testing parameters (i.e., human engineering design and user confidence), and the system proceeded into production. Later, as the users became more familiar with the system's operation, their opinions changed for the better.

In summary, tests may be reported in many different ways depending on size of the test, phase of system development, and the urgency of the report. In our multiphased approach, we describe preliminary test reports, which are important for conveying feedback to the development community early during development and developmental phases; interim test reports, which provide snapshots of testing in progress; and a final human factors test report, which documents the entire testing process and provides a final evaluation of the human-system interface data. The results section of the final report should summarize pertinent human-system interface data across all three dimensions, if possible: human engineering design criteria, user acceptance and opinion data, and human performance. The conclusions section should provide a succinct evaluation based on the three categories of human factors data. Comparisons are used to determine whether problems indicated by users or failures to comply with design standards have a significant effect on overall system performance. Human performance data might also identify areas in which human engineering design improvements may offer a high payoff in terms of improved system performance or reliability or in reduced training and support costs.

REFERENCES

AFOTECI-99-101. (1993). *Developmental test and evaluation.* Kirtland AFB, NM: U.S. Air Force Operational Test and Evaluation Center: Author.

American Society for Testing and Materials (ASTM)-1166. *Standard practice for human factors engineering design for marine systems, equipment, and facilities.* West Conshohocken, PA: Author.

Department of Defense, DOD Directive 5000.1. (1991). *Defense acquisition.* Author.

Department of Defense (DOD) Handbook (HDBK)-743. *Anthropometry of U.S. military personnel.* U.S. Army Soldier Systems Center, Natick, MA: Author.

Department of Defense, DOD Instruction 5000.2. (1991). *Defense acquisition management policies and procedures*. Author.

DOD-HDBK-759. *Human factors engineering design for army materiel*. U.S. Army Aviation and Missile Command. Huntsville, AL: Author.

International Organization for Standardization (ISO) 9241 Series. *Ergonomics of VDT workstations, Part II*. Geneva, Switzerland: Author.

Juran, J., & Gryna, F. (1980). *Quality planning and analysis* (2nd ed.). New York: McGraw-Hill.

Meister, D. (1986). *Human factors testing and evaluation*. New York: Elsevier.

Military Standard, MIL-STD-1472. (1989). *Human engineering design criteria for military systems, equipment and facilities*. Author.

Military Specification, MIL-HDBK-46855. (1994). *Human engineering requirements for military systems, equipment, and facilities*. Author.

National Air and Space Administration (NASA) Standard 3000. *Man-systems integration standard*. Washington, DC: Author.

Nuclear Regulatory (NUREG) 0070 Series. *Human-system interface design review guideline*. Washington, DC: Author.

NUREG/CR-5908. *Advanced human system interface design review guideline*. Washington, DC: Author.

O'Brien, T. (1982). *Critical human factors test issues for the M1E1 Abrams Main Battle Tank*. Aberdeen Proving Ground, MD: U.S. Army Materiel Systems Analysis Activity.

Parente, R., Anderson, J., Myers, D., & O'Brien, T. (1984). An examination of factors contributing to delphi accuracy. *Journal of Forecasting*, November.

Peacock, B., & Karwowski, W. (Eds.). (1993). *Automotive ergonomics*. Bristol, PA: Taylor & Francis.

U.S. Army Test Operating Procedure 1-2-610. Army Developmental Test Center, Aberdeen Proving Ground, MD: Author.

U.S. Department of Transportation, Federal Aviation Administration, Report CT-96/1. (1996). Human factors design guide for acquisition of commercial-off-the-shelf subsystems, non-developmental items, and developmental systems: Office of the Chief Scientific and Technical Advisor for Human Factors (AAR 100). National Technical Information Service, Springfield, VA: Author.

Chapter 5

Human Performance Testing

Douglas H. Harris
Anacapa Sciences, Inc.

The objective of this chapter is to provide a source of information and guidance for the development and use of human performance tests. The chapter was produced in parallel with the ongoing development and assessment of performance tests; as a consequence, it is based on actual experience in performance test design, administration, analysis, and evaluation. The chapter also encompasses earlier work on human performance testing, such as that reported by Guion (1979), Siegel (1986), and Smith (1991).

THEORY AND FOUNDATIONS

Definition

A performance test is any standardized, scorable procedure in which performance capability is demonstrated on tasks that relate directly to tasks performed on a job. Performance tests, also commonly called practical tests and work sample tests, are thus distinguished from other tests by their close relationship with job tasks. Within the relatively broad definition given above, performance tests might take a wide variety of different forms, limited only by the nature of jobs and the imagination of test developers.

Applications

The principal incentive for the development and implementation of human performance testing is to enhance the effectiveness of operations in systems and organizations. Applications of performance testing include

(1) qualifying personnel for assignment to critical tasks and for diagnosing individual strengths and weaknesses for the purpose of providing remedial skills development; (2) evaluating the design of equipment, procedures, and systems, particularly where human performance is critical to successful operation; and (3) assessing individual performance in training programs and evaluating the training programs themselves.

Theory and History

The history of human testing has been principally one of assessing individual differences through the development and administration of pencil-and-paper tests. For this reason, most of psychometric theory—the theory of tests and measures applied to people—has used the written test as the prototype. (A summary of psychometric theory based on written tests can be found in Gulliksen, 1987.) Although performance tests have had an equally lengthy history, there has not been a corresponding effort expended on the development of theory or practical guidelines for these tests. Thus, with the exception of a relatively small number of studies, the knowledge base for performance tests derives mainly from the extrapolation of research on written tests.

Performance testing has been used in one form or another for more than 50 years. Adkins (1947) provided an early description of the approach in his report on the construction and analysis of achievement tests. Asher and Sciarrino (1974) reviewed the history of performance testing through 1973 and concluded that this approach was consistently equal to or superior to other approaches used for predicting performance on jobs. Other approaches assessed by them included pencil-and-paper aptitude tests, tests of intelligence, personality tests, psychomotor tests, and biographical data. The state of knowledge of performance testing was reviewed and described more recently by a panel of the National Research Council in its report *Performance Assessment for the Workplace* (Wigdor & Green, 1991). A more extensive discussion of the theory and history of performance testing can be found in this reference.

Advantages

As suggested by the theory and history of the approach, performance testing has many advantages for evaluating the capabilities of personnel, equipment, procedures, and systems. These include the following:

1. Being superior predictors of human performance relative to other assessment approaches.

2. Possessing an undeniable theoretical basis for test validity by providing a point-to-point correspondence between test task and job task.

3. Reflecting the complex interactions among various capabilities required for successful performance.

4. Reflecting various noncapability factors that contribute to level of job performance, such as experience and work habits.

5. Possessing a high degree of face validity because they consist of a sample of the job—they look valid and, as a consequence, they are able to gain acceptance by those who take and administer them.

Test Components

The following eight components constitute and define a performance test for any given application.

1. The test objective is the explicit specification of the overall purpose of the test, such as evaluation of procedures or qualification of personnel.

2. The task breakdown is the specification of the steps or actions required to perform the test, typically in the order in which they are to be performed.

3. Test conditions include the physical environment, restrictions on communications, time limitations, and the resources provided for the period of testing.

4. Test procedures and instructions are rules that are designed to guide test taking and to minimize variability in test taking from one individual to another, such as rules covering test procedures, materials, recording of responses, and scoring.

5. The equipment and materials required for testing might include hardware, software, reference documents, supplies, specimens, and reporting forms.

6. Grading units are the essential elements of a test, such as test items, that provide the basis for measuring and scoring performance.

7. Scoring criteria and methods provide the system for obtaining responses from test takers, transforming responses to test scores, and applying criteria for determining test outcome, such as qualification for a task, passing a training course, or determining the usability of a system.

8. Administration procedures address the specification of instructions, practices, and procedures to assure their consistent application.

Characteristics of a Useful Test

A performance test must have several characteristics to provide assurance that it can be used with confidence for its intended purpose. Test validity refers to the extent to which the test measures what it was designed to measure. Reliability, a necessary requirement for validity, is the extent to which the test consistently measures the performance being assessed, minimizing the error of the measurement process. Face validity is realized when the test appears to measure what it purports to measure, thus providing assurance to test takers and gaining acceptance for the test. Finally, the test must be designed to be administered within the practical constraints on its use, such as cost, time, safety, and portability.

Test Validity.   There are three types of test validity—content, criterion related, and construct—as summarized in this paragraph. Each is defined in terms of the methods used to assess the extent to which a test measures what it purports to measure. Content validity employs expert judgment to assess how well the content of the test relates to whatever the test is designed to measure. Criterion-related validity is assessed by comparing test scores with one or more independent criteria, such as measures of job performance, using statistical correlation techniques. Construct validity is the relationship determined, usually by means of statistical factor analysis, between the test and some quality or construct that the test is designed to measure (pilot response to aircraft emergencies, for example).

The most meaningful and useful type of validity for tests of human performance is content validity, in which validity is assured by careful, point-by-point matching of test grading units to the critical tasks of the operation or job. The process of test-to-task matching should involve experts in the performance of the tasks. When this process is completed systematically, there is typically little question about the basis for test validity. For additional information and discussion on the concepts of and approaches to test validity see Schmitt and Landy (1991).

Test Reliability.   The upper limit of test validity is determined by test reliability—the extent to which the test is consistent in its measurement. Even if the match between test and task elements is perfect, test validity might be unacceptable because of low reliability. Because the test-to-task matching can typically be done satisfactorily through the application of expert judgment, the critical consideration in the development of performance tests is that they meet minimal standards of reliability. An operational definition of test reliability is given by the following expression:

$$\text{Test reliability} = \frac{\text{True variability} - \text{Error variability}}{\text{True variability}}$$

In this expression, True variability is the variability among test scores due to true differences among the capabilities being tested; Error variability is the variability among test scores due to all other factors. These other factors might include the differential effects of test instructions, physical conditions, motivation, distractions, and other test conditions on individual test scores—factors which may cause measured capability to deviate from true capability.

This definition serves as the basis for calculating an index of reliability that has a meaningful range of 0 to 1. At the extremes, if Error variability = 0, then Test reliability = 1, and if Error variability is as great as True variability, then Test reliability = 0. A performance test with a low reliability will not be useful even if, as discussed earlier, there were a perfect and complete match between test items and task requirements. Low reliability will degrade test validity, reduce the confidence that can be placed in the test, and reduce the accuracy of decisions based on test results. For most applications of performance testing, the index of reliability should be equal to or greater than 0.80.

There are two fundamental approaches to estimating the reliability of a performance test: (1) correlational and (2) analysis of variance, with several variations of each approach. These approaches and the procedures for their application are presented in detail elsewhere (Guilford, 1954; Gullikson, 1987; Zeidner & Most, 1992).

Standard Error of Measurement.   Measurement error figures significantly in the definition of test reliability and in some of the procedures employed for calculating test reliability. The statistic, standard error of measurement, although closely related to reliability, should be considered as a separate characteristic of a performance test because it provides information beyond that provided by the index of reliability. It also provides the basis for establishing confidence levels for test results. Standard error of measurement is a function of test reliability and the variability of test scores, and is both defined and calculated by the following expression (Zeidner & Most, 1992):

$$SE = SD \; [\text{Square Root} \; (1 - R)]$$

Where: SE = Standard error of measurement
SD = Standard deviation of test scores
R = Test reliability

As was explained earlier, test reliability is a function of the ratio of true variability to error variability—that is, the variability of test scores due to true differences between individuals relative to that of the error variability introduced by the testing process (measurement error). The greater the ratio of true to error variability, the greater will be the reliability of the test, and the more reliance we can have that the test will provide consistent results from one administration to another.

However, the reliability index based on this ratio says nothing about the variability of test scores. Within limits, test reliability can be the same whether there is high variability or low variability among the test scores. Thus, while the reliability index tells us about the consistency of the test, standard error of measurement tells us about the precision of the test. The smaller the standard error of measurement, the more precise the test.

Suppose that we administered equivalent tests, under identical conditions, to the same individual 100 times. Because we are testing the same individual, the mean score would be our best estimate of the person's true score; deviations of scores from the mean would be due to variations (error) contributed by the testing process. We measure the standard error of measurement directly by calculating the standard deviation of the 100 test scores about the mean score of 70, and we obtain a standard error of 11.70. This result is shown in the top part of Fig. 5.1. The distribution of scores around the mean is also illustrated graphically.

Because the shape of the distribution is normal, we know that 90% of the scores are to the right of the value $= 70 - 1.28 (11.70) = 55$. That is,

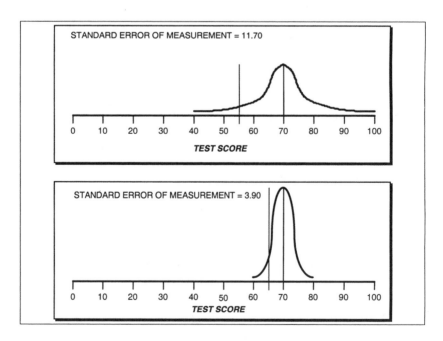

FIG. 5.1.   Effect of differences in standard error of measurement.

the point 1.28 standard errors to the left of the mean marks the 10/90 dividing line—10% of the scores are to the left of that point, and 90% of the scores are to the right. Because 90% of the scores are greater than 55, we are 90% confident that the individual's true score is greater than 55. This example illustrates the principal value of standard error of measurement—providing a basis for determining the degree of confidence one can place in test results.

The effect of standard error of measurement on the degree of confidence one can place in a test score is further illustrated by the illustration in the lower part of Fig. 5.1. In this example, the mean is also 70, but the standard error is only 3.90. As before, the 90% confidence limit is determined by calculating the test score that is 1.28 standard errors to the left of the mean: 70 – 1.28 (3.90) = 65. Because we have determined by this calculation that 90% of the obtained scores are equal to or greater than 65, we can say that we are 90% confident that the individual's true score is equal to or greater than 65.

Thus, the 90% confidence limit of the first example (55) was much lower than the 90% confidence limit of the second example (65) because of differences in the standard error of measurement. We can say, then, that the test employed in the second example had greater precision than the one employed in the first example because the lower-bound confidence limit was closer to the average value.

Face Validity and Acceptance. Performance tests enjoy an advantage over other types of tests in that they have the appearance of validity; that is, they have face validity. This is because the test usually looks very much like the tasks being addressed by the test. Thus, if the test matches the tasks well, face validity and acceptance are not likely to be a major problem. On the other hand, even minor discrepancies in the content of the test and in the procedures employed in testing sometimes raise concerns among those taking the test.

HOW TO DESIGN AND ASSESS
A PERFORMANCE TEST

To be useful, a performance test must be designed specifically for its intended purpose, satisfy certain fundamental requirements of tests and measures, be acceptable to those giving and taking the test, and meet the practical limitations of test administration. Because these issues provide the foundation and direction for test development, they should be addressed early in the development effort. Moreover, desired test characteristics should be formulated and stated explicitly prior to initiating the test construction effort.

Defining and Scoring Grading Units

Central to the design of any performance test is the way grading units are specified and the way responses to grading units are counted to arrive at a test score. In contrast to written tests, there are many possible combinations of grading units and counting schemes that might be employed. Written tests commonly consist of a set of questions, with each question serving as a grading unit. The test score is determined by counting the number of questions answered correctly. Performance tests, on the other hand, are like the actual job. Grading units and counting systems are patterned after successful job performance. As a consequence, a wide range of methods might be considered for defining and scoring grading units. The following examples cited by Guion (1979) illustrate the variety of possibilities.

For rifle marksmanship tests, a grading unit was defined by a specified number of rounds fired from a specified position at a specified distance from the target. The test score was obtained by counting the number of shots in the target and by weighting each relative to its distance from the center of the target.

A dental hygienist's performance test consisted of performing several task sequences. Each step of each sequence was considered as a separate grading unit, and was rated by an observer using a checklist. A test score was obtained by adding the observer ratings for the separate steps.

In testing lathe operator performance, each operator produced a taper plug in conformance to engineering specifications. Grading units consisted of the eleven characteristics specified for the taper-plug product. In scoring the test, one point was counted for each specification met or exceeded by the resulting product.

Performance of printing press operators was assessed by a test in which grading units were defined as violations of good printing practices and safety rules. Using a checklist, an observer recorded each violation as the operator performed a specified set of tasks. The test score was obtained by subtracting the number of violations from a perfect score.

As suggested by the preceding examples, there are two principal aspects of performance that can serve as the basis for measurement—the process and the product. In these examples, the tests for dental hygienist and printing-press operator were based on the work process. Grading units were defined by steps performed in representative tasks of hygienists and were defined by operator violations in performing printing tasks. In the tests for rifle marksmanship and lathe operators, grading units were defined and scores determined from the product produced—the pattern of holes in the target and the characteristics of the machined taper plug.

Attaining Adequate Test Reliability

As discussed earlier, the principal problem facing the developer of a performance test is that of attaining a satisfactory level of test reliability. Although reliability can be influenced by many different characteristics of a test, the most common solution to inadequate reliability is increasing the length of the test (i.e., increasing the number of grading units). This solution has been well studied, and mathematical formulas, such as the Spearman-Brown formula (Guilford, 1954), have been developed and used for decades to predict the effect on reliability of changing the length of a test. Assuming that the grading units added to a test are comparable to the grading units already in the test, this formula calculates the reliability of the longer test.

You might face some severe limitations, however, on providing a test of adequate length or in increasing the length of a test to increase its reliability. Test materials representative of conditions encountered on the job might be difficult to obtain or costly to manufacture, the amount of time and funds available for testing are likely to be limited, and the logistics required for testing could be burdensome and disruptive of normal operations and personnel assignments.

Increasing Reliability by Subdividing Grading Units

A potentially effective means of attaining useful levels of performance-test reliability is by subdividing grading units rather than increasing the number of grading units—the number of grading units is increased while other aspects of the test remain the same. Studies of operator performance in ultrasonic inspection (Harris, 1996) provided sufficient data to permit an evaluation of this approach. Results of that evaluation were positive and have significant implications for the design of performance tests.

The effect of subdividing grading units was assessed empirically using a performance test of the ultrasonic detection of defects in pipe welds in nuclear power plants. Performance tests involving the examination of 10 pipe-weld specimens were completed by each of 52 experienced ultrasonic operators as part of their qualification for performing tasks of this type on the job. Subdivision of grading units (each section of pipe was divided into 20 scorable units, increasing the number of grading units for the test from 10 to 200) was found to increase the reliability of the test from 0.28 to 0.92, to decrease the standard error of measurement of the test from 13.81 to 1.35, and to decrease the 90% confidence band around test scores from ±22.60 to ±2.20. Moreover, the increase in reliability was predicted by application of the Spearman-Brown prophecy formula, the method commonly employed for predicting the effect of increasing test length. In this manner, increased reliability can be accomplished without any additional testing burden—time, costs, materials, and so on.

Assignment and Interpretation of Test Scores

A key issue in the assignment of scores to individuals is whether to provide a binary score (pass or fail), a multilevel score (such as A, B, C, D, or F), or a quantitative score (such as probability of defect detection, time required to machine a satisfactory part, or average percentage deviation from designated flight path). If the objective is solely to qualify personnel for the performance of specified tasks, the binary score might be sufficient. However, even in this case, a multilevel or quantitative score might be useful in determining the action to be taken with those who do not pass. For example, an individual who barely misses a passing score, getting a 78 when passing is 80, might be given another opportunity to qualify right away. On the other hand, an individual who does not come close, scoring a 63, might be required to take additional training along with a specified amount of practice before being given another chance to qualify.

Testing the Test

The design and construction of a performance test involves addressing and resolving many technical issues. At this time, test technology is not sufficient to ensure that the initial product of such efforts will always be adequate in terms of test validity, reliability, and acceptability. For example, a test with low reliability will not serve the purpose for which it was designed. As a consequence, it is likely to have a negative, rather than positive, impact on individual careers, effectiveness of systems and organizations, and efficiency and safety of operations. Consequently, as in the case of most any newly designed system, every performance test should be evaluated and refined on a sample of the population for which it is intended before it is put into use. Testing the test involves the application, on a limited basis, of the procedures addressed in the remaining portion of this chapter.

HOW TO ADMINISTER A PERFORMANCE TEST

The essential strategy for administering a performance test is to match test conditions and procedures as closely as possible to job conditions and procedures. For this reason, designing the details of test administration requires in-depth knowledge of the job tasks, gained through experience or analysis. However, deviations from operating conditions and procedures are usually required to meet the controlled conditions needed to provide reliable, valid test results. Therefore, expertise in personnel testing is also required. This section provides principles and procedures for test administration based on the experience of testing specialists in the development and administration of performance tests.

Standardizing Testing Conditions

The conditions under which performance tests are administered can have a significant effect on the reliability, validity, acceptability, and precision of test results. Critical issues to be considered in the establishment of testing conditions include computer-aided test administration; facilities, equipment, and materials; test instructions; supporting information; provision of job aids; and time requirements. Overriding these specific issues is required to standardize the testing conditions for those being tested to minimize sources of variability other than true differences in performance. Standardization requires forethought in the planning and specification of testing conditions and in the discipline with which the test administrator adheres to these specifications.

Using Computer-Based Testing. The use of computer technology might help standardize testing conditions and generally facilitate the testing process. The extent to which this technology might be employed is, of course, dependent on the nature of tasks for which the testing is conducted. Even so, computer-based testing, including the use of synthetic environments, is now applicable to a wide range of tasks from typing to power-plant maintenance to spacecraft operation.

For those tasks in which computer-aided test administration in whole or part is feasible, shell software is available to ease the burden of standardizing testing procedures. As the name implies, shell software provides a predeveloped framework for the system. The test developer starts with the shell and fills it to meet the specific requirements and objectives of the specific testing system. Instructions, training, practice data, written tests, practical performance tests, test scoring, recording of results, and even the statistical analyses designed to evaluate and improve the system can be incorporated in the shell. Theoretically, the whole testing process can be self-administered. However, experience with computer-aided testing systems indicates that best results are obtained when a qualified test administrator is present to address problems that might arise.

Providing Appropriate Facilities, Equipment, and Materials. Facilities, equipment, and materials are major factors in the establishment of testing conditions and can greatly influence test results. In providing these needed elements, one is faced with making trade-offs among competing considerations. Under ideal conditions, the equipment and materials used by the test takers in performance testing would be identical for each individual. For example, ultrasonic instruments and search units from different manufacturers will add variability to performance measures. On the other hand, an individual's performance may be hindered if not permitted to employ a system that he or she is experienced in using or if not provided sufficient training and practice on a new system. Thus, standardization of test conditions must be traded off against the possibility of handicapping a person in demonstrating qualifications or in accurately reflecting system conditions.

Providing Test Instructions. Performance testing requires instructions to test takers regarding the specific steps involved in the testing process and how each of these steps is to be completed during testing. These instructions must also emphasize any differences that might exist between the way the task is performed on the job and the manner in which it is to be performed during the test. For example, if results are to be recorded in a manner that differs from how results are typically recorded, this difference needs to be highlighted in the test instructions. In designing and providing test instructions, the objective is to communicate precisely what is expected of the test taker and to provide exactly the same information to each test taker. Thus, the content of test instructions and the method of providing test instructions constitutes important conditions under which testing takes place.

Instructions to test administrators and test takers can be a significant source of measurement error. Instructions convey to administrators the rules under which testing is to take place, covering all aspects of test conditions, procedures, equipment, materials, reporting, and scoring methods. Test takers require instructions to understand the test objectives, what they are to do, and how they are to be evaluated. Research has shown that test instructions have the potential to contaminate the testing process if they are incomplete, unclear, or incorrect. Also, test results can be biased when instructions are provided somewhat differently to different individuals.

Providing Supporting Information. As part of the performance testing process, or as part of the preparation needed for testing, test takers will require some supporting information. The amount of information will range from small to large, depending on the objectives and nature of the testing. At the small end of the continuum, supporting information might consist of only a few instructions about the task to be performed, such as in a data-entry task to test a new data-entry device. At the large end would be the instructions to a flight crew in preparation for a mission-planning exercise.

Information can be provided in a variety of forms, from a simple data package containing text and drawings to a database implemented by computer that provides a complete course of preparatory instruction including practice tasks. As a consequence, the extent and nature of the supporting information provided to the test taker can be a significant condition under which the test is administered.

Using Performance Aids. Tasks are aided and supported on the job with various types of information. As discussed previously, drawings, photographs, maps, and radiographs might be required for task performance. Explicit instructions and other aids might also be available for system set up and calibration. The issue is to what extent this type of information should be provided in support of the test task. Will more error variability be added in providing or not providing these types of aids?

In a previous study of decision aiding (Harris, 1992), aids provided on worksheets for performance tests were effective, but they were not used all the time by all operators. Also, there was no consistent pattern among operators relative to when they did and did not use the aids. The results, however, left little doubt about the value of the aids. As shown in Fig. 5.2, when the aids were used, the percentage of successful outcomes was significantly greater than when aids were not used.

These results raise an issue to be addressed in deciding on the extent to which job aids should be provided for performance testing. Does the variability contributed by the potential differential use of job aids reflect true differences in performance capability, or does it reflect error introduced into the testing process? The demonstrated ability to use job aids effectively is likely to be component of job proficiency. Therefore, the effective use of job aids contributes to the true measure of performance rather than to measurement error. On the other hand, the inconsistent use of aids is likely to contribute to measurement error.

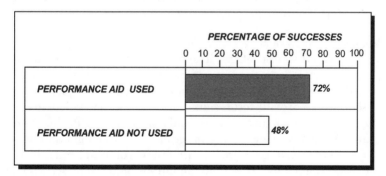

FIG. 5.2.   Percentage of successes when performance aids were used compared with when performance aids were not used.

Setting Testing Time Requirements. The amount of time required for testing will normally be greater than the time required for performing the same or similar tasks on the job. More time is needed to understand and comply with the added rules and procedures imposed by the testing process. In addition, because performance testing will serve as the basis for decisions likely to influence the design of new equipment and systems, or an individual's career progression and remuneration, it is advisable to assure that no one is put at a disadvantage by having too little time to properly complete test tasks. On the other hand, time limits are needed to assure that testing can be scheduled in advance and can be completed within prescribed budgets.

Time requirements for a new performance test can seldom be established accurately without a trial administration of the test to a representative sample of test takers. Obtaining time lines from a trial administration is part of the test assessment process discussed earlier in this chapter.

Applying Testing Procedures

Testing procedures should be designed with two objectives in mind: (1) to maximize the proportion of true performance represented in test scores and (2) to minimize the amount of error introduced into the test score by the testing process. In this section, procedures are introduced and discussed in support of these objectives.

Responding to Grading Units. Responses to grading units serve as the basis for assessing a test taker's capability for performing the task. They contain all the information that will be used to calculate a test score used in arriving at a decision, such as accepting Design A rather than Design B or to qualify or not qualify the individual to perform the task. Test administration procedures and instructions to test takers can help assure that appropriate responses are obtained. These procedures will depend specifically on how the grading units are defined and scored, as discussed earlier.

Providing Test Results to Test Takers. When and how should test results be provided to test takers? The answer, of course, depends on the purpose of the test, how the test is administered and scored, and on the specific limitations of the process employed. For example, computer-aided scoring can provide test results quickly and with a high degree of accuracy (assuming that there are no undiscovered bugs in the computer program). Manual scoring, on the other hand, takes longer and might be more subject to human errors of interpretation, counting, and calculation.

The answer depends also on the limitations imposed by security measures employed to assure the integrity of the test and to minimize cross talk among subjects in a system evaluation. Security measures are required to ensure that the test is not compromised through the release of sensitive information about the test. For example, knowledge of the correct responses to specific grading units could give some test takers an advantage over others or bias the results of the study. Thus, even though some information would be helpful to the test taker in understanding the basis for a test score, it might not be provided because of the possible compromise of the integrity of the test.

Establishing Rules for Borderline Failures.   A performance test score is seldom a pure measure of performance. The test score might also reflect factors that contribute errors and inaccuracies to the testing process. For this reason, in testing individual capabilities, individuals with test scores very close to but below the criteria for passing may fail because of errors inherent in measuring their capabilities rather than their lacking the required capabilities. The best remedy, of course, is to use tests in which the standard error of measurement is very small. As illustrated by the example in Fig. 5.3, a borderline region can be established below the passing score for the test. If a test score falls into this region, the candidate may examine additional grading units in an attempt to score above the passing criterion.

In the example, the borderline region encompasses test scores from 85 to 90. Because the obtained test score of 87.5 falls into this region, the candidate is given the opportunity to continue testing. The figure shows three possible outcomes that might result from additional testing. The individual (1) scores above the passing score and passes, (2) stays in the border-

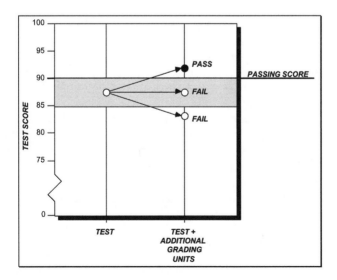

FIG. 5.3.   An approach to obtaining more accurate test results for borderline failures.

line region and fails, or (3) scores below the borderline region and fails. If testing is stopped at this point, a score that remains in the borderline region would be considered a failing score.

Continued testing has the effect of increasing the length of the test. Because increasing the length will increase the reliability of the test and reduce the standard error of measurement, continued testing will lead to a more accurate estimate of an individual's true capability. In other words, continued testing will increase the proportion of the test score that reflects true ability relative to measurement error. Thus, if an individual is capable of passing the test, continued testing will provide an additional opportunity to demonstrate that capability. If an individual is not capable of passing, continued testing will demonstrate that as well. In short, continued testing of borderline failures will provide greater confidence in the ultimate outcome, whatever that outcome is.

Setting Guidelines for the Participation of Support Personnel

Personnel other than the test taker might be required to support the testing process. Examples of the types of support personnel that might be employed include (1) technicians to set up, maintain, and troubleshoot equipment employed during testing, (2) computer programmers to modify, maintain, and debug software of computer-based systems in conjunction with testing, (3) craft workers to rig and move materials during testing, and (4) personnel to assist with data recording during the data acquisition phase of testing.

In general, the participation of support personnel will complicate the test environment and add opportunities for measurement error. In designing the test, designers should limit the participation of support personnel to situations where it is absolutely necessary. When support personnel are employed, their characteristics should be explicitly defined, and their role should be limited and explicitly described.

The general guideline for the utilization of support personnel during qualification testing is that these personnel should have no influence on the responses of the test taker or, if some influence is unavoidable, that it be the same for all test takers. The procedures provided for the participation of support personnel should preclude the possibility that the skills, knowledge, and general effectiveness of these personnel have any effect on test results.

Selecting the Test Administrator

The test administrator is the person formally assigned the responsibility for establishing the required testing conditions, assuring that specified testing procedures are followed, and safeguarding and disseminating testing results and records. Qualifications of the test administrator, or administrators, should match the test administration tasks. Although this guideline appears obvious, meeting it requires a detailed understanding of the tasks to be performed and the skills and knowledge required by each of the tasks. For example, we need to determine whether or not the test administrator must be proficient in the specific task being tested. If so, there is no alternative but to assign such a person; if not, there is little point in doing so. It will limit the number of personnel from which an administrator can be selected, it will add unnecessary costs to the testing process, and it will tie up a person who might be better used elsewhere.

SUMMARY

This goal of this chapter was to provide a source of information and guidance for the development and use of human performance tests—standardized, scorable procedures in which human performance capability is demonstrated on tasks that relate directly to tasks performed on a job. The chapter outlined the theory and foundations for performance testing, including history, applications, advantages, test components, and the characteristics of useful tests. Specific guidance was provided for designing and assessing performance tests, with special emphasis on the greatest challenge to the test developer—attaining adequate test reliability. The final part of the chapter addressed how to administer performance tests, particularly in applying the critical strategy of establishing standardized testing conditions. Guidelines were provided for computer-based testing, facilities and materials, instructions, supporting information, performance aids, time requirements, and testing procedures.

Although the number of considerations in the development and administration of performance tests might seem intimidating, the attempt here is to provide sufficient breadth and depth of coverage to be helpful in a wide range of test and evaluation applications. Also, as with so many other technical endeavors, the success of human performance testing is in the details of their development and application. Hopefully, some of the details provided in this chapter will help extend the success of the past 50 years of human performance testing into the coming century.

REFERENCES

Adkins, D. (1947). *Construction and analysis of achievement tests.* Washington, DC: U.S. Government Printing Office.
Asher, J. J., & Sciarrino, J. A. (1974). Realistic work sample tests: A review. *Personnel Psychology, 27,* 519–533.

Guilford, J. P. (1954). *Psychometric methods* (2nd ed.). New York: McGraw-Hill.

Guion, R. M. (1979). *Principles of work sample testing: III. Construction and evaluation of work sample tests* (NTIS AD-A072 448). Washington, DC: U.S. Army Research Institute for the Behavioral and Social Sciences.

Gulliksen, H. (1987). *Theory of mental tests*. Hillsdale, NJ: Lawrence Erlbaum Associates.

Harris, D. H. (1992). *Effect of decision making on ultrasonic examination performance* (EPRI TR-100412). Palo Alto, CA: Electric Power Research Institute.

Harris, D. H. (1996, September). Subdivision of grading units can increase the reliability of performance testing. *Proceedings of the 40th Annual Meeting of the Human Factors and Ergonomics Society*, 1032–1035.

Schmitt, N., & Landy, F. J. (1991). The concept of validity. In N. Schmitt & W. C. Borman (Eds.), *Personnel selection in organizations*. San Francisco: Jossey-Bass.

Siegel, A. I. (1986). Performance tests. In R. A. Berk (Ed.), *Performance assessment: Methods and applications*. Baltimore: Johns Hopkins University Press.

Smith, F. D. (1991). Work samples as measures of performance. In A. K. Wigdor & B. F. Green, Jr. (Eds.), *Performance assessment for the workplace, Volume II: Technical issues* (pp. 27–74). Washington, DC: National Academy Press.

Wigdor, A. K., & Green, B. F., Jr. (Eds.). (1991). *Performance assessment for the workplace: Volume I*. Washington, DC: National Academy Press.

Zeidner M., & Most, R. (1992). *Psychological testing*. Palo Alto, CA: Consulting Psychologists Press.

Chapter 6

Measurement of Cognitive States in Test and Evaluation

Samuel G. Charlton
Waikato University
and Transport Engineering Research New Zealand Ltd.

The measurement of cognitive states is an essential ingredient of HFTE. As one of the big three psychological research paradigms (i.e., psychophysiology, behavioral performance, and cognitive states), they provide an important link between laboratory theory and field application. There is, however, an important distinction to be made in the case of cognitive states in that they must be measured indirectly. Because cognitive states, such as workload, fatigue, and pain, are subjective psychological experiences, their momentary presence or absence cannot be readily quantified by objective means. Instead, they must be operationally defined in terms of indirect measures, such as verbal reports, or through objectively measurable behavioral and psychophysiological indices.

Although this inherent subjectivity led an entire school of psychology to abandon the study of cognition in favor of behaviorism, few researchers today would reject out of hand the importance of cognitive processes, such as perception, attention, memory, and states such as mental workload and fatigue. In HFTE, as well, the measurement of cognitive states has become ubiquitous. In part, it is because cognitive states such as mental workload and fatigue have considerable predictive and explanatory utility, such that we are able to use them to predict changes in task performance. The principal dangers associated with these measures arise from definitions that are inherently circular and from measurement techniques that result in biased or unreliable data. Although these are not trivial obstacles for the effective measurement of cognitive states, they are at least apparent to, and addressed by, most researchers using subjective measures.[1] (The inherent subjectivity pervading objective data-collection methodologies is much less readily recognized or controlled.) This chapter will describe the measurement issues and techniques surrounding two of the most widely used cognitive measures in HFTE today, mental workload and situation awareness. There are certainly many other cognitive states relevant to HFTE (boredom, psychological stress, and fatigue, to name a few), but coverage of all of them is beyond the scope of any single chapter. The measurement issues associated with these other states, to a large degree, are common to assessment of cognitive states in general, and thus we have elected to focus on the two with the most widespread use: mental workload and situation awareness.

DEFINING MENTAL WORKLOAD AND SITUATION AWARENESS

The issues surrounding mental workload and situation awareness have, in many important aspects, been cast from the same dye. Both subjects have their origins in the everyday vernacular of system operators, and yet both continue to puzzle and exercise human factors researchers and testers. Readers with even a passing familiarity with human factors testing will presumably not need to be convinced that the measurement of mental workload is important to test and evaluation. It is difficult to think of a human factors issue that has been the subject of as many research dollars, hours, and published pages. Situation awareness, the younger sibling of the pair, has followed much the same path. Garnering significant amounts of money, time, and pages in its own right, situation awareness is proving to be just as resistant to consensus on its relationship to cognitive theory, its implications for operator performance, and how best to measure it. Debates surrounding workload and situation awareness have, at times, taken on an almost religious character, with the participants no closer to agreement today than when the debates began. In this chapter, we will try to avoid the heat of these debates and simply cast light on the variety of measures available and some of the issues involved in their use.

Mental workload is often characterized as the amount of cognitive or attentional resources being expended at a given point in time. Situation awareness can be regarded as the momentary content of those resources, a subjective state that is afforded by the object or objects of one's attentional

---

[1]Recognizing that a great deal of other types of human factors data are gathered with subjective measures (principally questionnaires) a later chapter provides an overview of the salient issues in the design and use of questionnaires and interviews.

resources. Most people recognize what is meant by the term *mental work-load* and can provide numerous anecdotal examples of situations that lead to the subjective experience of high workload. Although the term *situation awareness* possesses less common currency, here too its meaning in every-day terms is clear. There is not, however, unanimous agreement in the psychological community as to the underlying cognitive mechanisms re-sponsible for either subjective state. Nor is there consensus on measures that can best quantify the level of workload experienced or the degree of situation awareness possessed by an individual.

There are many reasons for this lack of consensus, not the least of which is the very nature of the workload and situation awareness phenomena. As alluded to previously, workload and situation awareness are transitory, subjective states that have no obvious, direct external manifestations. As with research into many other psychological phenomena, theoretical in-terpretations serve to provide much of the direction for the research ques-tions and methodologies. In the area of cognitive workload, theorists con-tinue to debate the nature of attentional capacity with regards to its limiting effect on the overall cognitive system (Kantowitz, 1985): its interac-tion with the attributes of the objects and events being attended to (Kahne-man & Treisman, 1984), whether it reflects the interaction of individually constrained cognitive resources (Wickens, 1991), and whether or not atten-tion is the most appropriate theoretical model for cognitive workload (Moray, Dessouky, Kijowski, & Adapathya, 1991). Thus, the evolution of a single theory of cognitive workload and the development of one or even a few universally accepted workload metrics have been to some degree sty-mied by the profusion of theories and models of cognition and attention and the resulting proliferation of competing workload measures.

Situation awareness has also been marked by divergent theoretical con-ceptualizations and measurement approaches. Situation awareness has been hypothesized as a process that is separate from, but interdependent with, other cognitive processes, such as attention, perception, working memory, and long-term memory (Endsley, 2000), occurring sometime prior to decision making and performance (Endsley, 1995, 1996). Fur-ther, situation awareness has been characterized as three hierarchical stages or levels: the perception of elements in the environment, the com-prehension of the current situation, and the projection of the future status of events and elements in the environment.

Other researchers have pointed out that there is a danger in treating situation awareness as a process distinct from other cognitive activities (Sarter & Woods, 1991). Situation awareness viewed as a separate cogni-tive process stage, occurring between attention and decision making, de-nies the essential interdependence of cognitive processes that is the hall-mark of information processing in naturalistic environments (Flach,

1995). These authors also point out that situation awareness cannot be considered independently of the environment or external goal for achieving it. In other words, to evaluate the quality of one's situation awareness there must be an external standard or set of situation awareness criteria to compare human performance against (Adams, Tenney, & Pew, 1995; Smith & Hancock, 1995). This point of view emphasizes the inseparability between cognitive processes and the environment in which these processes function.

In considering these alternative views of situation awareness, a useful distinction can be made between situation awareness as a cognitive process and situation awareness as a product. As a process, situation awareness can be thought of as either the depth of involvement in the tripartite hierarchy of situation awareness stages or alternatively as the current state of the adaptive perception-decision-action cycle as directed toward some task or goal. As a product, situation awareness transcends the differences between these views and manifests itself as a measurable state of the perceiver, the momentary knowledge about the environment that is appropriate to the task or goal.

Another consideration in the development of the constructs of mental workload and situation awareness has been the nature of the relationship between workload and situation awareness, whether or not either of them can be considered unitary entities. It has been hypothesized, for example, that the subjective experience of workload is the collective result of several independent components, such as task-related input loads, operator effort and motivation, and operator performance (Jahns, 1973). Some researchers have found that operators can indeed discriminate different constituents of workload in their subjective experience (Hart, Childress & Hauser, 1982). One of the prevailing conceptualizations resulting from this work is that workload is a tripartite amalgam consisting of time load, mental effort, and psychological stress (Reid, Shingledecker, & Eggemeier, 1981). Other conceptualizations are more complex, involving as many as six to nine workload components (Bortolussi, Kantowitz, & Hart, 1986; Hart & Staveland, 1988). Similarly, a widely known view of situation awareness holds that it is composed of three distinct components: the perception of elements in the environment, the comprehension of the elements' meanings, and the projection of the future status of the elements in the environment (Endsley, 1995). Other views of situation awareness have included 6 or more context-dependent elements (Pew, 2000); 10 constructs ranging across situational demands, mental resources, and information availability (Selcon, Taylor, & Koritsas, 1991); 31 behavior elements (Bell & Waag, 1995); and the view of situational awareness as an indivisible condition reflecting a knowledge of the set of possible reactions that can be made to a changing situation (Klein, 2000). As noted in passing above, some researchers have included mental workload as a component of situation awareness, whereas others have viewed them as independent, albeit complementary, constructs (Endsley, 2000).

Human factors professionals have appropriately addressed this multidimensional aspect by developing measures that are themselves multidimensional. These different measures often produce different results—the individual measures being differentially sensitive to the various components of the workload and situation awareness constructs occasioned by the task or system environment under study. What we are left with, then, is a situation where there exists a multiplicity of workload and situation awareness measures, ranging from global to multidimensional scales and from methods with high sensitivity to those with greater diagnosticity. In spite of this potentially confusing situation, workload and situation awareness remain issues of obvious importance for HFTE.

In the first edition of this handbook, we described six attributes of workload metrics, advanced by a variety of researchers, that could be used to assess the relative suitability of the measures available for a given application (Charlton, 1996). We repeat them here as they are applicable not just to the measurement of mental workload but to cognitive states in general. The first feature is sensitivity, that is, the ability of the measure to detect changes in cognitive state (in mental workload or situation awareness) resulting from task demands or environmental changes. The second feature, intrusiveness, is the degree to which the measure interferes with operator tasks or introduces changes into the operational situation. Third is the diagnosticity of the measure, that is, the ability to discriminate the specific relationship between the task environment and the cognitive state of the operator (e.g., schedule demands and the experience of time pressure or task goals and display elements attended to). Fourth is the convenience of the measure, that is, the amount of time, money, and instrumentation required for measurement. Fifth, relevance or transferability, is concerned with the applicability of the measure to a variety of different operator tasks. Finally, operator acceptance refers to the operators' willingness to cooperate with, or be subjected to, the demands of measurement methodology. An ideal workload or situation awareness measure should be sensitive, nonintrusive, diagnostic, convenient to use, relevant to different tasks, and acceptable to operators.

The remainder of this chapter provides descriptions of several frequently used workload metrics, all of which fare well in meeting the six selection criteria described previously. The methods covered are in no way intended to provide the comprehensive detail available in the original citations. Instead, the goal of this chapter is to acquaint the reader with the techniques most frequently used in test and evaluation, along with some of the situations and conditions where they can be used to the greatest advantage.

MEASURES OF MENTAL WORKLOAD

Three fundamental techniques for collecting mental workload data, performance measures, psychophysiological methods, and rating scales, are presented in this section. Also included are analytic techniques that have been used to provide an a priori estimate of the amount of workload an operator might experience under specific conditions. Although these a priori or projective measures are properly considered task analysis techniques, their relationship to other workload measures and usefulness in early tests merit a brief discussion at the end of this section.

Performance Measures

Performance measures of mental workload involve collection of data from one or more subjects performing the task or tasks of interest. Two types of task performance metrics of mental workload have traditionally been used: primary task performance and secondary task performance. Operator performance on a primary task as a method of measuring mental workload suffers from several problems, not the least of which is a lack of generalizability of methods and results. Because the tasks are often unique to each system, nearly every situation requires its own measure of task performance (Lysaght et al., 1989).

More to the point, covariance of primary task measures with other measures of mental workload is not always reliable. For example, high workload levels cannot always be logically inferred from poor task performance. Increased response latencies, error rates, and decision times could be attributable to any number of other factors, such as training, motivation, communications, user interface problems, and so on (Lysaght et al., 1989). Similarly, levels of mental workload, as indicated by subjective reports, may rise without any degradation in task performance (Hart, 1986), a phenomenon difficult to separate from dissociation or insensitivity of the scales and operator tasks under examination (Tsang & Vidulich, 1989; Wierwille & Eggemeier, 1993). For some easy tasks, increased workload (i.e., effort) may actually lead to improved performance (Gopher & Donchin, 1986; Lysaght et al., 1989). Over a wide range of moderate workload levels, subjects are able to adjust their level of effort and maintain acceptable levels of performance (Lysaght et al., 1989). For moderately difficult tasks, subjects may not be able to increase their effort enough to meet the task demands, and thus increased workload is associated with poorer performance (O'Donnell & Eggemeier, 1986). For very difficult tasks, subjects may not continue to expend the extra effort in the face of what are perceived as unreasonable task demands; instead, they reduce their effort and allow task performance to deteriorate so they can return to normal levels of workload (Charlton, 1991). In short, task demands do not always drive a subject's workload and effort in predictable ways; users actively manage their time and effort to be as comfortable as practically possible.

The other performance-based approach to workload assessment has been to measure performance on a secondary task as an indication of spare mental capacity. The logic behind the secondary task paradigm is that if the subject is only partially loaded by the requirements of the primary task, performance on a secondary task (e.g., a simple memory, mental arithmetic, time estimation, or reaction-time task) should remain efficient. As the processing requirements of the primary task increase, performance on the secondary task is expected to suffer. Some researchers, however, have argued that the introduction of a secondary task can change the nature of the primary task and therefore contaminate workload measures thus obtained (O'Donnell & Eggemeier, 1986). An alternative to introducing an arbitrary secondary task that may intrude on primary task performance and workload is to identify an embedded task, an existing concurrent operator task, or one that might plausibly coexist with typical operations (Wierwille & Eggemeier, 1993).

In addition to plausibility and nonintrusiveness, the multiple resource account of mental workload (Wickens, 1991) would maintain that the secondary task must be similar enough to the primary task so as to be sensitive to the capacity demands of the cognitive resource of interest; that is, if there is little or no overlap in the attentional demands of the primary and secondary tasks, secondary task performance will not be sensitive to the workload demands imposed by the primary task (Tsang & Vidulich, 1989; Wierwille & Eggemeier, 1993). A final consideration for use of secondary task measures is whether or not they are perceived as secondary by test participants (Lysaght et al., 1989). In our research on attention and mental workload in driving, we have used a radio tuning task (reaction time to retune a car radio once the signal changes to static) and a cellular phone answering task (again reaction time) to estimate momentary mental demands of various driving situations (see chapter 15 of this book). In these tests, we have found that instructions to participants must be carefully worded to ensure that good driving performance is perceived as the overriding goal. Although these secondary task measures may lack global sensitivity and transferability, we have found them to be reasonably diagnostic and nonintrusive. In addition, we have found them to covary fairly well with rating scale measures taken at the end of the driving scenarios.

Psychophysiological Methods

Psychophysiological approaches have long been of interest in the history of workload measurement (Wierwille & Williges, 1978). The reason for their appeal is that they offer the potential for objective measurement of a physiological correlate of mental workload and release the human factors tester from reliance on self-reports and rating scales. For example, eye-blink rate, heart rate variability, electrodermal activity (skin conductance), and electroencephalograph (cortical activity) recordings have all been used successfully as indicators of mental workload. Unfortunately, many of the psychophysiological measures thus far developed in the laboratory have not always lived up to their potential in offering a suitable field test methodology. The reasons for this vary from the cost, complexity, and intrusiveness of some of the required instrumentation to a lack of agreement on the validity of the physiological data collected in relatively uncontrolled field settings. Nonetheless, psychophysiological measurement of mental workload continues to offer considerable promise for HFTE and is covered in some detail in chapter 7 of this book.

Rating Scales

By far, the most frequently used measures of mental workload are rating scales. As with many other psychological phenomena, subjective self-reports (typically questionnaires) are the measure closest to a direct measure (Lysaght et al., 1989; Moray, 1982; O'Donnell & Eggemeier, 1986; Sheridan, 1980; Wierwille & Williges, 1978). The number of different mental workload rating scales available, however, can be rather daunting to the human factors tester. The methods range from simple, unidimensional scales that provide a single value for overall workload to multidimensional scales that measure various components of workload, as described earlier. In our coverage, we describe four common scales representing the historic development of workload measurement: The Cooper-Harper and its derivatives, the Subjective Workload Assessment Technique, the NASA Task Load Index, and the Crew Status Survey. Each of these measures have been in widespread use in HFTE for some time and are generally found to be sensitive to changes in workload levels, minimally intrusive, diagnostic, convenient, relevant to a wide variety of tasks, and possessive of a high degree of operator acceptance (Gawron, Schiflett, & Miller, 1989; Hill et al., 1992; Wierwille & Eggemeier, 1993).

Cooper-Harper Scales. The Cooper-Harper Scale is perhaps the earliest standardized scale used for measuring mental workload. Originally developed for the assessment of aircraft handling characteristics (Cooper & Harper, 1969), it has a long tradition of use for the workload associated with psychomotor tasks (O'Donnell & Eggemeier, 1986). As shown in Fig. 6.1, the scale is a decision tree that leads the pilot to one of 10 ordinal ratings. The primary advantages of the scale are that it is well-known in the testing community, easy to use, and the resulting ratings correlate highly ($r = .75$ to $.79$) with other, more sophisticated workload scales (Lysaght et al., 1989). Its applicability, however, is limited to workload associated with

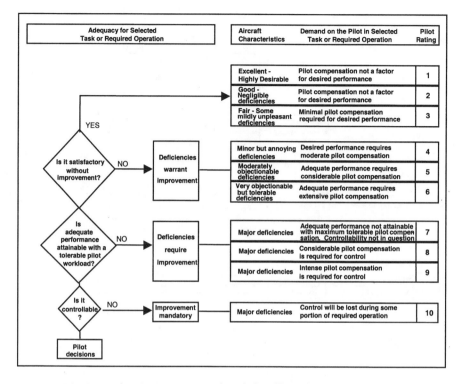

FIG. 6.1. The Cooper-Harper aircraft handling characteristics scale (Cooper & Harper, 1969).

piloting aircraft. The original Cooper-Harper Scale has been adapted for use by other researchers over the years under a number of different names: Honeywell Cooper-Harper (Wolf, 1978), the Bedford Scale (Roscoe, 1987), and others. The most widely used Cooper-Harper derivative, however, is known as the Modified Cooper-Harper, or MCH, scale (Wierwille & Casali, 1983).

The Modified Cooper-Harper scale was developed to extend the workload assessment capabilities of the Cooper-Harper Scale to situations and tasks outside the psychomotor-aircraft piloting domain. As shown in Fig. 6.2, Wierwille and Casali (1983) retained the basic structure of the Cooper-Harper but modified the scale descriptors to capture the workload associated with a variety of perceptual, cognitive, and communications tasks. The resulting scale, known as the MCH, provides a sensitive measure of overall mental workload for a wide variety of operator tasks, aircraft handling included (Wierwille & Casali, 1983). Subsequent investigators (Warr, Colle, & Reid, 1986) confirmed that the MCH was at least as sensitive as the Subjective Workload Assessment Technique (SWAT) to task difficulty and mental workload across a wide range of tasks.

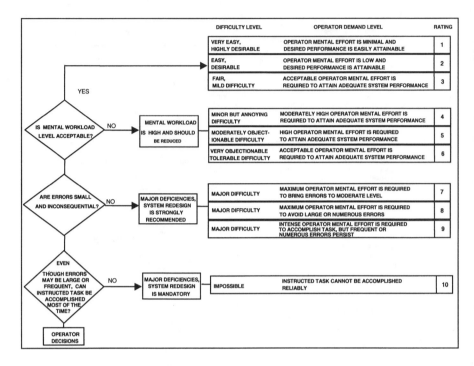

FIG. 6.2. The MCH rating scale. *Note.* From W. W. Wierwille and J. G. Casalli, 1983, *Proceedings of the Human Factors Society 27th Annual Meeting.* Reprinted with permission from *Proceedings of the Human Factors and Ergonomics Society.* Copyright 1983 by the Human Factors and Ergonomics Society. All rights reserved.

Like the original Cooper-Harper, the MCH is typically administered to subjects at the end of a test event, such as a flight, task sequence, or duty shift. This is in contrast to some other workload measures that can be collected during the test event with little or no interruption to task performance. As with other types of subjective data collection instruments, such as questionnaires, the quality of the MCH workload data will depend on the memorability of the event of interest and how soon after task completion the MCH is administered. On the whole, subjective measures of workload have been found to be fairly insensitive with respect to delays in obtaining ratings and even intervening tasks (Lysaght et al., 1989). Because of its good sensitivity, wide applicability, and ease of use, the MCH has become a popular and frequently used workload metric for many field test situations (Charlton, 1991).

Subjective Workload Assessment Technique. SWAT (Reid & Nygren, 1988) is perhaps the most frequently cited methodology in the workload literature. SWAT was developed by the U.S. Air Force Armstrong Aeromedical Research Laboratory (AAMRL) to provide a quantitative workload instrument with interval scale properties. SWAT is theoretically grounded in a multidimensional view of mental workload comprising time load, mental effort, and psychological stress. Methodologically, SWAT is based on an application of conjoint measurement and scaling procedures, which result in a normalized, overall workload score based on ratings of time load, mental effort, and stress for each subject. Before SWAT can be used to collect workload ratings, the scale must be normalized for each subject. During this first phase, called scale development, subjects rank order 27 possible combinations of 3 different levels of time load, mental effort, and stress by means of a card-sorting technique. The resulting rankings are then subjected to a series of axiom tests to identify a rule for combining the three dimensions for each subject. Once the rule has been established, conjoint scaling is applied to develop an appropriate unidimensional workload scale that ranges from 0 (*no workload*) to 100 (*highest workload possible*). If a Kendall's coefficient of .79 or greater is obtained when the scale is compared back to the original rank orderings, then the subject's individualized workload scale is deemed valid and can be used to collect data. An example of the definitions used in scale development is shown in Fig. 6.3.

During the data collection phase, subjects provide time load, mental effort, and stress levels (e.g., 2, 2, 1), either verbally or by means of a checklist, at predetermined times throughout the test activity. For some applications, particularly in aircraft testing, there can be resistance to collecting workload ratings real-time during the mission for fear of introducing an additional task. In these cases SWAT ratings are collected at the end of the test flight or duty shift, using a structured debriefing or in conjunction with viewing a mission videotape. At the end of the test, the ratings are converted to the normalized workload scale using the rule identified during scale development. In addition, the individual time, effort, and stress ratings can also be used for workload assessment, representing 3 separate, 3-point scales.

SWAT is widely used in test and evaluation because it is generally regarded to be a well-developed, reliable, and valid metric of workload. It carries with it the background of extensive laboratory research and field use. Further, it produces a quantitative result with which to compare different systems and, if desired, the diagnosticity inherent in its three subscales. The power of SWAT, however, comes at a cost. SWAT requires a fair amount of preparation and pretraining of subjects prior to use. Further, the analysis required after data collection does not lend itself to immediately available quick-look results. There have been efforts to establish an appropriate limit or red-line value for SWAT scores, and although a score of 40 has been proposed for some tasks, it has still not been acknowledged by the test community (Reid & Colle, 1988; Reub, Vidulich, & Hassoun, 1992; Wierwille & Eggemeier, 1993).

SUBJECTIVE WORKLOAD ASSESSMENT TECHNIQUE

NAME	DATE AND TIME

TASK BEING RATED

(MARK AN X IN ONE CHOICE FOR EACH OF THE THREE AREAS BELOW THAT BEST DESCRIBES WHAT YOU BELIEVE THE TASK WORKLOAD TO BE.)

I. TIME LOAD

1 Often have spare time. Interruptions or overlap among activities occur infrequently or not at all.

2 Occasionally have spare time. Interruptions or overlap among activities occur frequently.

3 Almost never have spare time. Interruptions or overlap among activities are frequent, or occur all the time.

II. MENTAL EFFORT

1 Very little conscious mental effort or concentration required. Activity is almost automatic requiring little or no attention.

2 Moderate conscious mental effort or concentration required. Complexity of activity is moderately high due to uncertainty, unpredictability, or unfamiliarity. Considerable attention required.

3 Extensive mental effort and concentration are necessary. Very complex activity requiring total attention.

III. PSYCHOLOGICAL STRESS

1 Little confusion, frustration or anxiety exists and can be easily accomodated

2 Moderate stress due to confusion, frustration, or anxiety. Noticeably adds to workload. Significant compensation is required to maintain adequate performance.

3 High to very intense stress due to confusion, frustration, or anxiety. High to extreme determination and self-control required.

AFOTEC FORM JUN 83 81

FIG. 6.3.   SWAT rating card.

NASA Task Load Index. At first blush, the NASA Task Load Index (Hart & Staveland, 1988), or TLX, appears similar to SWAT. Like SWAT, it is based on a multidimensional approach to the workload phenomenon and uses an adjustment to normalize ratings for each subject. The assumptions and procedural details of the two methods, however, are substantially different. Instead of using the three workload components assessed by SWAT, TLX divides the workload experience into six components: mental demand, physical demand, temporal demand, performance, effort, and frustration. Instead of 3-point subscales for each component, TLX measures each component subscale with 20 levels. Figure 6.4 shows the TLX subscales and their definitions. Instead of using the conjoint measurement procedures used in SWAT's scale development phase, TLX uses a somewhat simpler postdata collection weighting procedure for combining information from the six subscales.

During data collection participants are asked to rate each task component of interest on the six subscales. After completing data collection, the participants make 15 paired comparisons of each subscale to provide a weighting of each workload aspect's importance to the just-completed operational task. Totaling the number of times a subscale is chosen over another gives the weighting for that subscale. A subscale that is never chosen over another is assigned a weight of 0. Following data collection, overall workload scores that range from 0 to 100 are computed for each test participant by multiplying the rating obtained for each subscale by its task weighting, then adding up all of the subscale scores and dividing by the total number of paired comparisons used to obtain the weights (i.e., 15). Like SWAT, the resulting score can be used to evaluate overall workload, and the individual subscale scores can be used to diagnose the various workload components associated with a given task or event. As compared to SWAT, the chief advantage of TLX is that data collection can begin without the complexities of pretest scaling. On the other hand, participants must still complete the pair-wise rating procedure at the conclusion of the test.

Crew Status Survey. The Crew Status Survey (CSS) was developed at the U.S. Air Force School of Aerospace Medicine to assess dynamic levels of workload and fatigue throughout a crewmember's shift (Gawron et al., 1988; Rokicki, 1982). It was designed to be easily understood by the test participants, easy to administer and complete in the field, and readily analyzed by the tester. Although used somewhat less frequently than the other scales described in this section, the CSS has been found to be quite sensitive across a wide variety of tasks and to correlate well with other workload measures (Charlton, 1991; Gawron, Schiflett, & Miller, 1989). As shown in Fig. 6.5, the CSS requires three ratings collected on a single form: subjective fatigue, maximum workload, and average workload. The survey is typically

NAME	TASK	DATE

TASK LOADING INDEX

Mental Demand — How mentally demanding was the task?

Very Low — Very High

Physical Demand — How physically demanding was the task?

Very Low — Very High

Temporal Demand — How hurried or rushed was the pace of the task?

Very Low — Very High

Performance — How successful were you in accomplishing what you were asked to do?

Perfect — Failure

Effort — How hard did you have to work to accomplish your level of performance?

Very Low — Very High

Frustration — How insecure, discouraged, irritated and annoyed were you?

Very Low — Very High

AFOTEC FORM JUL 94 SAH-10

FIG. 6.4. The NASA TLX rating scale.

110

NAME	DATE AND TIME

SUBJECTIVE FATIGUE

(Circle the number of the statement which describes how you feel RIGHT NOW.)

1	Fully Alert; Wide Awake; Extremely Peppy
2	Very Lively; Responsive, But Not At Peak
3	Okay; Somewhat Fresh
4	A Little Tired; Less Than Fresh
5	Moderately Tired; Let Down
6	Extremely Tired; Very Difficult To Concentrate
7	Completely Exhausted; Unable To Function Effectively; Ready To Drop

COMMENTS

WORKLOAD ESTIMATE

(Circle the number of the statement which best describes the MAXIMUM workload you experienced during the past work period. Put an X over the number of the statement which best describes the AVERAGE workload you experienced during the past work period.)

1	Nothing to do; No System Demands
2	Little to do; Minimum System Demands
3	Active Involvement Required, But Easy To Keep Up
4	Challenging, But Manageable
5	Extremely Busy; Barely Able To Keep Up
6	Too Much To Do; Overloaded; Postponing Some Tasks
7	Unmanageable; Potentially Dangerous; Unacceptable

COMMENTS

SAM FORM 202
APR 81

CREW STATUS SURVEY

FIG. 6.5.   The Crew Status Survey.

administered at regular intervals throughout the test participants' duty shifts or flight missions. The survey can be completed very quickly with minimal experience and thus produces little disruption or additional operator tasking (Charlton, 1991). As with SWAT, the three scale values can easily be reported verbally by operators in situations where handling paper forms is difficult. Data analysis as well is very simple; the ratings are plotted over time and compared to performance measures, significant task sequences, or the occurrence of other mission events of interest. In addition, for test situations where a pass-fail workload score is needed, the CSS provides an inherent test criterion by virtue of the absolute wording of the scale.

The primary advantages of the CSS are its inherent simplicity and its good agreement with other workload scales, such as SWAT and TLX. The ratings obtained with the survey, however, had not been demonstrated to possess the equal interval scale properties reported for SWAT and TLX. To correct this situation, researchers at the Air Force Flight Test Center (AFFTC) undertook a significant effort to revise and validate the psychometric properties associated with the workload portion of the survey (Ames & George, 1993). The wording of CSS was mapped onto four distinct workload components: activity level, system demands, time loads, and safety concerns. Scale analysis and development then proceeded through several iterations of revision and examination in an attempt to produce subjectively equal intervals between scale steps without sacrificing the understandability of the original scale.

The resulting revision of the CSS workload scale, called the AFFTC Revised Workload Estimate Scale (ARWES), was then subjected to verification testing using paired comparison and rank-order estimation tests. The revised scale descriptors are shown in Fig. 6.6. The results of the verification testing showed excellent concordance and a nearly ideal straight line function for the revised scale. Analysis of confusability of scale descriptors showed that the new scale preserved the understandability of the original response alternatives. Although the scale is not as diagnostic as SWAT or TLX, it can provide the tester with interval data without the card-sorting or paired comparisons required by the multidimensional workload scales. Finally, the terminology inherent in the revised scale provides the tester with an easily interpretable result and lends itself to wording absolute, rather than relative, workload standards for those situations where a criterion is needed for evaluation.

Projective Techniques

As described earlier, projective techniques have been developed to predict levels of workload at the very beginning of a system's development process. Typically, they involve analysis of the proposed tasks required to oper-

NAME		DATE AND TIME
WORKLOAD ESTIMATE		
(Circle the number of the statement which best describes the MAXIMUM workload you experienced during the past work period. Put an X over the number of the statement which best describes the AVERAGE workload you experienced during the past work period.)		
1	Nothing to do; No system demands.	
2	Light activity; Minimum demands.	
3	Moderate activity; Easily managed; Considerable spare time.	
4	Busy; Challenging but manageable; Adequate time available.	
5	Very busy; Demanding to manage; Barely enough time.	
6	Extremely busy; Very difficult; Non-essential tasks postponed.	
7	Overloaded; System unmanageable; Essential tasks undone; Unsafe.	
COMMENTS		

AFOTEC FORM JUL 94 SAH-20

FIG. 6.6. The AFFTC Revised Workload Estimate Scale.

ate the developmental system and comparisons to similar systems for which empirical workload data exist. While these techniques are not a substitute for assessing workload while a user actually operates the system, they can provide valuable indications of potential workload problems. In approximate order of complexity, from least to most complex, the more common projective techniques follows.

Time-line Analysis. Time-line analysis is used to identify how long tasks and task components will take and if they can be accomplished in the allotted time (Charlton, 1991; DOD-HDBK-763, 1987; Parks, 1979). It is assumed that if the sum of all of the task times is less than the time available, then there will be some operator slack time and therefore less potential for operator overload. Task times are based on a combination of empirical data from predecessor or baseline systems and expert judgment. In addition to time estimates for each task, skill levels, difficulty ratings, repetition frequencies, and perceptual-motor channels can be identified for each task. Tasks that occur infrequently, have relatively high difficulty ratings, or involve several perceptual motor-channels simultaneously receive special attention when estimating workload from the time lines. The time-line analysis proceeds by organizing all of the required tasks into a linear sequence using a milestone format, operational sequence diagram, flow process chart, or another formalism. Task times are simply summed for each time period of interest (typically at intervals of 5 to 15 minutes) with the goal of arriving at a total of no more than 75% of the time available for an acceptable level of workload. Cases where there are significant numbers of high-difficulty tasks at 75% loading, or where task times total more than 100% of the time available, should be regarded as unacceptable. The resulting workload estimates, however, represent an approximation of only about ±20% (DOD-HDBK-763, 1987) and are only as good as their constituent task time and difficulty estimates.

Scheduling Theory. The primary tenet of scheduling theory as applied to workload analysis is that in real-world tasks, time pressure is the major source of cognitive workload (Moray et al., 1991). In a manner similar to time-line analysis, scheduling theory compares task times to the time available for completion. The primary advantage of scheduling theory over time-line analysis is the identification of optimal task sequences. Experiments with the scheduling theory approach have shown good correspondence between subjective workload ratings and the number of tasks completed in the allotted time (i.e., a greater number of tasks completed were associated with lower workload ratings). It was not the case, however, that workload was lower for subjects who were informed of, and followed, the optimal task sequences. In fact, the optimal task sequence was often contrary to the intuitively selected task sequence, was perceived as more demanding, and resulted in higher ratings of perceived workload. The lesson for human factors testers and system designers is that the sheer number of tasks assigned within a specific time period is not a good predictor of workload. The ordering of those tasks within a prescribed time frame is at least as important in determining subjects' perceptions of high workload.

Pro-SWAT. Projective application of the Subjective Workload Assessment Technique (Pro-SWAT) is essentially a role-playing exercise in which subjects project themselves into the system tasks one at a time and complete SWAT ratings (Kuperman & Wilson, 1985; Reid, Shingledecker, Hockenberger, & Quinn, 1984). The procedure usually involves giving a detailed briefing on each task function and topography, creating some level of equipment mock-up, and providing an extensive debriefing in which tasks receiving high workload ratings are discussed in detail. The higher the fidelity of the mock-up and the task procedures used, the better the predictive utility of the workload results. The technique also provides a means of direct comparison to existing systems or fielded versions of the same system for which SWAT ratings are collected. In short, Pro-SWAT offers a relatively low-cost method for assessing the workload of developmental systems, provided that functional tasks are well-defined and that an equipment mock-up is available.

MEASUREMENT OF SITUATION AWARENESS

As with mental workload, most of the techniques for collecting situation awareness data can be categorized as being either performance measures, psychophysiological methods, or rating scales. In addition, situation awareness is often assessed by procedures testing participants' knowledge of the situation. Knowledge-based procedures differ from rating scales in that participants' scores are matched against the true state of system and its likely future state. We describe situation awareness measures representing all four techniques in the sections to follow.

Performance Measures

Performance measures of situation awareness use objective response data from one or more subjects performing the task or tasks of interest. As with mental workload performance measures, two types of task performance metrics can be used: primary task performance and secondary task performance. Primary task performance measures rely on situations with specific test points that, if the operator possesses good situation awareness, would elicit action of a particular sort. For example, experimenters may introduce specific events, such as mechanical failures or air traffic control transmissions, into an aircraft simulation and measure the participants' time to respond appropriately (Pritchett & Hansman, 2000; Sarter & Woods, 1991). Similarly, in a driving simulation task, experimenters have measured participants' time to override an autopilot to avoid other traffic (Gugerty, 1997). In both cases, participants' situation awareness was inferred from their production of a response appropriate to the current situation or event. It can be argued, however, that appropriate performance in these cases may be the result of many factors other than situation awareness (e.g., participant skill, differing goals or motivations, and even mental workload). To make the link between task performance and situation awareness somewhat less tenuous, researchers have recommended that the test responses involve only obvious, highly trained or proceduralized actions where minimal decision making or interpretation by the participants is required (Pritchett & Hansman, 2000).

Another approach is to measure performance on a secondary task to determine the momentary focus (or foci) of operators' attention. In an ongoing series of tests of drivers' attention, we have introduced a range of embedded situation awareness stimuli into various aspects of a simulated driving environment (Charlton, 2000). Similar to embedded secondary task measures of workload, the drivers press a button on the dashboard when they detect one of the prescribed stimuli. The essential difference is that these stimuli occur across a range of sources in the driving environment. For example, stimuli occur inside or outside the cab (e.g., on the instrument panel or on the roadway), can be visual or acoustic (e.g., a dashboard light or an auditory warning), and can be either relevant or irrelevant to the driving task (e.g., traffic control signs vs. arbitrary roadside signs). In one recent experiment, drivers were told to respond as soon as they could whenever they detected a threesome (e.g., three lights on the dashboard, a road sign with three dots, three stripes on the road, three car horns, three church bells). The driving scenarios contained many such stimuli, sometimes containing the target threesomes, other times containing single, double, or quadruple-element distractor stimuli. The goals of the exercise were to assess the proportion of the drivers' attention that was directed inside the vehicle versus outside and the capacity available to process stimuli irrelevant to the driving task. To date, this approach has provided useful information on drivers' allocation of attention in tests of different road conditions (Charlton, 2000).

Performance measures of situation awareness do not have a long history of use, and, as a result, it is difficult to offer any definitive conclusions on their utility. It is clear that both primary and secondary task methods suffer from an inherent lack of generalizability as the measures are defined in terms of participants' reactions to specific task or situation elements. Although the performance measures are objective, they can be influenced by a range of response-related factors and thus cannot be said to provide a pure indication of the operator's internal state of awareness, making their sensitivity suspect as a result. However, because they are reasonably diagnostic, they can be used without substantially altering the task environment, are generally convenient to use, and meet with good operator acceptance. Performance measures hold considerable promise in assessing situation awareness.

Psychophysiological Methods

Like performance measures, psychophysiological measures of situation awareness have a fairly short history. In one of the few studies to explore situation awareness using psychophysiological measures, Vidulich, Stratton, Crabtree, and Wilson (1994) found that electroencephalographic activity and eye-blink durations paralleled display types associated with either good or poor situation awareness. In a recent review (Wilson, 2000) it has been proposed that it may be possible to use psychophysiological measures, such as changes in heart rate and blink activity, to monitor operators' detection of environmental stimuli (level 1 situation awareness) and brain activity to indicate their general level of expectancy (contributing to level 2 situation awareness). The challenge will be to identify measures that unambiguously associate with situation awareness as opposed to cooccurring cognitive factors, such as workload and surprise. Another approach has been to use eye-scanning behavior as an indication of what aspects of the situation are being attended to (Isler, 1998). With this approach, there is no guarantee that what is looked at is receiving attention (although it is certainly a necessary precursor), and there is certainly no indication of the later stages of situation awareness, comprehension, or projection of future states.

Rating Scales

Situation Awareness Rating Technique. The Situation Awareness Rating Technique (SART) has received the greatest attention among the situation awareness rating scales commonly in use. Originally developed by interviewing aircrews regarding the elements of situation awareness, SART was conceptualized as 10 rateable constructs (Taylor & Selcon, 1991). The 10 constructs included the instability, variability, and complexity of the situation; the arousal, spare mental capacity, and concentration of the operator; the division of attention in the situation; the quantity and quality of information in the situation; and the familiarity of the situation. In practice, each of these 10 constructs could be rated on a 7-point scale ranging from *low* to *high* using a 1-page question form (see Fig. 6.7). A simplified version of SART was also developed based on the finding that the 10 constructs could be subsumed under 3 headings (subscales): demand on attentional resources, supply of attentional resources, and understanding. The simplified version of SART asks operators to rate a just-completed task by placing a mark on each of three 100-mm lines corresponding to each subscale category. The answers are scored by measuring the location of the mark on each line and converting it to a score ranging from 0 (*low*) to 100 (*high*). The results of either method (10-question or 3-question SART) can be used to calculate an overall metric of situation awareness by combining subscale ratings according to the following formula: SA = Understanding – (Demand – Supply).

SART has been used across a fairly wide range of application areas and has proven to be a robust measure with good sensitivity to the experimental manipulations (Jones, 2000; Vidulich, 2000). There has been some question, however, as to what SART ratings are indicating. The SART method contains a large mental workload component, and analysis of

		Low				High		
		1	2	3	4	5	6	7
DEMAND	Instability of Situation							
	Variability of Situation							
	Complexity of Situation							
SUPPLY	Arousal							
	Spare Mental Capacity							
	Concentration							
	Division of Attention							
UNDERSTANDING	Information Quantity							
	Information Quality							
	Familiarity							

FIG. 6.7. The Situation Awareness Rating Technique (SART) data collection form. *Note*. From S. J. Selcon, R. M. Taylor, and E. Koritsas, 1991, *Proceedings of the Human Factors Society 35th Annual Meeting*. Reprinted with permission from *Proceedings of the Human Factors and Ergonomics Society*. Copyright 1991 by the Human Factors and Ergonomics Society. All rights reserved.

subscale data has suggested that SART scores may be reflecting participants' workload rather than their situation awareness (Endsley, 1996; Vidulich et al., 1994). Further, SART scores have shown a significant correlation with TLX scores, with the exception that SART is sensitive to differing experience levels in the participants whereas TLX is not (Selcon et al., 1991). Although the cognitive states of workload and situation awareness may be positively correlated in many situations, some authors have argued that these two constructs must be kept distinct (Pew, 2000). Nonetheless, because of its relative ease of use, sensitivity, and apparent ecological validity, SART continues to enjoy widespread acceptance in the test community (Jones, 2000).

Situation Awareness Rating Scale.   The Situation Awareness Rating Scale (SARS) represents another approach to a multidimensional situation awareness metric (Waag & Houck, 1994). SARS, however, differs from SART in that participants rate other test participants, as well as themselves, in regards to their situation awareness. SARS is based on 31 behavior elements of experienced fighter pilots, each of which is rated on a 6-point scale ranging from *acceptable* to *outstanding*. The 31 elements are arranged in 8 operational categories, such as general traits, system operation, communication, and information interpretation. Subscale scores are calculated by averaging the ratings in each of the eight categories, or an overall situation awareness score is taken using the overall mean rating.

SARS scores have shown good interrater reliability and consistency (Bell & Waag, 1995), but there is some question regarding their utility for test and evaluation. SARS appears to measure individual differences in situation awareness more readily than the effects of a given event, display, or scenario (Endsley, 1996; Jones, 2000). Further, the 31 behavior elements are drawn exclusively from the fighter pilot domain and thus not applicable to other systems or crews.

SA-SWORD.   Another promising rating scale technique was developed from the Subjective Workload Dominance (SWORD) methodology by changing the instructions to participants (Vidulich & Hughes, 1991). The SA-SWORD data-collection procedure is essentially a variant of Saaty's Analytic Hierarchy Process (Saaty, 1980) and requires participants to complete a series of pair-wise comparisons between tasks, events, or interfaces, identifying which one of each pair is associated with the highest workload. When used to assess situation awareness, the method is known as SA-SWORD. Participants rate the display or interface in each pair that produced the best situation awareness, using a 19-point scale (corresponding to scores from 2 to 9 for the stronger member of each pair, ½ to ⅑ for the weaker member, and a score of 1 when the members are equal). Ratings are calculated by constructing a matrix of the individual scores, calculating the geometric mean for each row, and normalizing the means accordingly (Vidulich, Ward, & Schueren, 1991). SA-SWORD has been very limited, but the initial report indicates that it was sensitive, consistent, and showed a markedly different pattern than workload ratings collected in the same scenarios (Vidulich & Hughes, 1991).

Knowledge-Based Procedures

Knowledge-based procedures provide an objective measure of situation awareness by comparing participants' perceptions of the situation to the actual state of the world. The best-known of these techniques is the Situation Awareness Global Assessment Technique, or SAGAT (Endsley, 1996). SAGAT consists of a series of questions posed to subjects engaged in a medium- or high-fidelity simulation of some activity (e.g., piloting an aircraft). The simulation is interrupted at random times, the display screens are blanked, and a series of questions (called probes) about the current status of elements in the situation are presented to the participant. The answers to the questions are scored as *correct* or *incorrect* depending on their concordance with the actual states of the various elements at the time the simulation was halted. Following presentation of the questions, the simulation is typically resumed where it stopped, to be frozen again periodically with more questions presented as the simulation progresses through the planned scenario.

The SAGAT procedure has been employed across a wide range of dynamic tasks, including aircraft control, air traffic control, driving, and nuclear power plant control (Endsley, 2000). The element common to all of them is that, by necessity, they must be tasks that can be simulated with reasonable fidelity. For each task, SAGAT questions must be developed to fully probe the situation awareness construct. Thus, probe questions should be directed at the presence or absence of elements in the situation (level 1 situation awareness), the participants comprehension of their meaning (level 2), and the anticipated future state of the elements (level 3). Question construction requires careful thought and crafting on the part of the tester. Examples of situation awareness questions presented during one SAGAT freeze in a simulated air traffic-control task are shown in Fig. 6.8 (Luther & Charlton, 1999). The number of questions presented during each freeze should be kept to a small number to minimize interference effects in working memory. The questions presented during any one freeze are thus drawn from a larger set of potential questions. Scoring participants' answers to the questions may require establishing a tolerance band around the actual value (e.g., airspeed within 10 kn), depending on the task tolerances. As the resulting scores reflect a binomial distribution of correct/incorrect answers, a chi-square or other binomial test of significance will be required.

Another practical aspect is knowing when and how often to freeze the simulation. Endsley recommends that the timing of the freezes and probe questions be randomly determined so that test participants cannot prepare for them in advance (Endsley, 2000). Endsley also recommends that no freezes occur earlier than 3 to 5 minutes into the simulation, and no 2 freezes occur within 1 minute of each other to allow participants to build up a mental picture of the situation. Combining data from different freeze points to form an overall situation awareness score may also compromise the sensitivity of the method because momentary task demands will change throughout the simulation (Endsley, 2000).

Although the SAGAT method is not without its drawbacks, particularly in regard to its dependence on the degree to which the subjects are consciously aware of all the information actually involved in their decisions and actions (Nisbett & Wilson, 1977), it does have the advantages of being easy to use (in a simulator environment), possessing good external indices of information accuracy, and possessing well-accepted face validity (Endsley, 1996). A criticism of SAGAT and related techniques has been that the periodic interruptions are too intrusive, effectively contaminating any performance measures gathered from the simulation. A related concern has been the effect of the probe questions themselves: setting up an

---

*Please answer the questions about each aircraft
by writing in the space provided or by circling an answer*

Aircraft [1]: Speed – Under 100 knots
                     100-150 knots
                     150-200 knots
                     Over 200 knots
What is this aircraft's   Altitude _____
Is this aircraft   ascending/descending/level
Is this aircraft   taking off/ landing/overflight
.
.
.

Aircraft [n]: Speed – Under 100 knots
                     100-150 knots
                     150-200 knots
                     Over 200 knots
What is this aircraft's   Altitude _____
Is this aircraft   ascending/descending/level
Is this aircraft   taking off/ landing/overflight

Where will the next aircraft to enter your zone appear from?
                     Auckland airport
                     Whenuapai Airport
                     Ardmore Airport
                     Passed off from Centre

Will Aircraft [1] get to Ardmore before Aircraft [3] gets to
   Whenuapai airport? _____

FIG. 6.8.   Example situation awareness questions from a simulated air traffic control task.

expectancy for certain types of questions and affecting what aspects of the environment the participants attend to (Pew, 2000; Sarter & Woods, 1991). Endsley has attempted to address these concerns by showing that participants' performances on probe trials were not statistically different than performances on nonprobe trials (Endsley, 2000). Another approach has been to include the use of control or distractor questions that probe aspects of the situation that are irrelevant to the primary task to mitigate any effects of expectancy or attentional focusing (Charlton, 1998).

The knowledge-based technique SAGAT is a valuable contribution to the study of situation awareness and is certainly well worth considering for tests where the focus of operator attention is of interest. For many application domains it has been found to possess good sensitivity, diagnosticity, and relevance. The user does need to be prepared, however, for questions and concerns inevitably raised regarding operator acceptance and intrusiveness of the technique.

SUMMARY

Given the variety of mental workload and situation awareness metrics available, there is one or more methodology suitable for nearly every field test application. In the case of mental workload measurement, the TLX scale has seen the widest use because of its good sensitivity and excellent diagnosticity. In situations involving continuous or extended operations where a real-time inventory of workload that does not adversely affect task performance is needed, the revised Crew Status Survey, the ARWES scale, may be used to good advantage. Both the TLX and the ARWES are sensitive and reliable, and they meet with good operator acceptance. Another indication of momentary workload can be gathered with secondary task measures, where such tasks are plausible, minimally intrusive, or preexisting. If a detailed analysis of the constituent elements of task-specific workload is necessary, powerful but more cumbersome methods, such as SWAT or psychophysiological metrics, may be appropriate. Where a simple indicator of potential workload is needed early on in system development, time-line analysis or scheduling theory will be useful.

The choice of a situation awareness metric will usually be a trade-off between the type of results desired against constraints on test resources. For example, in a field testing environment, where data collection opportunities may be restricted to a noninterference basis, knowledge-based measures such as SAGAT typically will be impractical. In these situations, a simple, practical measure of situation awareness, such as SART, can provide a reasonably diagnostic, overall indicator of situation awareness. Where more controlled test conditions are available, SAGAT and secondary task measures can enable moment-by-moment analysis of the situation awareness afforded by a given interface or the focus of users' attention during a task scenario. The underlying message for the human factors tester is that measurements of cognitive states, such as mental workload and situation awareness, are likely to remain an important component of test design for the foreseeable future. A favorite methodology, however, should never drive the test design. The nature of the system and the issues to be resolved will determine the nature of the test, which in turn will determine the appropriate mental workload and situation awareness methods.

REFERENCES

Adams, M. J., Tenney, Y. J., & Pew, R. W. (1995). Situation awareness and the cognitive management of complex systems. *Human Factors, 37,* 85–104.

Ames, L. L., & George, E. J. (1993, July). *Revision and verification of a seven-point workload scale* (AFFTC-TIM-93-01). Edwards AFB, CA: Air Force Flight Test Center.

Bell, H. H., & Waag, W. L. (1995). Using observer ratings to assess situational awareness in tactical air environments. In D. J. Garland & M. R. Endsley (Eds.), Experimental analysis and measurement of situation awareness (pp. 93–99). Daytona Beach, FL: Embry-Riddle Aeronautical University Press.

Bortolussi, M. R., Kantowitz, B. H., & Hart, S. G. (1986). Measuring pilot workload in a motion base simulator. *Applied Ergonomics, 17,* 278–283.

Charlton, S. G. (1991, July). *Aeromedical human factors OT&E handbook* (2nd ed.) (AFOTEC Technical Paper 4.1). Kirtland AFB, NM: Air Force Operational Test and Evaluation Center.

Charlton, S. G. (1996). Mental workload test and evaluation. In T. G. O'Brien & S. G. Charlton (Eds.), *Handbook of human factors testing and evaluation.* Mahwah, NJ: Lawrence Erlbaum Associates.

Charlton, S. G. (1998). Driver perception and situation awareness. *Proceedings of the Eighth Conference of the New Zealand Ergonomics Society, 1,* 51–54.

Charlton, S. G. (2000, July). *Driver vehicle interactions: Maintenance of speed and following distances.* (Technical Report contracted by the Foundation for Research Science and Technology). Hamilton, NZ: Transport Engineering Research NZ Ltd.

Cooper G. E., & Harper, R. P. (1969). *The use of pilot rating in the evaluation of aircraft handling qualities* (NASA TN-D-5153). Moffett Field, CA: NASA Ames Research Center.

DOD-HDBK-763. (1987). *Human engineering procedures guide.* Author.

Endsley, M. R. (1995). Toward a theory of situation awareness in dynamic systems. *Human Factors, 37,* 32–64.

Endsley, M. R. (1996). Situation awareness measurement in test and evaluation. In T. G. O'Brien & S. G. Charlton (Eds.), *Handbook of human factors testing and evaluation.* Mahwah, NJ: Lawrence Erlbaum Associates.

Endsley, M. R. (2000). Direct measurement of situation awareness: Validity and use of SAGAT. In M. R. Endsley & D. J. Garland (Eds.), *Situation awareness analysis and measurement.* Mahwah, NJ: Lawrence Erlbaum Associates.

Flach, J. M. (1995). Situation awareness: Proceed with caution. *Human Factors, 37,* 149–157.

Gawron, V. J., Schiflett, S. G., & Miller, J. C. (1989). Cognitive demands of automation in aviation. In R. S. Jensen (Ed.), *Aviation psychology* (pp. 240–287). Brookfield, VT: Gower.

Gawron, V. J., Schiflett, S. G., Miller, J. C., Slater, T., Parker, F., Lloyd, M., Travele, D., & Spicuzza, R. J. (1988, January). *The effect of pyridostigmine bromide on inflight aircrew performance* (USAF-SAM-TR-87-24). Brooks AFB, TX: USAF School of Aerospace Medicine.

Gopher, D., & Donchin, E. (1986). Workload: An examination of the concept. In K. R. Boff, L. Kaufman, & J. Thomas (Eds.), *Handbook of perception and human performance, Vol. 2: Cognitive processes and performance* (pp. 41.1–41.49). New York: Wiley.

Gugerty, L. (1997). Situation awareness during driving: Explicit and implicit knowledge in dynamic spatial memory. *Journal of Experimental Psychology: Applied, 3,* 1–26.

Hart, S. G. (1986). Theory and measurement of human workload. In J. Zeidner (Ed.), *Human productivity enhancement, Vol. 1: Training and human factors in systems design* (pp. 396–456). New York: Praeger.

Hart, S. G., Childress, M. E., & Hauser, J. R. (1982). Individual definitions of the term "workload." *Proceedings of the Eighth Symposium on Psychology in the Department of Defense,* 478–485.

Hart, S. G., & Staveland, L. E. (1988). Development of the NASA Task Load Index (TLX): Results of empirical and theoretical research. In P. A. Hancock & N. Meshkati (Eds.), *Human mental workload* (pp. 139–183). Amsterdam: North-Holland.

Hill, S. G., Iavecchia, H. P., Byers, J. C., Bittner, A. C., Zaklad, A. L., & Christ, R. E. (1992). Comparison of four subjective workload rating scales. *Human Factors, 34,* 429–439.

Isler, R. B. (1998). Do drivers steer in the direction they look? *Proceedings of Eighth Conference of the New Zealand Ergonomics Society, 1,* 55–57.

Jahns, D. W. (1973). Operator workload: What is it and how should it be measured? In K. D. Gross & J. J. McGrath (Eds.), *Crew system design.* (pp. 281–288). Santa Barbara, CA: Anacapa Sciences.

Jones, D. G. (2000). Subjective measures of situation awareness. In M. R. Endsley & D. J. Garland (Eds.), *Situation awareness analysis and measurement.* Mahwah, NJ: Lawrence Erlbaum Associates.

Kahneman, D., & Treisman, A. (1984). Changing views of attention and automaticity. In R. Parasuraman & D. R. Davies (Eds.), *Varieties of attention* (pp. 29–61). New York: Academic Press.

Kantowitz, B. H. (1985). Channels and stages in human information processing: A limited analysis of theory and methodology. *Journal of Mathematical Psychology, 29,* 135–174.

Klein, G. (2000). Analysis of situation awareness from critical incident reports. In M. R. Endsley & D. J. Garland (Eds.), *Situation awareness analysis and measurement.* Mahwah, NJ: Lawrence Erlbaum Associates.

Kuperman, G. G., & Wilson, D. L. (1985). A workload analysis for strategic conventional standoff capability. *Proceedings of the Human Factors Society 29th Annual Meeting,* 635–639.

Luther, R. E., & Charlton, S. G. (1999). Training situation awareness for air traffic control. *Proceedings of the Ninth Conference of the New Zealand Ergonomics Society,* 52–56.

Lysaght, R. J., Hill, S. G., Dick, A. O., Plamondon, B. D., Linton, P. M., Wierwille, W. W., Zaklad, A. L., Bittner, A. C., & Wherry, R. J. (1989, June). *Operator workload: Comprehensive review and evaluation of operator workload methodologies* (Technical Report 851). Alexandria, VA: U.S. Army Research Institute.

Moray, N. (1982). Subjective mental load. *Human Factors, 23,* 25–40.

Moray, N., Dessouky, M. I., Kijowski, B. A., & Adapathya, R. (1991). Strategic behavior, workload, and performance in task scheduling. *Human Factors, 33,* 607–629.

Nisbett, R. E., & Wilson, T. D. (1977). Telling more than we can know: Verbal reports on mental processes. *Psychological Review, 84,* 231–259.

O'Donnell, R. D., & Eggemeier, F. T. (1986). Workload assessment methodology. In K. R. Boff, L. Kaufman, & J. Thomas (Eds.), *Handbook of perception and human performance, Vol. 2: Cognitive processes and performance* (pp. 42.1–42.49). New York: Wiley.

Parks, D. L. (1979). Current workload methods and emerging challenges, In N. Moray (Ed.), *Mental workload its theory and measurement* (pp. 387–416). New York: Plenum.

Pew, R. W. (2000). The state of situation awareness measurement: Heading toward the next century. In M. R. Endsley & D. J. Garland (Eds.) *Situation awareness analysis and measurement.* Mahwah, NJ: Lawrence Erlbaum Associates.

Pritchett, A. R., & Hansman, R. J. (2000). Use of testable responses for performance-based measurement of situation awareness. In M. R. Endsley & D. J. Garland (Eds.), *Situation awareness analysis and measurement.* Mahwah, NJ: Lawrence Erlbaum Associates.

Reid, G. B., & Colle, H. A. (1988). Critical SWAT values for predicting operator overload. *Proceedings of the Human Factors Society 32nd Annual Meeting,* 1414–1418.

Reid, G. B., & Nygren, T. E. (1988). The Subjective Workload Assessment Technique: A scaling procedure for measuring mental workload. In P. A. Hancock & N. Meshkati (Eds.), *Human mental workload* (pp. 185–218). Amsterdam: North-Holland.

Reid, G. B., Shingledecker, C. A., & Eggemeier, F. T. (1981). Application of conjoint measurement to workload scale development. *Proceedings of the Human Factors Society 25th Annual Meeting,* 522–525.

Reid, G. B., Shingledecker, C. A., Hockenberger, R. L., & Quinn, T. J. (1984). A projective application of the Subjective Workload Assessment Technique. *Proceedings of the IEEE 1984 National Aerospace Electronics Conference-NAECON 1984*, 824–826.

Reub, J., Vidulich, M., & Hassoun, J. (1992). Establishing workload acceptability: An evaluation of a proposed KC-135 cockpit redesign. *Proceedings of the Human Factors Society 36th Annual Meeting*, 17–21.

Rokicki, S. M. (1982, October). *Fatigue, workload, and personality indices of air traffic controller stress during an aircraft surge recovery exercise* (Report SAM-TR-82-31). Brooks AFB, TX: USAF School of Aerospace Medicine.

Roscoe, A. H. (1987). In-flight assessment of workload using pilot ratings and heart rate. In A. H. Roscoe (Ed.), *The practical assessment of pilot workload, AGARDograph No. 282* (pp. 78–82). Neuilly Sur Seine, France: AGARD.

Saaty, T. L. (1980). *The analytic hierarchy process*. New York: McGraw-Hill.

Sarter, N. B., & Woods, D. D. (1991). Situation awareness: A critical but ill-defined phenomenon. *The International Journal of Aviation Psychology, 1*(1), 45–57.

Selcon, S. J., Taylor, R. M., & Koritsas, E. (1991). Workload or situational awareness? TLX vs. SART for aerospace systems design evaluation. *Proceedings of the Human Factors Society 35th Annual Meeting*, 62–66.

Sheridan T. B. (1980). Mental workload—What is it? Why bother with it? *Human Factors Society Bulletin, 23*, 1–2.

Smith, K., & Hancock, P. A. (1995). Situation awareness is adaptive, externally directed consciousness. *Human Factors, 37*, 137–148.

Taylor R. M., & Selcon, S. J. (1991). Subjective measurement of situation awareness. *Designing for everyone: Proceedings of the 11th Congress of the International Ergonomics Association*, 789–791.

Tsang, P. S., & Vidulich, M. A. (1989). Cognitive demands of automation in aviation. In R. S. Jensen (Ed.), *Aviation psychology* (pp. 66–95). Brookfield, VT: Gower.

Vidulich, M. A. (2000). Testing the sensitivity of situation awareness metrics in interface evaluations. In M. R. Endsley & D. J. Garland (Eds.), *Situation awareness analysis and measurement*. Mahwah, NJ: Lawrence Erlbaum Associates.

Vidulich, M. A., & Hughes, E. R. (1991). Testing a subjective metric of situation awareness. *Proceedings of the Human Factors Society 35th Annual Meeting*, 1307–1311.

Vidulich, M. A., Stratton, M., Crabtree, M., & Wilson, G. (1994). Performance-based and physiological measures of situation awareness. *Aviation, Space, and Environmental Medicine, 65* (Suppl. 5), 7–12.

Vidulich, M. A., Ward, G. F., & Schueren, J. (1991). Using the subjective workload dominance (SWORD) technique for projective workload assessment. *Human Factors, 33*(6), 677–691.

Waag, W. L., & Houck, M. R. (1994). Tools for assessing situational awareness in an operational fighter environment. *Aviation, Space, and Environmental Medicine, 65* (Suppl. 5), 13–19.

Warr, D., Colle, H. A., & Reid, G. B. (1986). A comparative evaluation of two subjective workload measures: The Subjective Workload Assessment Technique and the Modified Cooper-Harper Scale. *Proceedings of the Tenth Symposium on Psychology in the Department of Defense*, 504–508.

Wickens, C. D. (1991). *Engineering psychology and human performance* (2nd ed.). Colombus, OH: Merrill.

Wierwille, W. W., & Casali, J. G. (1983). A validated rating scale for global mental workload measurement application. *Proceedings of the Human Factors Society 27th Annual Meeting*, 129–133.

Wierwille, W. W., & Eggemeier, F. T. (1993). Recommendations for mental workload measurement in a test and evaluation environment. *Human Factors, 35*, 263–281.

Wierwille, W. W., & Williges, B. H. (1978, September). *Survey and analysis of operator workload analysis techniques* (Report No. S-78-101). Blacksburg, VA: Systemetrics.

Wilson, G. F. (2000). Strategies for psychophysiological assessment of situation awareness. In M. R. Endsley & D. J. Garland (Eds.), *Situation awareness analysis and measurement*. Mahwah, NJ: Lawrence Erlbaum Associates.

Wolf, J. D. (1978). *Crew workload assessment: Development of a measure of operator workload* (AFFDL-TR-78-165). Wright Patterson AFB, OH: Air Force Flight Dynamics Laboratory.

Chapter 7

Psychophysiological Test Methods and Procedures

Glenn F. Wilson
Air Force Research Laboratory

Psychophysiological measures are considered by many people to be one of the three main methods of monitoring operator states. The other two are operator performance measurement and measures of cognitive states. Taken together these three methods provide a well-rounded approach to human factors testing. The multidimensional approach is especially appropriate when dealing with how operators function when interacting with the current, complex systems that are prevalent in most areas of human endeavor. Any one method will probably not provide enough information to understand the full scale of human skills and abilities required to operate and interact with these complex systems. Although the full suite of methods may not be required or feasible in every instance of test and evaluation, one should consider using as complete a slate of methods as possible and reasonable.

Psychophysiological measures can provide insight into the basic responses and functional state of operators. Because they do not depend on verbal report or performance measures derived from the operated system, they are sometimes easier and less intrusive than other methods. Heart rate, for example, provides continuous information about how a person is coping with a task. This can be accomplished in situations where it is not possible or feasible to acquire performance data. For the most part, the psychophysiological data can be collected nonintrusively. This means that not only is the operator unaware of the data collection, but also he or she does not have to interrupt work to provide opinions about the nature of the task. Although electrodes are applied to operators and recorders may be worn in order to collect electrophysiological data, operators quickly adapt to this equipment. This is because of the same phenomenon that makes us unaware of wearing wrist watches and clothing. Because amplifiers and records are small and light weight, they can be worn in almost any environment, including in space and under water.

Psychophysiological measures are capable of apprising the tester of the functional state of an operator; the operator's response to new equipment, procedures, or both; and the operator's level of workload and vigilance (Wilson & Eggemeier, 1991). They can also be used to determine if task stimuli are perceived and how efficiently the brain is able to process relevant information from the task. They probably won't be able to tell you what the person is thinking about. They may be able to discern biases when the physiological data differ from subjective reports. Covariation with performance data, which is also at variance with subjective reports, is especially helpful in these situations. Psychophysiological measures can be used to determine the operator's overall state. For example, you can determine if their mental workload is high or if they are about to fall asleep. It is possible to determine their level of attention by monitoring the pattern of electrical brain activity, for example.

AN EXPLANATION OF PSYCHOPHYSIOLOGICAL MEASURES

Psychophysiology is literally the joining of psychology and physiology to understand how they interact. This interaction goes both ways. The psychology part can cause changes in the person's physiology, and the physiological state influences a person's behavior. Ackles, Jennings, and Coles (1985) defined psychophysiology as "a scientific discipline concerned with the theoretical and empirical relationships among bodily processes and psychological factors" (p. ix). Surwillo's (1986) definition states, "In general terms, psychophysiology is the scientific study of the relationship between mental and behavioral activities and bodily events" (p. 3). In test and evaluation we are typically concerned with how an operator's interaction with a system influences their physiology. By monitoring their physiology we are able to infer the cognitive and emotional demands that the job places on the person. For example, cardiac measures were used to evaluate the utility of a new ambulance dispatching system and showed that the new system, compared with the older system, reduced systolic blood pressure during times of high workload (Wastell & Newman, 1996). This result, along with performance improvements, were seen as validating implementation of the new system.

Anatomically the nervous system is divided into two main divisions, central nervous system (CNS) and peripheral nervous system (PNS). The peripheral portion is further divided into somatic and autonomic. The CNS consists of the brain and spinal cord, and the autonomic system is divided into the sympathetic and parasympathetic. Basically, the CNS receives input from the senses, processes this information in the context of past experience and inherent proclivities, and initiates actions upon the external world. The PNS provides the sensory input and carries out the actions initiated by the CNS. The autonomic portion of this complex is responsible for the internal milieu, thus providing an overall environment that supports the well-being of the person. CNS measures of interest to test and evaluation include the electroencephalograph (EEG) and event-related potentials (ERP). The EEG provides information concerning the electrical activity of the CNS. The EEG is typically decomposed into several bands of activity that originated in clinical medicine but have been found to have application to many fields. Using 20 or more EEG channels permits the construction of topographic maps that provide additional spatial information. The topography shows the areas of the scalp where relevant frequency changes occur. ERPs represent the brain activity associated with the brain's processing of discrete stimuli, information, or both. ERPs are thought to represent the several stages of information processing involved in cognitive activity related to particular tasks. Other measures including positron emission tomography (PET), functional magnetic resonance imaging (fMRI) and the magnetioencephalogram (MEG) provide fantastic insights into CNS functioning. However, they are presently limited to laboratory environments. They will no doubt provide the impetus for field applications, but in their current state their use is not at all feasible for test and evaluation applications with most systems. The restraints involve the recording equipment that requires subjects to remain immobile in the recording apparatus and does not permit ambulation. Also, high levels of shielding from external magnetic and electrical fields are required, which limit real-world application at present.

PNS measures of interest to test and evaluation include cardiac functions and respiration, eye, and electrodermal activity. Cardiac activity, particularly heart rate, has been used for the longest time in applied situations. Heart rate was monitored in Italian pilots by Gemelli as early as 1917 (Roscoe, 1992). It is a robust measure that is easy to collect and is widely used. Heart rate has been employed in the certification of commercial aircraft (Blomberg, Schwartz, Speyer, & Fouillot, 1989; Wainwright, 1988). The U.S. Air Force has used heart rate in several operational test and evaluation programs. It has been used in the testing of the B-1 bomber, C-17 transport, C-135 tanker cockpit upgrade, and the C-130J transport. The variability of the heart rhythm has also been used for a number of years. Theoretically, this rhythmic activity of the heart is influenced by higher brain centers. Cognitive activity is hypothesized to reduce the rhythmicity of the heart. Increased cognitive activity, such as mental workload, causes reduction in the rhythmic activity. One difficulty with this measure is the lack of consensus on the most appropriate procedure of quantifying the variability. In 1973, Opmeer reported finding 26 different methods for calculating heart rate variability (HRV) in the literature; no doubt the number is even higher now. Eye activity has been used in a number of investigations to monitor the visual demands of a task. Increased visual demands by the task usually result in decreased blinking. Because we are functionally blind during the short period of the eye lid closure, reduced blinking will improve the likelihood of detecting important visual events. Respiration has been used to measure operator responses to work situations. Typically the rate and depth of breathing are measured. Electrodermal activity (EDA) monitors the activity of the eccrine sweat glands. These sweat glands are enervated by the sympathetic branch of the autonomic nervous system (ANS). Several ways to measure the activity of these glands exist. EDA is closely related to emotional responses, a fact that led to including EDA as a measure of the lie detector. EDA has been found useful in test and evaluation as well. These measures will be described more fully in the following sections.

Hormones produced in the body in response to environmental changes can also be measured. These include cortisol and epinephrine levels in the blood, urine, and saliva. Changes in hormone levels can take 5 to 30 minutes to become evident in body fluids following significant environmental events. This is a handicap in situations where the task is associated with rapidly changing events. However, in other situations, where overall effects on operators over longer periods of time is of concern, hormonal measures have value.

ADVANTAGES OF PSYCHOPHYSIOLOGICAL MEASURES

Psychophysiology provides a number of measures that can be used by the human factors practitioner to determine the impact of system operation on the operator. With these tools, it is possible to nonintrusively and continuously monitor the state of the human operator. Tasks do not have to be interrupted to obtain self-reports. Operators do not have to suspend task performance or break their train of thought to provide an immediate perspective on their interaction with the system that they are operating. For the most part, psychophysiological measures do not require modification of operated systems, which can be the case with performance measures that require data from the system. The continuous nature of these measures provides several advantages to test and evaluation. In real-world testing, unplanned events can provide important data. Because the physiological data is continuously recorded, it is possible to analyze the data preceding, during, and following these unplanned events. Because a time line accompanies the physiological data, it is easy to locate the relevant portion of the data for analysis. Rokicki (1987) used continuously recorded electrocardiogram (ECG) data as a debriefing tool during aviation operational test and evaluation flights. During postflight debriefs, the heart-rate data were examined to locate episodes for further discussion. Using these methods, it was possible to note unexpected increases in heart rate. This information was used to help crew members recall important or forgotten relevant events during the flights. Currently, most psychophysiological equipment involves recording of the data during task performance and playing it back for analysis at a later time. However, online monitoring is possible because of the continuous nature of the physiological data. This online capability means that the operator's physiological responses to system operation are immediately available. Based on this information, the test scenario can be modified at once and more data can be collected to determine the effects of the new manipulation. One such system has been developed by the U.S. Air Force (Wilson, 1994). Physiological data could also be used as input to adaptive aiding systems that monitor the functional state of the operator and alert the operator and system when the operator's state is not optimal. Modifications to the task can be implemented immediately to adjust to the current needs of the operator. This should result in improved overall system performance. Laboratory research has demonstrated the feasibility of this approach (Freeman et al., 1999; Wilson, Lambert, & Russell, 2000).

Performance measures may not be good gauges of operator capacity because of the ability of human operators to maintain performance up to the point of breakdown. The correlation between psychophysiological measures and performance means that the physiological data can show changes that precede performance breakdown. This may permit the prediction of performance breakdown and allow for the avoidance of errors.

In the wide range of test and evaluation situations, no one measure will be able to provide all of the required data. Psychophysiological data can be used to augment subjective and performance methods to acquire a better evaluation of the effects of system operation upon the operator. In situations where it is not possible to collect subjective or performance data, or both, it may be possible to collect physiological data. There are, of course, times when it is not feasible to collect physiological data.

IMPLEMENTATION OF PSYCHOPHYSIOLOGICAL METHODS

As with subjective and performance measures, there are issues that must be addressed prior to implementing psychophysiological measures. Their proper use requires knowledge and experience; however, this is the case with all of the methods used in test and evaluation. Although the skills might be more specialized, the principles of physiological data collection and analysis are not difficult to learn and apply. Several good handbooks, guides, and journal articles on the proper collection and analysis procedures for these data exist and are listed at the end of this chapter. With the inclusion of psychophysiology into college courses, many people have some familiarity with these methods.

One of the issues that must be addressed when using psychophysiological methods is the requirement of specialized equipment. Although subjective methods use only paper and pencils, physiological data collection requires more specialized equipment. If the operators are free to move about their work space, then ambulatory physiological recorders will be needed. Small, lightweight recorders are readily available. Several manufacturers provide excellent equipment that can be worn by operators as they go about their daily routines. Examples of ambulatory recorders that can be easily worn by operators while performing their jobs are shown in Fig. 7.1.

If operators do not move about during testing, then it is possible to use larger and more readily available laboratory-type equipment. Several manufacturers provide amplifiers, filters, and analysis systems that can be used when operators do not need to move about their work space. This is similar to the familiar laboratory situation where subjects remain seated during testing. Operators can be quickly and conveniently disconnected from the recording equipment during breaks. Amplifiers for recording all of the described measures are readily available in a wide price range. Off-the-shelf data analysis software that runs with standard analog-to-digital converters is also available.

Another option is to use telemetry. Small amplifier/transmitter sets are worn by the operators while on the job. These units are connected to the electrodes and perform the same function as the amplifiers in other systems. The amplified data is sent to a receiving unit. This permits the operator to be totally free to move about the work space and be unencumbered by attached wires. The range of transmission varies from several feet to several miles.

Physiological measures provide relative, not absolute, data. Although there are physical limits and guidelines for the physiology of physical work, no guidelines exist for cognitively related psychophysiological data.

FIG. 7.1.   Three representative physiological recording devices. The recorder on the right is a clinical Holter monitor that records three channels of ECG data on a microcassette tape. The tape is played back to analyze the data. The device in the center is a Holter monitor that detects ECG R waves online and stores the interbeat intervals in memory. After data collection the interbeat intervals are transferred to a computer for further analysis. The device on the left is a multipurpose recorder that can record up to eight channels of data on either a PCMCIA disk or flash memory card. Each channel is separately programmable with regard to gain, filter settings, and digitization rate. All three devices can hold several hours of data and can be easily worn by ambulatory operators.

Because of individual differences in responding, repeated-measures experimental designs are often used. For example, resting heart rate varies a great deal from person to person, depending upon their level of physical fitness, general health, and other factors. Further, the reliability of individuals' data varies. That is, some people do not exhibit large increases in heart rate when performing a difficult task, whereas others will show large increases. By using each operator as their own control, one can avoid some of the problems inherent with these measures. For example, determining change from baseline or resting periods is a useful procedure to account for individual differences. This problem is not unlike differences in the subjective workload reports from different operators; some restrict their ratings to a small portion of the range, and others use the full range.

APPLICATION AREAS

Psychophysiological measures have been used in a number of areas for test and evaluation. The following discussion provides examples, organized by the type of system evaluated. Transportation testing makes up the bulk of this work and ranges from automobile to space transportation.

Trains, Planes, and Automobiles

Automobiles. Brookhuis and de Waard (1993) used psychophysiological measures to validate performance measures of driver impairment. ECG and EEG measures were recorded while drivers drove with blood-alcohol levels just under the legal limit and separately for 2.5 hours to monitor their vigilance. HRV was also used but because it did not differ from HR it was not included in the total analysis. The psychophysiological measures correlated with performance measures of lane deviations, steering wheel movements, and car following distance. The psychophysiological changes preceded the performance impairments. Using this positive correlation between the physiological and performance measures, the authors felt that it was possible to develop performance-only based measures of driver impairment. The psychophysiological measures were used to validate the adequacy of the performance measures.

Richter, Wagner, Heger, and Weise (1998) used psychophysiological measures to ergonomically design roads to improve driver performance. Road curvature on older roads needed to be improved to make the roads safer. To accomplish this task the worse sections of the roads had to be identified so they could be fixed first. HR, EDA, and eye blinks were recorded while drivers drove on roads with varying curvature change rates. The strongest results were found for HR. HRV effects were not found for all road conditions, only with driving loads at lower and intermediate curvature levels. Blink rate exhibited stable correlations, with fewer blinks correlating with the highest curvature-change rates and lowest blink rates with straight road driving. The authors felt that EDA was more influenced by situationally dependent events, such as oncoming traffic, than the other psychophysiological measures. They suggest that an applied psychophysiological approach to ergonomic design of new roadways should use cardiovascular and blink rate measures for biopsychological assessment.

HR and HRV were used to measure the effects on truck driving with different numbers of trailers and connection types (Apparies, Riniolo, & Porges, 1998). Twenty-four professional truck drivers drove pulling one or three trailers over a standard route that took 8 to 10 hours to complete. Only HR showed significant changes among the trailer configurations. Increased HRs were found when driving with the more demanding configurations of trailer and connections. HR was found to increase over the driving session, whereas HRV decreased.

The effects of overnight driving on drivers and passengers was measured using blink frequency and duration, steering wheel inputs, and subjective reports (Summala, Häkkänen, Mikkola, & Sinkkonen, 1999). The drivers drove 1,200 km during one night and rode as a passenger on the same route during another night. Eye blinks were monitored via video camera. Blink frequency increased with driving time, which indicated tiredness, and was higher when driving than when riding. Further, blink frequency decreased when the driving task became more difficult, such as when approached by oncoming trucks. Steering wheel inputs did not vary systematically. Zeier (1979) evaluated the merits of driving automobiles with manual or automatic transmissions in city conditions. Each of the 12 operators served as driver and passenger. Acting as driver and driving with the manual transmission were associated with greater adrenaline secretion and higher HR and HRV and more EDAs. These results were interpreted to show that automobiles with automatic transmissions were safer and less stressful to operate in city traffic than those with manual transmissions.

To reduce traffic speed on rural roads changes were made to the roadways to reduce driver comfort and raise driver workload (De Waard, Jessurun, & Steyvers, 1995). Psychophysiological, subjective, and performance measures were used to examine the effect of this new road design on driver performance and mental load. The statistical variability of the HR and the spectrally determined 0.1-Hz band decreased on the modified roadway, and heart rate increased. All of the cardiac measures showed changes when compared to baseline data. Lateral position on the roadway and speed also showed significant effects. Forehead corrugator EMG did not show significant effects. It was concluded that the changes to the roadway were an effective means to reduce driver speed and increase safety on rural roads.

Taylor (1964) reported that higher rates of EDA activity were significantly correlated with sections of roadway that had a higher accident rate. This suggests that EDA could be used to determine which sections of new roadways will be more prone to higher accident rates. Helander (1978) ranked traffic events on the basis of EDA data and found that the rankings had a very high correlation with braking activity (.95) and had high face validity. He suggested that highways should be designed to minimize braking, thereby lowering drivers' stress and increasing safety.

Casali and Wierwille (1980) used psychophysiological measures to determine the causes of automobile simulator sickness when using motion-based simulators. HR, skin pallor, forehead perspiration, and respiration rate were used to monitor subjects' reactions to the simulator conditions.

All psychophysiological measures, except HR, showed effects caused by the simulator manipulations. Casali and Wierwille concluded that simulator sickness could be reduced by not rotating the simulator platform to simulate translation, avoiding system delays, and not enclosing the subjects inside the simulator.

Buses. Bus driving is a very stressful job and is associated with high rates of illness (Mulders, Meijman, O'Hanlon, & Mulder, 1982). In an effort to improve the traffic environment and reduce driver stress, Stockholm, Sweden, made a number of changes to bus routes, bus stops, and passenger information systems (Rydstedt, Johansson, & Evans, 1998). These changes resulted in reduced HR and systolic blood pressure in the drivers, thus validating the changes. It was felt by the authors that the bus drivers' health environment would be made better by such changes. In another effort to reduce stresses on bus drivers, a project was undertaken to isolate problem areas in the drivers' workstation, design a new workstation and evaluate the new workstation, using psychophysiological and subjective measures in all phases (Göbel, Springer, & Scherff, 1998). Based on the identified factors, a new workstation was designed and implemented. Among the changes was a reduction in the number of elements on the instrument panel from 64 to 30. Testing of the new workstation revealed lowered HR, decreased task execution times, and 84% drivers preference for the new configuration. The authors felt that the psychophysiological measures identified important design factors that would have been overlooked with subjective data alone. Though novel and slightly time-consuming, they felt that the use of psychophysiological measures yielded a better design and evaluation of the new workstation than that yielded by conventional methods.

Trains. Sleepiness in train drivers poses safety hazards to drivers, other crew, passengers, and the general public. To identify the level of alertness among night and day train drivers, Torsvall and Akerstedt (1987) studied the psychophysiology of 11 train drivers. EEG, ECG, electrooculogram (EOG), and subjective sleepiness were recorded during approximately 4.5 hours of driving, during either the night or the day. Night driving was associated with increased subjective sleepiness, increased levels of slow eye movements, and alpha-, theta-, and delta-band EEG activity. Four night drivers admitted dozing off, and two of them missed signals during times of large alpha bursts. On the other hand, Myrtek et al. (1994) reported lower heart rates during automated high-speed train driving than when stopping and starting the train at stations. These psychophysiological results were at odds with subjective and observer data, indicating their utility. As expected, drivers on a mountain route showed decreased HRV and T-wave changes compared with drivers on a high-speed, level route. Taken together, these results show that psychophysiological measures can provide unique data not available by other methods and that they can be used to evaluate human systems.

Airplanes. Psychophysiological measures have been used for years in aviation. HR has had the most use with the more recent addition of EEG and ERPs. EDA and respiration have had limited use. Notable of the earlier studies are the works of Roman in the 1960s and Roscoe during the 1970s and 1980s. (See Roscoe [1992] for an excellent review of the work through the 1980s.) Using HR, Roscoe (1979) demonstrated that Harrier ski-jump takeoffs were no more difficult than conventional, short takeoffs from a runway. Roscoe (1975) evaluated the effects on pilots of steep gradient approaches that were to be used for noise abatement. His results demonstrated that the steeper approaches did not involve higher pilot workload than the customary approaches. One question regarding the value of psychophysiological data had to do with the relative contributions of cognitive activity, such as mental workload versus stress and fear. The data in Fig. 7.2 show that the mental demands of landings are associated with increased heart rates, predominately in the pilot in command of an aircraft. Increased heart rates are not found in the pilots who are not in command (i.e., not performing the landing) but who are in the same aircraft and facing the same danger.

Wilson (1993) used multiple psychophysiological measures to evaluate the workload of pilots and weapon system officers during air-to-ground training missions with F4 fighter crews. Workload differences between pilots and weapon system officers were clearly distinguished, as were the demands of the various segments of the flights. HR was the best at discriminating among the flight segments. Blink rate was sensitive to the high visual demands of landing and bombing range activity. Wilson, Fullenkamp, and Davis (1994) further reported the first known ERPs to be collected during actual flight. The P2 component of the ERP was reduced in amplitude during flight relative to ground-recorded ERPs. Using general aviation aircraft, Hankins and Wilson (1998) determined the workload of a multiple-segment flight scenario using HR, EOG, EEG, and subjective measures. HR was sensitive to the cognitive demands of the flights, though not diagnostic. That is, heart rate increased almost equally during both takeoff and landing. Eye blink rate was found to be sensitive and diagnostic, decreasing during the visually demanding instrument flight-rules portions of the flights. Theta-band EEG showed increased activity during mentally demanding portions of the flights. More recently, Wilson (in press) used topographic EEG data, from 29 channels, to evaluate visual and instrument flight effects on general aviation pilots. The pilots exhib-

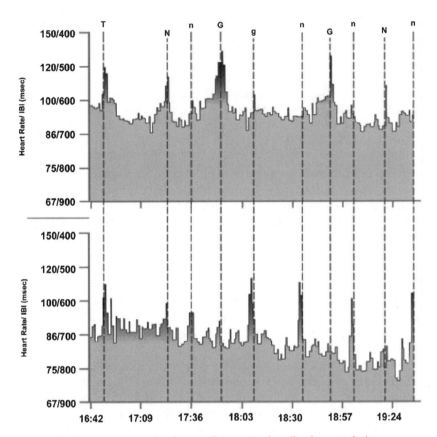

FIG. 7.2. Heart rate data from a pilot (top) and copilot (bottom) during a portion of an operational test and evaluation flight assessing a new transport aircraft. Mission time of day is shown on the abscissa. The figure demonstrates the well-known phenomenon in which the pilot in command has the higher heart rate during high workload segments. The events shown are takeoff (T), go arounds (G), and touch and goes (N). Uppercase letters designate that the pilot was in command and lowercase letters represent the copilot in command. Higher heart rates in the person responsible are associated with maneuvers. Even though the same level of danger was experienced by both flyers, only the one responsible showed increased heart rates. The level of heart rate increase is sizable. Both pilots show this response pattern, even though there are individual differences in their basic levels of heart rate. The pilot's heart rate was approximately 10 beats/minute higher than the copilot's.

ited changes in the scalp-recorded patterns of alpha activity that were consistent with the mental demands of the various segments of the flights. Topographic EEG methods permit the identification of the EEG bands affected by flight parameters and further localize the affects to specific brain regions. This enhances the power of EEG measures in test and evaluation by increasing the level of information about brain function during testing. An example of the merits of multiple measures is shown in Fig. 7.3, which depicts changes in heart rate and eye blink rate to the differing demands of general aviation flight.

Hunn and Camacho (1999) used HR, pilot performance, and subjective ratings during tests of an automatic ground collision avoidance system in high performance military aircraft. To test the system, pilots intentionally flew at the ground with various incidence angles. The pilots activated the automatic system as close to the ground as they felt comfortable. HR showed significant increases with each test run. HR and subjective ratings of anxiety were good measures of pilot anxiety. HR was a good indicator of higher-risk flight maneuvers.

In a helicopter simulation study designed to test the relative cognitive demands of verbal versus digital communication formats, Sirevaag et al. (1993) found blink-closure duration and ERP P300 amplitude to discriminate between the two modes. The P300s were elicited to nonattended stimuli. HRV also exhibited significant differences; however, further analysis indicated that these differences were probably due to respiration artifacts. Using ERP methodology, Fowler (1994) found that the latency of the P300 component increased with the difficulty of the simulated landing task. The P300s were elicited using stimuli of a secondary task. Veltman and Gaillard (1996) also reported respiration contamination of HRV data in a flight simulator study. HR and blood pressure showed significant effects of the task difficulty manipulations. The gain between blood pressure and HRV was sensitive to mental effort and not influenced by respiration. Eye blink duration was also found to be sensitive to the visual but not the cognitive demands of the simulated flights. In a later study, Veltman and Gaillard (1998) investigated the effects of flying a "tunnel in the sky" and aircraft following in a simulator. HR showed increases with heightened difficulty of the tunnel task. Blink rate decreased with increased visual demands. HRV, blood pressure, and respiration did not show reliable differences. Svensson, Angelborg-Thanderz, Sjöberg, and Olsson (1997) analyzed the effects of information complexity on pilot mental workload in a fighter simulator. Increased complexity of aircraft display was associated with impaired performance and increased HR. HR variability also showed decreased activity with increased complexity.

In a comparison of electromechanical versus glass cockpits, Itoh, Hayashi, Tsukui, and Saito (1990) used B747 and B767 simulators to have pi-

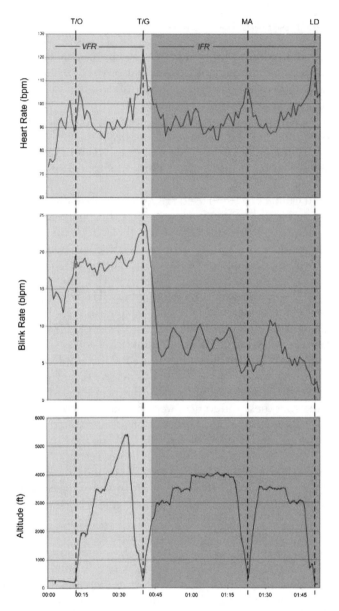

FIG. 7.3. Heart rate, blink rate, and altitude from a general aviation pilot demonstrating the value of multiple measures. Mission duration is shown on the X axis. Vertical lines denote takeoff (T/O), touch and go (T/G), missed approach (MA), and landing (LD). The light and dark shading designate visual flight rule (VFR) and instrument flight rule (IFR) flight segments, respectively. During IFR the pilot wore goggles that restricted vision to the aircraft instrument panel, which simulated night or adverse weather conditions. Note the increased heart rates preceding and following the landings. The smaller heart rate increases were associated with navigation and other mentally demanding portions of the flight. The increased visual demands of IFR flight result in a dramatic decrease of blink rate during the entire IFR segment. The smaller perturbations during the IFR segment were related to changes in visual demand associated with flight maneuvers such as performing holding patterns.

lots fly takeoffs and landings. Differences in gaze patterns were reported between the two groups. The gaze differences were due to the type of displays in the two simulators. No differences in workload were found. However, in abnormal situations, fewer gaze changes were found for B767 pilots, indicating that they were able to more quickly find needed information than the B747 pilots. Differences in the cockpit instrumentation between the two simulators resulted in more out-of-window gazing during landings by the B747 pilots. HRV decreased during the abnormal maneuvers. Lindholm and Cheatham (1983) used HR, HRV, and EDA to determine the effects of learning a simulated carrier-landing task. HR and EDA increased during the landings as the subject got closer to the touchdown on the carrier. Only HR showed decreased rates over the 30 trials while performance was improving.

Backs, Lenneman, and Sicard (1999) used the concept of autonomic space to reanalyze simulator data from a previous investigation of simulator workload in commercial aviation pilots. Principal components analysis was used to derive estimates of the relative influences of parasympathetic and sympathetic activity in response to the flight tasks. Different patterns of autonomic activity were associated with the various aspects of the flight scenario. This new approach yields greater insight into cardiac dynamics than simple HR analysis and may prove fruitful in the future.

The effects of simulated air refueling during straight and level flight versus a 10-degree turn and landing under visual or instrument conditions was studied by Sterman et al. (1993). Refueling during the 10-degree turn was associated with decreased EEG alpha band power over parietal scalp sites. Instrument landings showed decreased alpha power over right and left temporal scalp sites. These results suggest that multiple EEG recording sites are useful to detect significant changes in brain electrical activity that is related to different flying tasks. Electrical brain activity was also recorded in a low-fidelity simulation task by Sterman, Mann, Kaiser, and Suyenobu (1994) to determine changes in topographically recorded EEG between performing simulated landings, with hand movement only and eyes-closed conditions. The main finding was reduction of alpha-band activity over central and parietal scalp sites during landing. The authors felt that the reduction at the parietal scalp sites was related to cognitive processing. Application of EEG topographic procedures may yield a great deal of information about how pilots process the various types of information encountered during flying. As cited previously, Wilson (in press) used these procedures during actual flight.

The effects of space flight on cosmonaut sleep was investigated using EEG and body temperature during a 30-day mission (Gundel, Polyakov, & Zulley, 1997). Normal sleep patterns, determined by preflight ground data, were disrupted by space flight. The circadian phase was delayed by 2 hours, the latency to rapid eye movement (REM) sleep was shorter, and more slow wave sleep was found in the second sleep cycle. Further, the heart period was found to increase by about 100 ms during sleep in space (Gundel, Drescher, Spatenko, & Polyakov, 1999). HRV in the high-frequency band also increased during sleep in space. Although space is an extreme environment, psychophysiological measures can be used there to determine the effect on humans.

Psychophysiological measures have been used to evaluate air traffic control (ATC) environments. Zeir, Brauchli, and Joller-Jemelka (1996) evaluated busy compared with slow periods of actual ATC enviroments and found increases in salivary cortisol and subjective reports of difficulty after 100 minutes of work. The authors felt that salivary cortisol was a valid indicator of ATC workload. To study long-term stress effects on air traffic controllers, Sega et al. (1998) measured blood pressure and HR in controllers and a matched control group during a 24-hour period that included work shifts. Blood pressure and HR did not differ between the controllers and the control group, thus leading to the conclusion that controllers were adequately coping with the stresses of the job.

The effects of three different manipulations of workload during simulated ATC was evaluated using multiple psychophysiological measures (Brookings, Wilson, & Swain, 1996). Eye blink, respiration, EEG spectra, performance, and subjective reports reliably discriminated between task difficulty effects. Only the EEG discriminated between the different manipulations of workload. In further analysis of the EEG from this study, Russell and Wilson (1998) used a neural network classifier to discriminate between the manipulation of traffic volume, traffic complexity, and the overload condition. They reported a mean correct classification of 84% with a 92% classification accuracy between the overload condition and other conditions. This suggests the level and nature of mental workload can be accurately determined online using psychophysiological data.

Automation is suggested as a means to reduce workload and improve controllers' performance. However, the nature and means of application of the automation may actually interfere with the controllers' performance. Hilburn, Jorna, and Parasuraman (1995) used HRV and performance and subjective measures to determine the effects of three levels of automation. HRV demonstrated that the levels of workload could be distinguished and that the automation reduced workload. The subjective measure only discriminated between the workload levels and did not show an automation effect. Hilburn, Jorna, Byrne, and Parasuraman (1997) took a further step and used psychophysiological measures to determine the utility of adaptive aiding in contrast to the static aiding scheme used in the first study. The physiological data showed that mental workload could be reduced with the use of adaptive aiding in the air traffic controller task. Again, the physiological and subjective data dissociated, and the authors caution designers of automated systems not to rely entirely on operator opinion.

Workstation Evaluation

In an evaluation of new and conventional telephone switchboards, Brown, Wastell, and Copeman (1982) used ECG and EEG measures. Performance and subjective reports both favored the conventional system over the newer one. The physiological measures supported these findings, even though the effects were not strong.

In a simulation study in which operators controlled an electrical network, Rau (1996) used HR and blood pressure to determine the effects of decision latitude. HR and blood pressure increased with increasing task demands and with increased decision latitude. The increased stress responses with increased decision latitude were unexpected because it was predicted that the increased latitude would permit the operators more control over their jobs. However, these increased physiological responses are similar to aviation data from two pilot aircraft where the pilot in charge has been shown to have the higher HR (Hart & Hauser, 1987; Roscoe, 1978; Wilson, 1992). The influence of externally versus internally paced work has been investigated by Steptoe, Evans, and Fieldman (1997) using a battery of physiological measures. They found that externally paced tasks were associated with more deleterious effects than the self-paced conditions. Systolic blood pressure and EDA increased during the externally paced tasks.

Computer system response time is an important system feature for the operator to enhance performance and reduce stress. Kohlisch and Kuhmann (1997) evaluated the effects of either 2 s or 8 s system response times on HR, EDA, blood pressure, performance, and pain responses. The longer response time was judged to be better because it was associated with lower systolic blood pressure, more EDAs, and fewer errors in their performance. Ten years later, Kohlisch and Kuhmann (1997) investigated three system response times: 1s, 5s, and 9s. The short response time was associated with increased cardiovascular activity and poorer performance, whereas the long response time was associated with decreased cardiovascular activity and more reports of headaches. The medium response times were best in terms of performance and moderate levels of HR and blood pressure changes. The psychophysiological data imply that computer system response times that are optimal in terms of performance and operator health are in the range of 5s.

In an interesting use of ERP methods, Fukuzumi, Yamazaki, Kamijo, and Hayashi (1998) evaluated visual-display terminal colors using the latency of the ERP P100 component. The color evoking the shortest latency P100 was the most readable. Fukuzumi et al. concluded that the ERP technique could be used to quantify display-color readability. Yamada (1998) recommends the use of EEG theta-band activity and eye blinks to determine workers' attention and mental load and also to evaluate the attractiveness of video games.

EQUIPMENT

Some Basics of Electrophysiological Recording

The electrical signals from the body that are used for psychophysiological analysis are quite small. Eye blinks and heart beats are about 1 mV or 0.001 V. EEG activity is smaller yet and is in the range of 5 to 70 μV or 0.000005 V. Amplifiers are necessary to enlarge these signals so that they can be recorded and analyzed. Fortunately, the recent advances in electronics has provided good, small amplifiers. These devices require little power, so they can be operated by batteries. Electrophysiological recordings often contain unwanted signals or noise. This noise can come from several sources and will be discussed more fully. Filters can be used to reduce many sources of noise. Filters are of three general types: high pass, low pass, and band pass reject. High-pass filters attenuate activity below a certain frequency while passing unattenuated frequencies above that frequency. Low-pass filters attenuate frequencies above a certain frequency and pass those below the cutoff frequency. Band-pass filters combine the characteristics of high- and low-pass filters. Band-reject filters attenuate only small portions of the frequency spectrum and pass all the rest. These are useful to eliminate electrical mains noise at 50 or 60 Hz. In addition to amplifiers, small general purpose computer processor chips are available, making it possible to conduct some of the signal processing at the recording device itself. This includes digitization and filtering of the signals. Another necessary feature of physiological recording devices is data storage. The processed data can be stored either on the recording device or telemetered to a receiving station for recording. Tapes, disks, and flash memory are used to store the data on the recorder. Some devices use standard dictating equipment, such as microcassettes, to store the data. These tapes are robust and relatively immune from electrical noise and shock damage. Other devices use small PCMCIA hard disks that are currently capable of storing up to 500 MB of digitized data. The flash cards cannot store as much data as the PCMCIA disks, but they are not as sensitive to environmental factors, such as vibration and G forces. The disks must not be dropped, and both the disks and flash memory cards should be protected from electrical fields, which can damage the stored information. One drawback to the analog tapes is that they must be replayed to digitize the stored data for analysis. Depending on the system, this can be a time-consuming process and adds another step to the analysis. If the data are digitized and stored at the recorder, then the data only need to be transferred to a computer for analysis.

A further concern for aviation test and evaluation and environments containing sensitive electronic equipment is the electromagnetic interference (EMI) emitted by the recording device. Units with computer processors are especially prone to radiate EMI that could interfere with aircraft instruments. If EMI noise is found to be present, it is usually a fairly simple matter to provide shielding to the recorder so that it passes the standards for use in aircraft. Analog tape-recording devices typically emit very low levels of EMI and have no trouble meeting the EMI standards. In some electronically noisy environments the effects of EMI on the recorder is of concern and must be checked.

Characteristics of Commercial Psychophysiological Recording Equipment

The different categories of recording equipment were discussed previously. There were four categories of recorders listed: general purpose ambulatory, specialized ambulatory, stationary laboratory, and telemetry systems. The general purpose ambulatory recording devices provide several channels of data, typically 8 to 64. They have amplifiers whose gain and filter settings are programmable. This allows simultaneous recording of different types of physiological data. For example, ECG, EDA, EEG, and respiration can be recorded at the same time. Units are available that weigh about 1 lb and are capable of recording eight or more channels of data continuously for many hours. Because the gain and filter settings of each channel can be individually determined, any number of different measures can be simultaneously collected. This flexibility permits one to take advantage of the unique information provided by the various physiological measures. They also permit recording of other information from the environment, such as temperature, acceleration, motion, light conditions, altitude, and event markers from the operator, the controlled system, or both. Units are available that record the data in analog or digital format. The storage media can be tape, disk, or flash memory. These units are battery powered, so they are electrically isolated.

A second type of ambulatory recorder is the special purpose unit, which is more restricted in the types of data that can be recorded. An example is the Holter monitor that is designed to record only ECG for clinical evaluation of cardiac patients. Typically, from one to three independent channels are provided. Most of these devices have at least one channel for backup in case of channel failure. One advantage of these units is that they are capable of recording for up to 24 hours at a time. It is possible to have the Holter monitors modified so that EOG can be recorded on one of the channels. Another avenue is the use of inexpensive sports heart-rate monitors. They record heart rate on a wristwatch device that receives interbeat interval signals from a strap worn around the chest. One disadvantage of these devices is that they usually provide only averaged data. They average over several seconds or minutes because they do not have the storage capacity to store the actual ECG. However, in many situations this is perfectly adequate, although this will preclude the use of HRV methods and hinder artifact detection and correction. Ambulatory EEG recorders are also available that are designed to record only EEG. These devices are typically intended for clinical neurological recording and may permit the recording of 32 or more channels of EEG. Additionally, they may record EOG data for purposes of artifact detection and correction. These EOG data can be used for other analyses, such as blink rate.

The third class of ambulatory recorder is actually a hybrid. They can record several different types of physiological signals, but the types of data that can be recorded is fixed. For example, a four-channel recorder might be set to record only ECG and three EEG channels. The amplifier settings do not have the flexibility to permit recording other types of data, such as EDA or respiration. These units typically include an event-marker channel so that the operator or an observer can indicate when significant events occur.

The fourth type of recorder is the telemetry unit. The physiological data are amplified and filtered by a small unit that is worn on the operator and contains a radio transmitter. A separate receiving device can be located more than 10 ft away to pick up the transmitted data and decode it. If space is available, it is possible to use larger transmitters that are capable of transmitting several miles. The output data from the receiver may be recorded in analog or digital form. Telemetry devices permit operators to be unencumbered and yet provide data that can be analyzed online. As described earlier, online monitoring of operator functional state and adaptive aiding can be implemented using telemetry systems.

To find sources of information about currently available equipment search the Internet and communicate with people using these techniques. Sources for ambulatory recorders, telemetry equipment, laboratory amplifiers, electrodes, electrolyte, and other supplies can be found on the Internet.

RECORDING BASICS

Electrodes: Getting Good Data

To reduce environmental electrical noise (e.g., from electrical wiring) differential amplifiers are used. Two inputs are provided to the amplifier and summed together, which has the result of canceling noise common to each input. The common noise is the electrical noise from the environment. Signals that are not common to both inputs (e.g., ECG, EOG, EEG) are amplified. The weak electrical signals from the body are picked up via small metal sensors. These electrodes are attached to the surface of the body and held in place with an adhesive or bands. Low impedance contacts between the sensors and skin are necessary to provide clean signals. To reduce the levels of environmental noise a low impedance interface between the electrode and skin is desirable. This is important to remember because a great deal of expense and time are involved in the collection of test and evaluation data. If electrode application is not done properly, you may have to spend much time with signal processing later. In the heat of battle during data collection, there may be an emphasis to quickly apply the electrodes. If the physiological data are worth collecting, then sufficient time must be allotted for proper electrode application.

The two sources of high electrode impedance are dead skin and the natural oils in the skin. An effective way to reduce the impedance is to swab the skin with alcohol to remove the oils and abrade the skin with gauze or other mild abrasive. The conduction of the biopotentials from the skin to the electrode is enhanced by using a conductive electrolyte. These are commercially available as ECG and EEG gels or pastes. These electrolytes are typically hypertonic, so they are good conductors of the electrical signals. In the case of EDA a normotensive interface is used because a hypertonic medium would interfere with data collection. This is also available commercially or can be readily made using published recipes.

Attachment of the electrodes to the skin is accomplished in several ways. Electrodes are either disposable or reusable. The disposable variety come with an adhesive collar and are used in clinical ECG recording. They are prejelled with electrolyte so that immediately after skin preparation and drying they can be placed on the skin and are ready for recording. The reusable type are usually the metal electrode itself or the electrode encased in a hard plastic case. They are held in place with double-sided adhesive collars or, in the case of EEG, they can be held in place with the electrode paste. If the paste is used and the recordings will go on over several hours, petroleum jelly applied over the electrode and covered with a piece of gauze will keep the paste from drying out. Another type of reusable electrode system that is used to record multiple EEG channels is the electrode cap or net. The electrodes or their holders are attached to a cap or net that fits on the head and is held in position with a chin strap or other arrangement. These are available with permanent or disposable electrodes. Reusable electrodes must be sterilized to prevent passing diseases from one operator to another. Guidelines for the proper methods to use have been published and are listed at the end of this chapter.

If data are collected over several hours and away from the hook-up room, then it is a good idea to prepare a traveling kit containing extra electrodes, leads, electrolyte, and tape. This can be used to quickly replace or repair a lost electrode and prevent loss of data. Although this rarely happens, the traveling kit does not take up much room and can prevent the loss of valuable data.

Artifacts and Other Things That Go Bump in the Night

Artifacts are signals that are not wanted and contaminate the physiological data of interest. Artifacts in the laboratory are usually not as much of a problem as in the field. Laboratory experiments can be cheaply rerun or extra data can be collected to replace any contaminated data. With test and evaluation the expenses are usually much higher, so repeating data collection may not be feasible. Further, in most test and evaluation situations, operators are naturally free to move about and cannot be restrained from talking and doing other artifact-causing behaviors. Also, the data of most interest can be accompanied by the sorts of activities that produce artifacts. Therefore, artifacts cannot be avoided, and methods must be used to either reduce or remove these artifacts from the data of interest. Cutmore and James (1999) provide an overview of the various artifacts encountered in psychophysiological research with suggestions on how to avoid and correct them.

There are two main types of artifacts encountered in test and evaluation. One type is physical artifacts from the environment, such as electrical interference. The second type is biological interference, such as EOG contamination of EEG signals. Some types of physical signals that seem to be important usually aren't. For example, people are often concerned about the effects of aircraft radio and radar signals on physiological data. The frequency range of these signals is much higher than the physiological signals of interest, and they are of no concern, unless the operator is very close to the source of these signals or they are extremely strong. However, electronic noise from power mains is in the range of the physiological signals, 50 or 60 Hz. They are often large enough to be picked up and become superimposed on the biological signals of interest. Some recording equipment comes with 50- or 60-Hz notch filters to reduce their impact. Low electrode impedance helps. If this does not eliminate the noise, changing the orientation of the operator or the mains cabling may help. Shielding of the offending signal sources may be required. With current high impedance amplifiers, this type of artifact is less of a problem than a few years ago. Earlier, shielded rooms had to be used; now it is common to record excellent quality biological signals in open rooms with no shielding.

The most problematic form of noise is usually biological in origin. This noise includes other unwanted biological signals with the desired ones. Muscle activity (EMG) can be fairly large and, when included with ECG or EEG signals, cause a problem. In test and evaluation operators are free to move about, action which is associated with EMG activity. Because the electrodes and amplifiers do not know the difference between EEG and EMG, they pick up and amplify both signals. Other types of interfering signals include eye activity in the EEG and EMG artifacts in ECG and EOG signals. Movement of the body also can cause artifacts caused by the electrode leads moving over the body and changes at the electrode/skin interface due to moving the skin and electrodes. Talking and chewing can also produce EMG and movement artifacts.

Movement-related artifacts can be reduced by taping the electrode leads to the body or clothing so that they are prevented from moving across the skin. Providing stress-reducing loops that are taped next to the electrodes can also help. Because the biological artifacts cannot be prevented by shielding, we are left with three options: Filters can be used, artifacts can be detected and removed, or we may have to remove the contaminated sections of data from analysis. Electronic filters are useful in removing unwanted signals when their frequency does not overlap with those of the wanted signals. For example, EOG activity is fairly slow, below 10 Hz. The frequency of the EMG activity is much higher, up to several hundred hertz, but does go as low as 10 Hz. By using filters that greatly reduce activity above 10 Hz, we can effectively eliminate the effects of EMG on the EOG. Sometimes we are not so fortunate. EEG activity is commonly analyzed up to 30 Hz or higher, which includes the lower end of the EMG spectra. Using filters set to filter out activity above 10 Hz would eliminate a large part of our wanted EEG signals. In this case it may be wise to analyze the EEG only to 15 Hz or so to avoid the EMG contamination. In other cases, such as EOG contamination of EEG, it is possible to identify the contaminating EOG signals and remove them from the EEG portion of the signal. Eye blinks and eye movements have characteristics that can be used to separate them from the EEG. Several published algorithms exist to detect and remove eye contaminates from EEG signals. Commercially available EEG systems often include these routines as part of their analysis packages.

ECG contamination of EEG data can often be corrected by moving the reference electrode. Its position may be responsible to the ECG that appears in the EEG. By moving the reference electrode, a more isopotential location can be found that eliminates the ECG.

Artifacts in ECG data can be corrected if one is only interested in interbeat intervals (IBIs). Because the cardiac rhythm is regular, observing the pattern of IBIs preceding and following missed or extra beats permits their correction. This works if only a few beats are missed or artifactually added. Mulder (1992) and Berntson and Stowell (1998) describe such a method. Artifact-free IBI data is crucial when determining HRV, which is very sensitive to artifacts—only one artifact can produce large errors.

If artifacts cannot be avoided, and because of their nature, cannot be removed from the data of interest, then the only option is to remove that data from analysis. This may mean the loss of critical data, but its inclusion into the test results would invalidate them. Good electrode application, artifact avoidance, and artifact removal takes care of most of the artifacts encountered in test and evaluation.

Practical Considerations

Of paramount importance, after getting the test director's approval to use psychophysiological measures, is convincing the operators of their utility in the testing. Until the use of psychophysiological measures is routine, their utility should be explained to the operators. Because they wear the electrodes, which take extra time to apply, they may be concerned about the value of the data. It can be useful to have someone that the operators know and trust be part of the test team. This often makes the use of psychophysiological measures more acceptable to the operators.

A typical and valid concern of operators who are required to pass a medical examination to be certified is that of privacy—who will see the ECG and EEG data. They are concerned that a physician may see something in the data that would cause the operator to lose certification. This is the case with military pilots who are required to undergo routine physical examinations that may include ECG testing. Because the ECG data is analyzed by computer, the vast majority of the data are never even seen by humans. Also, the test and evaluation staff almost never include physicians who are trained to interpret the ECG for clinical abnormalities. However, if something is seen in the record that looks suspicious, it should be brought to the attention of the operator so that he or she can take any necessary steps.

Safety issues include communicable diseases, electrical shock, and tripping or catching of electrode wires. In special cases, other safety considerations must be considered. In aircraft the safe egress from the aircraft in case of emergency must be considered. Ambulatory recorders can be worn in the pockets of flight suits, with any cables covered by tape or worn inside the flight suit. In the case of ejection seats safety personnel should be consulted for the need for quick disconnect cables or proper placement of the recorder on the operator or seat.

As mentioned, good electrode application is important. Although there may be pressure from other test team members and operators to rush the application, it is important to take enough time to properly apply the electrodes and recording equipment.

Because of individual differences in physiological response patterns, resting baseline periods should be included in the test plan. If time is especially precious, it may be possible to use pretest briefings for baselines, but a few minutes of resting data is optimal. Time pressures may make baseline periods difficult to schedule; however, they should be considered as part of the test sequence.

To correlate the psychophysiological data with the other test data, a coordinated time line is essential. Because testing is usually accomplished using a time line, all that is needed is synchronization of the physiological data with the time line being used during data collection. This may be local time, simulator time, or Zulu time. Further, notes taken during the actual testing with exact times noted are essential. Unplanned events can be identified from the physiological data and, using their time of occurrence, performance, subjective, video, and audio data can be examined to determine the cause. Notes should include equipment failures so that they can be correlated with the physiological data. Times when operators move from place to place should be noted if ECG data are being collected. The physical activity associated with getting up, walking around, and climbing stairs produces increased heart rate. Identification of these episodes allows eliminating them from consideration as unplanned events.

THE FUTURE

Further reduction in the size of electronics will mean that more powerful, smaller recording and analysis systems will be constructed. Increased density of storage devices will reduce the size of recording equipment, and make it possible to record data for longer periods, and increase the number of channels that can be recorded. With no limit in sight to the increased speed and storage capacity, very powerful complete analysis systems will be worn by operators. With this equipment, it will be possible to telemeter already analyzed data to the evaluators. It will be possible to modify test plans to provide better analysis of the systems being tested. With more powerful computer processors, more sophisticated signal processing will be possible. New developments in signal processing will be rapidly incorporated into test and evaluation systems. The entire psychophysiological data collection, storage, and analysis system may be worn by operators. Downloading or telemetering this data to the test center will permit swift system evaluation. An extension of this capability will be to provide the information about the operator's state to the system so that it can adapt itself to meet the current needs of the operator. In this fashion, total system performance will be improved by incorporation the operator's state into total system operation.

Sensor application will be improved with the development of rapid application sensors. These sensors will not require skin preparation prior to application. They will be dry electrodes; that is, no electrolyte will be required. This will greatly improve operator enthusiasm regarding acceptance of psychophysiological measures. One seemingly minor advantage that will be very important to operators is the lack of residual electrolyte left on the skin after electrode removal. They will be clean electrodes. Current projects show great promise toward the development and commercialization of these sensors. Coupled with the small and extremely powerful analysis systems discussed previously, many more areas of test and evaluation will be able to make use of psychophysiological measures.

REFERENCES

Ackles, P. K., Jennings, J. R., & Coles, M. G. H. (1985). *Advances in psychophysiology* (Vol. 1). Greenwich, CT: JAI Press.

Apparies, R. J., Riniolo, T. C., & Porges, S. W. (1998). A psychophysiological investigation of the effects of driving longer-combination vehicles. *Ergonomics, 41,* 581–592.

Backs, R. W., Lenneman, J. K., & Sicard, J. L. (1999). The use of autonomic components to improve cardiovascular assessment of mental workload in flight simulation. *The International Journal of Aviation Psychology, 9,* 33–47.

Berntson, G. G., & Stowell, J. R. (1998). ECG artifacts and heart period variability: Don't miss a beat? *Psychophysiology, 35,* 127–132.

Blomberg, R. D., Schwartz, A. L., Speyer, J. J., & Fouillot, J. P. (1989). Application of the Airbus workload model to the study of errors and automation. In A. Coblentz (Ed.), *Vigilance and performance in automatized systems* (pp. 123–137). Dordrecht, NL: Kluwer Academic Publishers.

Brookhuis, K. A., & de Waard, D. (1993). The use of psychophysiology to assess driver status. *Ergonomics, 36,* 1099–1110.

Brookings, J. B., Wilson, G. F., & Swain, C. R. (1996). Psychophysiological responses to changes in workload during simulated air traffic control. *Biological Psychology,* 361–378.

Brown, I. D., Wastell, D. F., & Copeman, A. K. (1982). A psychophysiological investigation of system efficiency in public telephone switch rooms. *Ergonomics, 25,* 1013–1040.

Casali, J. G., & Wierwille, W. W. (1980). The effects of various design alternatives on moving-base driving simulator discomfort. *Human Factors, 22,* 741–756.

Cutmore, T. R. H., & James, D. A. (1999). Identifying and reducing noise in psychophysiological recordings. *International Journal of Psychophysiology, 32,* 129–150.

De Waard, D., Jessurun, M., & Steyvers, F.J.J.M. (1995). Effect of road layout and road environment on driving performance, drivers' physiology and road appreciation. *Ergonomics, 38,* 1395–1407.

Fowler, B. (1994). P300 as a measure of workload during a simulated aircraft landing task. *Human Factors, 36,* 670–683.

Freeman, F. G., Mikulka, P. J., Prinzel, L. J., & Scerbo, M. W. (1999). Evaluation of an adaptive automation system using three EEG indices with a visual tracking task. *Biological Psychology, 50,* 61–76.

Fukuzumi, S., Yamazaki, T., Kamijo, K., & Hayashi, Y. (1998). Physiological and psychological evaluation for visual display color readability: A visual evoked potential study and a subjective evaluation study. *Ergonomics, 41,* 89–108.

Göbel, M., Springer, J., & Scherff, J. (1998). Stress and strain of short haul bus drivers: Psychophysiology as a design oriented method for analysis. *Ergonomics, 41,* 563–580.

Gundel, A., Drescher, J., Spatenko, Y. A., & Polyakov V. V. (1999). Heart period and heart period variability during sleep on the MIR space station. *Journal of Sleep Research, 8,* 37–43.

Gundel, A., Polyakov, V. V., & Zulley, J. (1997). The alteration of human sleep and circadian rhythms during space flight. *Journal of Sleep Research, 6,* 1–8.

Hankins, T. C., & Wison, G. F. (1998). A comparison of heart rate, eye activity, EEG and subjective measures of pilot mental workload during flight. *Aviation, Space, and Environmental Medicine, 69,* 360–367.

Hart, S. G., & Hauser, J. R. (1987). Inflight application of three pilot workload measurement techniques. *Aviation, Space, and Environmental Medicine, 58,* 402–410.

Helander, M. (1978). Applicability of drivers' electrodermal response to the design of the traffic environment. *Journal of Applied Psychology, 63,* 481–488.

Hilburn, B., Jorna, P. G. A. M., Byrne, W. A., & Parasuraman, R. (1997). The effect of adaptive air traffic control (ATC) decision aiding on controller mental workload. In M. Mouloua & J. M. Koonce (Eds.), *Human-automation interaction: Research and practice* (pp. 84–91). Mahwah, NJ: Lawrence Erlbaum Associates.

Hilburn, B., Jorna, P. G. A. M., & Parasuraman, R. (1995). The effect of advanced ATC automation on mental workload and monitoring performance: An empirical investigation in Dutch airspace. *Proceedings of the Eighth International Symposium on Aviation Psychology , 1,* 387–391.

Hunn, B. P., & Camacho, M. J. (1999). Pilot performance and anxiety in a high-risk flight test environment. *Proceedings of the Human Factors and Ergonomics Society 43rd Annual Meeting, 1,* 26–30.

Itoh, Y., Hayashi, Y., Tsukui, I. A., & Saito, S. (1990). The ergonomic evaluation of eye movement and mental workload in aircraft pilots. *Ergonomics, 33,* 719–733.

Kohlisch, O., & Kuhmann, W. (1997). System response time and readiness for task execution—The optimum duration of inter-task delays. *Ergonomics, 40,* 265–280.

Lindholm, E., & Cheatham, C. M. (1983). Autonomic activity and workload during learning of a simulated aircraft carrier landing task. *Aviation, Space, and Environmental Medicine, 54,* 435–439.

Mulder, L. J. M. (1992). Measurement and analysis methods of heart rate and respiration for use in applied environments. *Biological Psychology, 34,* 205–236.

Mulders, H. P. G., Meijman, T. F., O'Hanlon, J. F., & Mulder, G. (1982). Differential psychophysiological reactivity of city bus drivers. *Ergonomics, 25,* 1003–1011.

Myrtek, M., Deutschmann-Janicke, E., Strohmaier, H., Zimmermann, W., Lawerenz, S., Brügner, G., & Müller, W. (1994). Physical, mental, emotional, and subjective workload components in train drivers. *Ergonomics, 37,* 1195–1203.

Opmeer, C. H. J. M. (1973). The information content of successive RR interval times in the ECG, preliminary results using factor analysis and frequency analyses. *Ergonomics, 16,* 105–112.

Rau, R. (1996). Psychophysiological assessment of human reliability in a simulated complex system. *Biological Psychology, 42,* 287–300.

Richter, P., Wagner, T., Heger, R., & Weise, G. (1998). Psychophysiological analysis of mental load during driving on rural roads—A quasi-experimental field study. *Ergonomics, 41,* 593–609.

Rokicki, S. M. (1987). Heart rate averages as workload/fatigue indicators during OT&E. *Proceedings of the 31st Annual Meeting of the Human Factors Society, 2,* 784–785.

Roscoe, A. H. (1975). Heart rate monitoring of pilots during steep gradient approaches. *Aviation, Space, and Environmental Medicine, 46,* 1410–1415.

Roscoe, A. H. (1978). Stress and workload in pilots. *Aviation, Space, and Environmental Medicine, 49,* 630–636.

Roscoe, A. H. (1979). Handling qualities, workload and heart rate. In B. O. Hartman & R. W. McKenzie (Eds.), *Survey of methods to assess workload.* Paris: AGARD.

Roscoe, A. H. (1992). Assessing pilot workload. Why measure heart rate, HRV and respiration? *Biological Psychology, 34,* 259–288.

Russell, C. A., & Wilson, G. F. (1998). Air traffic controller functional state classification using neural networks. *Proceedings of the Artificial Neural Networks in Engineering (ANNIE'98) Conference, 8,* 649–654.

Rydstedt, L. W., Johansson, G., & Evans, G. W. (1998). The human side of the road: Improving the working conditions of urban bus drivers. *Journal of Occupational Health Psychology, 2,* 161–171.

Sega, R., Cesana, G., Costa, G., Ferrario, M., Bombelli, M., & Mancia, G. (1998). Ambulatory blood pressure in air traffic controllers. *American Journal of Hypertension, 11,* 208–212.

Sirevaag, E. J., Kramer, A. F., Wickens, C. D., Reisweber, M., Strayer, D. L., & Grenell, J. F. (1993). Assessment of pilot performance and mental workload in rotary wing aircraft. *Ergonomics, 36,* 1121–1140.

Steptoe, A., Evans, O., & Fieldman, G. (1997). Perceptions of control over work: Psychophysiological responses to self-paced and externally-paced tasks in an adult population sample. *International Journal of Psychophysiology, 25,* 211–220.

Sterman, M. B., Kaiser, D. A., Mann, C. A., Suyenobu, B. Y., Beyma, D. C., & Francis, J. R. (1993). Application of quantitative EEG analysis to workload assessment in an advanced aircraft simulator. *Proceedings of the Human Factors and Ergonomics Society 37th Annual Meeting, 1,* 118–121.

Sterman, M. B., Mann, C. A., Kaiser, D. A., & Suyenobu, B. Y. (1994). Multiband topographic EEG analysis of a simulated visuo-motor aviation task. *International Journal of Psychophysiology, 16,* 49–56.

Summala, H., Häkkänen, H., Mikkola, T., & Sinkkonen, J. (1999). Task effects on fatigue symptoms in overnight driving. *Ergonomics, 42,* 798–806.

Surwillo, W. W. (1986). *Psychophysiology: Some simple concepts and models.* Springfield, IL: Charles C. Thomas.

Svensson, E., Angelborg-Thanderz, M., Sjöberg, L., & Olsson, S. (1997). Information complexity—Mental workload and performance in combat aircraft. *Ergonomics, 40,* 362–380.

Taylor, D. H. (1964). Driver's galvanic skin response and the risk of accident. *Ergonomics, 7,* 439–451.

Torsvall, L., & Akerstedt, T. (1987). Sleepiness on the job: Continuously measured EEG changes in train drivers. *Electroencephalography and Clinical Neurophysiology, 66,* 502–511.

Veltman, J. A., & Gaillard, A. W. K. (1996). Physiological indices of workload in a simulated flight task. *Biological Psychology, 42,* 323–342.

Veltman, J. A., & Gaillard, A. W. K. (1998). Physiological workload reactions to increasing levels of task difficulty. *Ergonomics, 41* 656–669.

Wainwright, W. A. (1988). Flight test evaluation of crew workload for aircraft certification. In A. H. Roscoe & H. C. Muir (Eds.), *Workload in transport operations* (Report No. IB 316-88-06, 54-67). Cologne, Germany: DFVLR.

Wastell, D. G., & Newman, M. (1996). Stress, control and computer system design: A psychophysiological field study. *Behaviour and Information Technology, 15,* 183–192.

Wilson, G. F. (1992). Applied use of cardiac and respiration measures: Practical considerations and precautions. *Biological Psychology, 34,* 163–178.

Wilson, G. F. (1994). Workload assessment monitor (WAM). *Proceedings of the Human Factors Society,* 944.

Wilson, G. F. (in press). An analysis of mental workload in pilots during flight using multiple psychophysiological measures. *International Journal of Aviation Psychology*.

Wilson, G. F., & Eggemeier, F. T. (1991). *Physiological measures of workload in multi-task environments* (pp. 329–360). In D. Damos (Ed.), *Multiple-task performance*. London: Taylor and Francis.

Wilson, G. F., Fullenkamp, P., & Davis, I. (1994). Evoked potential, cardiac, blink, and respiration measures of pilot workload in air-to-ground missions. *Aviation, Space, and Environmental Medicine, 65*, 100–105.

Wilson, G. F., Lambert, J. D., & Russell, C. A. (2000). Performance enhancement with real-time physiologically controlled adaptive aiding. *Proceedings of the Human Factors Society, 44th Annual Meeting, 3*, 61–64.

Yamada, F. (1998). Frontal midline theta rhythm and eyeblinking activity during a VDT task and a video game: Useful tools for psychophysiology in ergonomics. *Ergonomics, 41*, 678–688.

Zeier, H. (1979). Concurrent physiological activity of driver and passenger when driving with and without automatic transmission in heavy city traffic. *Ergonomics, 22*, 799–810.

Zeier, H., Brauchli, P., & Joller-Jemelka, H. I. (1996). Effects of work demands on immunoglobulin A and cortisol in air traffic controllers. *Biological Psychology, 42*, 413–423.

SELECTED BIBLIOGRAPHY

Ackles, P. K., Jennings, J. R., & Coles, M.G.H. (1985). *Advances in psychophysiology* (Vol. 1). Greenwich, CT: JAI Press.

Andreassi, J. L. (2000). *Psychophysiology: Human behavior and physiological response* (4th ed.). Mahwah, NJ: Lawrence Erlbaum Associates.

Backs, R. W., & Boucsein, W. (2000). *Engineering psychophysiology: Issues and applications*. Mahwah, NJ: Lawrence Erlbaum Associates.

Cacioppo, J. T., Tassinary, L. G., & Berntson, G. (2000). *Handbook of psychophysiology*. Cambridge, England: Cambridge University Press.

Special Journal Issues Devoted to Psychophysiology in Human Factors

Human Factors. Cognitive Psychophysiology. (1987). *29*, 2.

Biological Psychology. Cardiorespiratory Measures and Their Role in Studies of Performance. (1992). *34*, 2–3.

Ergonomics. Psychophysiological Measures in Transport Operations. (1993). *36*, 9.

AGARD Advisory Report (AGARD-AR-324). *Psychophysiological Assessment Methods*. (1994). Paris, France: AGARD.

Biological Psychology. EEG in Basic and Applied Settings. (1995). *40*, 1–2.

Biological Psychology. Psychophysiology of Workload. (1996). *42*, 3.

Ergonomics. Psychophysiology in Ergonomics. (1998). *41*, 5.

International Journal of Aviation Psychology. Flight Psychophysiology. (in press).

Published Guidelines for Recording and Analysis of Measures

Fowles, D., Christie, M. J., Edelberg, R., Grings, W. W., Lykken, D. T., & Venables, P. H. (1981). Publication recommendations for electrodermal measurements. *Psychophysiology, 18*, 232–239.

Fridlund, A. J., & Cacioppo, J. T. (1986). Guidelines for human electromyographic research. *Psychophysiology, 23,* 567–589.

Jennings, J. R., Berg, K. W., Hutcheson, J. S., Obrist, P. L., Porges, S., & Turpin, G. (1981). Publication guidelines for heart rate studies in man. *Psychophysiology, 18,* 226–231.

Shapiro, D., Jamner, L. D., Lane, J. D., Light, K. C., Myrtek, M., Sawada, Y., & Steptoe, A. (1996). Blood pressure publication guidelines. *Psychophysiology, 33,* 1–12.

Chapter 8

Measurement in Manufacturing Ergonomics

Brian Peacock
National Space Biomedical Research Institute

To understand the purpose of measurement in manufacturing ergonomics it is instructive to consider a traditional definition of the profession: the scientific analysis of human characteristics, capabilities and limitations applied to the design of equipment, environments, jobs, and organizations. This definition clearly articulates that the profession is interested in both human capabilities and their limitations. It also insists that ergonomics is an applied science or technology ("applied to the design of"). Analysis (without design) is the primary purpose of science. Design without sufficient analysis can lead to a failure to appreciate the human requirements. Ergonomics involves both sufficient analysis and design. The definition does not explicitly describe a scope or purpose of ergonomics. It requires some extrapolation to appreciate the great variety of systems and processes to which ergonomics can be applied. Also, it does not indicate when, in the design and use processes, ergonomics can be applied. Finally, it does not identify a particular human or group of humans. A brief assessment of these omissions will set the scene for a discussion of measurement in manufacturing ergonomics.

The scope of ergonomics is virtually unlimited; it can be applied to any area of human experience. The ergonomist is only limited by his or her special knowledge of ergonomics techniques and the practice domain. For example, analysis can cover many techniques of physiology, medicine, and psychology, although it is clear that specialists in these professions may not necessarily consider themselves to be ergonomists. Similarly, ergonomics

157

can be applied to any area of design of the technological or organizational worlds and to modification of the natural world. Here again, the broad professions of engineering, medicine, law, and business commonly offer more specialized intervention or design skills. Commonly, ergonomists limit their applications to analysis, design, test, and evaluation of the human interfaces with the technological, organizational, and natural worlds.

MANUFACTURING ERGONOMICS PURPOSES

Manufacturing and manufacturing ergonomics have the purposes of quality, productivity, safety, health, and motivation. The quality purpose in manufacturing is to ensure that the human operator can perform his or her job correctly so that the end product will be to the customer's liking. The ergonomics tools that can be applied to this purpose include selection, job assignment and training (changing the operator), and error proofing (changing the components or manufacturing tools and equipment). The productivity objective may be achieved by the application of lean manufacturing principles that reduce nonvalue-added activity, such as walking, and by applying work measurement principles to the design of tasks. Another productivity intervention may be through the choice between automation and manual operations, based on some long-term analysis of relative costs. The safety purpose of manufacturing ergonomics is the avoidance of acute incidents (e.g., slips, falls, or wrong decisions) that may cause injuries to the human operators or damage to the hardware systems or products.

Work-related musculoskeletal disorders have taken precedence as the principal purpose of manufacturing ergonomics in the past decade. Epidemiological studies have shown that the interaction between inappropriate postures, movements, high forces, and high repetition rates can give rise to musculoskeletal disorders of various severities. However, not everyone doing the same job has the same outcome. This observation highlights both the raison d'être and the Achilles heel of ergonomics—human variability. In earlier times productivity, safety, product quality, and job satisfaction each had their periods of emphasis and also faced the same challenges of uncertain outcomes. The emphasis on these objectives is highlighted respectively by such people as Frederick Taylor, Ralph Nader, W. Edwards Deming, and A. H. Maslow. The multiple purposes of manufacturing ergonomics present both a dilemma and a major opportunity. The intrinsic nature of many contemporary manufacturing processes—with repeated short cycle work—may be productive, safe, and relatively error free, but it is not intrinsically motivating, and it may be associated with work-related musculoskeletal disorders. It should be noted that one of the earliest articulations of this side effect of physical work is attributed to Ramizini in the 18th century.

Consideration of these five purposes of ergonomics in manufacturing indicates that ergonomics may be applied simply to the maximization or minimization of any one purpose. Ergonomics is neutral. There are clear trade-offs—greater demands for productivity may result in lowered quality, health, and motivation. Conversely, greater concentration on quality and safety may result in lower productivity. Ironically some of the greatest applications of ergonomics may be found in the military and the health services, which have very different objectives. These trade-offs, per se, are not the province of ergonomists but rather of their employers. The ergonomist has the responsibility to measure, analyze, articulate, and evaluate the trade-offs and explore ways of optimizing the multiple outcomes or maximizing a particular outcome, if that is the purpose of his or her activity.

THE ERGONOMICS CYCLE

The ergonomics cycle describes the major activities that contribute to the application of ergonomics in manufacturing (see Fig. 8.1). In its simplest form, the job is a process involving the interaction between an operator, components, equipment, and tools in a production environment, such as assembly or component manufacturing. Reactive ergonomics is practiced when some history of unwanted outcomes precipitates a call for analysis, which will lead to some intervention (design). In proactive ergonomics there may be no actual operator or job, rather there may be a physical mock-up of the task, or a laboratory abstraction, a similar job, a computer model, or just a drawing. Under these circumstances the operator is represented by population data, such as anthropometric or biomechanical ta-

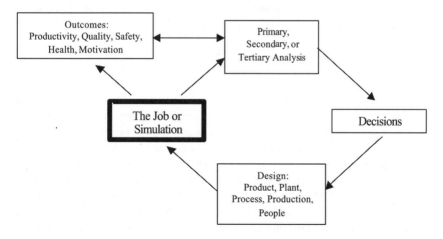

FIG. 8.1. The ergonomics cycle.

bles, an anthropomorphic model, or an experimental subject cohort that is assumed to be representative. In both the reactive and the proactive cases, the desirable or undesirable outcomes may be more or less direct. There may be a history of musculoskeletal disorders or quality or productivity concerns on similar jobs in the past. Quite often, however, the causal relationship between the characteristics of the job and the outcome may not be clear. Typically, company quality and medical records will indicate general problems in a wide variety of jobs, and the ergonomist is faced with the difficult task of making some link with changeable characteristics of jobs and justifying some intervention.

Outcome Measurement

Outcome measurement and analysis is generally complex. Typically, the quality, safety, productivity, health, or motivational measures will be counts of incidents that pass certain severity thresholds. For example, a quality incident may result in the repair or replacement of a component or the scrapping of the whole product. Similarly, the safety or health outcome may result in a bandage or aspirin and immediate return to work, some days off, or a permanent disability. Another challenge in outcome assessment is the accurate description of the denominators necessary for calculating meaningful rates. Conventionally, the denominator is the number of people working in that or a similar job multiplied by the number of hours worked. Commonly, a denominator of 200,000 hr or 100 employees per year is used in government statistics. Incidents per thousand products is a common index in the quality world. Of course, these numbers are meaningful at a gross level, but they often represent data from a wide variety of people and jobs. The challenge is to relate these general outcome measures to analysis of the changeable characteristics of a specific job.

Job Measurement and Analysis

Job, or simulation, measurement and analysis are the common foci of manufacturing ergonomics and will be discussed in depth later. However, at this time, it is appropriate to note the wide variety of levels of measurement and analysis, from checklists to detailed analysis and simulation to controlled experiments with sophisticated instrumentation. Typically this measurement and analysis will address the adjectives of the component systems, such as the height of a workbench, the force required to connect a hose, the duration of an awkward posture, or the number of repetitions of an activity per hour. Human measures may range from anthropometry to strength to eyesight, motor skill, and knowledge of fault types.

If the primary objectives of ergonomics are modification of the technological, physical, and organizational environments rather than the human operator, then it behooves the ergonomist to measure those things that will eventually be changed. Thus, measures of human characteristics may be of little use to the practitioner, unless there is some direct link to an engineering change. Ergonomics practitioners should generally focus on the measurement of the work itself and use published data on human interrelationships with that type of situation. These data generally have been collected under conditions conducive to accuracy, precision, and reliability. Measurement of human operators in practice generally leads to nongeneralizable conclusions because of sampling and methodological errors. Engineering measures, such as work heights, forces, repetition rates, and so on are generally more reliable and more useful.

Job Analysis—Outcome Relationships

The most difficult task for the ergonomist is the linking of these input system measures to the process outcome measures, particularly in light of the time lags and the mess of human variability that surrounds the investigation. In the laboratory, the research-trained ergonomist will identify independent, concomitant, and dependent variables and proceed according to his or her training. The independent variables are the system adjectives of interest; the concomitant variables are either unchangeable system or environmental adjectives, or human characteristics. The dependent variables are the multiple outcomes that must be optimized. A problem for measurement and analysis is that there may be a complex sequence of events. The adjectives of one system variable (the height of a work surface) may affect the back angle of the human system, which, in turn, may cause some outcome variables to change (e.g., the operator's back may be damaged or the time to move a component may be unnecessarily long). This problem of a probabilistic relationship between job characteristics and multiple outcomes is the greatest challenge to ergonomics practitioners. Although ergonomics researchers can deal with well-controlled experimental conditions and statistical tests of significance, the practitioner has to make a leap of faith in his or her design or intervention activities. Apart from the probabilistic relationships there is often a considerable time lag before the effect is seen.

Intervention and Design

The design element of the ergonomics cycle is the key to ergonomics and engineering practice. It involves the change of the attributes of the constituent systems. We could change the height of the work surface, put in a platform, select a taller operator, or require fewer repetitions of the activity. There are five general opportunities for intervention: product, plant, process, production, and people.

First, the product or component may be changed so that it is easier to manufacture or assemble. Second, there may be a major change to the manufacturing plant by the substitution of, for example, a robotic process. Third, the microprocess may be changed by attention to the workplace or tooling. Fourth, the production demands or staffing level may be changed to reduce the amount of repetition of a manual process. Finally, operator selection and training may be applied to address the human-system contribution. Generally speaking, these intervention or design opportunities should be investigated in order, with attention first to the product and finally to the people. However, the ever-present constraints on design and intervention may preclude certain approaches. For example, the weight of a component, such as a television set, may be due to the demands on the product design by the eventual user. Automation may be precluded for investment or versatility reasons. Workplace and tooling interventions may have progressed as far as possible. The production requirements may be set by the market and the competitors at a level of 1,000 items per day, and the lean manufacturing processes of measured and standardized work may have removed all the possible inefficiencies. Finally, the operators may have a union contract that mandates seniority-based job assignment rather than a management selection process. The confluence of these constraints is the usual domain of the practicing manufacturing ergonomist.

Ergonomics Decisions

These constraints lead naturally to the final element of the ergonomics cycle, the decision. This element is generally the responsibility of management. Given the historical outcome evidence, analysis of the job or process, and the set of alternative design interventions, the choice of intervention should be based on a rational cost-benefit assessment (assuming a capitalistic model of the organization where cost is all-important). This approach is based on key ratios relating the cost of alternative interventions to the benefits or cost of predicted outcomes. Unfortunately, the rationality of such decisions is not always perfect. Managers will weigh the evidence subjectively, according to their own and the company's priorities and policies. For example, popular company slogans regarding quality, safety, and efficiency may be reflected in the decision weighting for the choice of intervention. In many situations, ergonomics operates in a less-constrained decision-making environment, in which perceived health and safety outcomes receive high weightings.

MEASUREMENT AND PREDICTION

The purpose of measurement is to describe an individual or composite engineering (independent) variable in sufficient detail to communicate to the engineer what should be changed. These engineering values will be mapped into the simple or complex outcome (dependent) variable range by some predefined, predictive model. For example, the engineering variable may be lift moment and the outcome variable may be probability of damage to the lumbar spine. There are many complications of this simple mapping concept. First, individual variability will result in considerable error in the prediction model—robust populations will exhibit a shallower slope, and vulnerable populations will exhibit a steeper slope. Second, there may be no direct evidence of the relationship between the engineering variable and the outcome variable of interest because large-scale destructive tests cannot be performed on human subjects. Consequently it is necessary to obtain a surrogate measure, such as the response of cadavers in laboratory tests or, more conveniently, the assumption that human performance (e.g., strength) is inversely proportional to failure (L5/S1 damage.) Also, where composite or optimized combinations of outcome variables are of interest, there may be no empirical or theoretical science that can precisely describe the mapping. For example, how can the practicing ergonomist reliably combine quality, productivity, health, safety, and motivational outcomes into a single measure? Finally, although it is desirable to communicate one-dimensional variable values to the engineer, the well-known ergonomics problem of interactions may make the prediction unreliable.

For example, in manufacturing engineering it is common to refer to the OSHA logs for historical morbidity evidence and to similar data for product quality reports from customers or end-of-line audits. There may be anecdotal evidence of complaints, employee turnover, and defects from a particular operation. Job analysis of the actual operations may indicate poor postures that are held for a good portion of the job cycle, high forces, and high levels of repetition. The leap of faith that ergonomists may make, based on their training, is that interventions to correct the postures may resolve the multiple outcome problems.

Decision Thresholds

It must be noted that the mapping function may not be as monotonic as the one shown in Fig. 8.2. In some cases too little may be as bad as too much, as in the case of reach and fit, light and heavy, large and small, and loud and quiet. In these cases, it is convenient to involve two mapping functions—for minimization and maximization—for the separate ends of

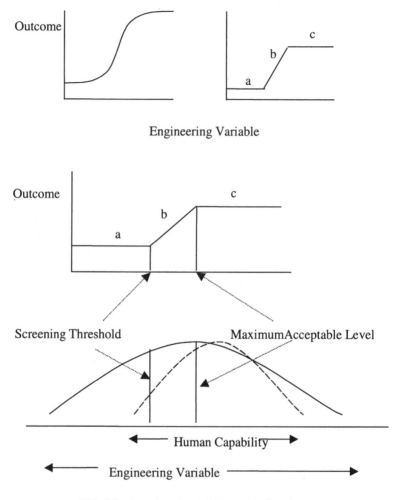

FIG. 8.2. Decision threshold mapping function.

the engineering variable range. Another important issue is the choice of decision points on the mapping curve. Whether the outcome variable is performance (e.g., productivity) or failure (e.g., musculoskeletal damage) there must be some policy input to the mapping curve that articulates the various decision thresholds. In general, it is convenient to consider two thresholds in these mapping curves: A low (screening) threshold is one below which the variable of interest will have no or minimal effect on the outcome variable. For example, in a lifting task, loads less than 5 or 10 lb will generally be acceptable to most people, under most conditions; a high threshold is that value beyond which the engineering variable on its own will be predicted to have an undesirable outcome. For example, the 1981 National Institute of Occupational Safety & Health (NIOSH) equation articulated an action limit and a maximum permissible limit. The region of the independent variable between these two thresholds is a region of uncertainty, where there may be an interaction between the engineering variable of interest and another variable. For example, if lower and upper limits of weight are considered to be 10 lb and 40 lb, then the upper region between these two values *will probably interact* with both the spatial lifting conditions and the lifting frequency, and the lower region *may interact*.

Resolution and Precision

The resolution of measurement is a very important consideration for manufacturing ergonomics. Ergonomic outcome (dependent) variables will generally be measured on an ordinal scale, which may be linear or nonlinear. It is popular in contemporary quantum risk analysis methods to use orders of magnitude, such as 1 in 1,000 or 1 in 1,000,000. Where the resolution of this scale is too coarse, there may be insufficient differentiation between different situations. For example, it is common to define green, yellow, and red zones to describe the risk associated with an engineering variable, with a practical result that most common work situations occur in the yellow zone, thus requiring further investigation of possible interactions. A general rule of thumb is that the resolution of the outcome scale (and also of the independent variable scale) should rarely involve more than 10 zones, and often 3 may be sufficient for practical purposes.

Precision and resolution are also concerns in the measurement of the independent (engineering) variables. In general, it is convenient to apply a law of round numbers, which sets thresholds at such levels as 5 or 10 lb, rather than 4.275 or 11.68 lb. This law of convenience can be justified because of the inherent effect of human variability in these mapping functions. The ergonomics (probabilistic) science of populations is not sufficiently precise to warrant the fine resolution offered by engineering science. Even the tailored fitting of clothes or shoes to an individual is not an exact science. The difference between accommodating 99%, 95% or 90% of the population may be of practical significance to the engineer, but error in the outcome variable may be such that the relationships are not precise, especially where the design is to affect only small cohorts of operators. Another example of this issue is seen in the design of vehicular speed limits, where a precise, round number is mapped into a much less precise measure of safety, but the rule is generally clear and appropriate. It should be noted that at the lower end of the speed-limit scale single digit differences may be suggested, but on the freeways differences are usually measured in 10s. Because human variability creates a wide band of uncertainty around the mapping function, an assumption of linearity over the critical region is probably warranted.

Interactions and Complex Stress-Exposure Measures

Ergonomists are well aware of the occurrence of interactions among independent and concomitant variables regarding their effects on outcome variables. Engineering science is also very cognizant of complex relationships between multiple dimensions. For example, the dynamics of a slip will be affected by the interaction between mass, speed, angle of incidence, and friction. Similarly, the stress of a lifting task has been articulated in lifting equations as being a complex interaction between weight, horizontal distance, vertical distance, distance moved, coupling, asymmetry, lift frequency, and shift length. European ergonomists have also included the effects of one-handed lifting, two-person lifting, and lifting under awkward conditions to their calculation of a lift index. However, a particular engineer can usually only affect one variable at once; indeed different engineers may be responsible for different variables. For example, the weight of a load may be dictated by other requirements of the product, such as the material and size of a door or television set. The weight variable will thus be the responsibility of the product engineer, who is constrained by more important criteria. The spatial conditions of lifting will typically be the responsibility of a manufacturing engineer, who sets up the conveyor systems, workplace fixtures, and handling aids. The frequency and shift duration factors will be the province of the production or industrial engineer. Finally, the selection of the person performing the task will be a complex decision of the individual, the supervisor, the union, and perhaps the medical department. The ergonomist must therefore decompose the complex lift index, or the findings of other complex analysis tools, into values of single variables that can be changed by these different engineers.

The interaction model shown in Fig. 8.3 has a number of practical uses. First, the values Y_L and X_L are those values of independent design variable, such as weight and horizontal reach, below which there is unlikely to be any adverse outcome, whatever the effect of other variables. Similarly, the upper values—X_U and Y_U—represent levels of the variables that are likely to be related to some unwanted outcome independent of other variables. The intermediate regions represent areas of uncertainty, where there may be adverse interactions. The resulting square, or box, contains the major area of interest for the ergonomist, given that the design, for other reasons, must be beyond the lower or screening level. Clearly the

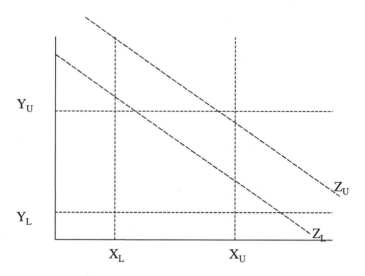

FIG. 8.3. The interaction model.

conditions where this model is most likely to fail, fall in the top right-hand corner of the box, because of the effects of other variables, such as frequency or population characteristics. Hence, the diagonal lines Z_U and Z_L represent the design cutoffs given the possibility of higher-order interactions. The choice of values for X, Y, and Z will be dependent on local knowledge of the processes and populations. For example, a materials-handling analysis could have X values of 1 and 6 lifts per minute, Y values of 10 and 40 lb, and Z values that cut off half or one quarter of the box, depending on the spatial conditions of the lifting task and the expected lifting population.

This model is linear and simple. It is of course possible to add non-linearities and develop more complex nomograms, such as those that were popular in thermal environment assessment. The value of this approach is that it provides an easy-to-understand and implementable method, which is sufficiently accurate, to analyze and design tasks.

PRIMARY AND SECONDARY ANALYSES

Given this dilemma of interactions and practicalities of individual variable measurement and communication, it is convenient to set up a sequence of assessments of increasing precision regarding interactions. For example, a screening checklist may identify levels of one-dimensional independent variables below and above certain thresholds (i.e., a primary analysis). Those variables above the high threshold will be immediately drawn to the attention of the appropriate engineer for rectification, and those below the lower threshold will be excluded from further investigation. The variables with values between the two thresholds should be investigated for interaction effects (with new composite thresholds) (i.e., a secondary analysis). A pass or failure at this secondary level (two-dimensional or multidimensional) will lead to decomposition and solution investigation. This is where the skill of the ergonomist comes into play in the trade-off between individual dimensions in the engineering resolution of complex interactions.

The NIOSH Lift Equation (Waters, Putz-Anderson, Garg, & Fine, 1993) is a good example of a secondary analysis. It has a multiplicative form in which the load constant is modulated by discounting factors, the values of which indicate the relative contribution of the individual components of the Lift Index. Other similar tools include Liberty Mutual's psychophysical tables (Snook & Cirrello, 1991), the University of Michigan's energy expenditure prediction model (Garg, 1976), the rapid upper limb assessment (RULA) model (McAtamny & Corlett, 1993), the Physical Work Strain Index (Chen, Peacock, & Schlegel, 1989), the New Production Worksheet (Schaub, Landau, Menges, & Grossman, 1997) and the Strain Index (Moore & Garg, 1995). The world of physical ergonomics has experienced an explosion in the development of such models and analysis tools. It is beyond the scope of this chapter to do a full review of all the alternatives; rather the chapter addresses the theoretical basis behind the tools and provides examples where appropriate.

The Science

Because of the inherent human, situational, and temporal variations in the relationship between simple or complex engineering variables and outcomes, there will never be a simple scientific relationship. Consequently, the development of mapping statements and the associated weightings and thresholds must involve a process that makes the best use of all the available evidence. Naturally, the most heavily weighted evidence will be a scientific assessment of the relationship, such as is available through the laws of physics and physiology. For example, it is understood that the lifting moment around the lumbar spine will interact with the back extensor muscles to create a compression of the intervertebral discs, which have been shown to be susceptible to failure under high loads. Similarly, the physiological processes of musculoskeletal hypertrophy and fatigue, the psychophysical data of human acceptability of various task conditions, and the epidemiological evidence of reported injuries may all contribute to a decision regarding the relationship between the engineering variables associated with lifting and the possible outcomes.

Policy

This scientific and other empirical data alone is not sufficient for the establishment of ergonomics mapping statements and design thresholds. There may be constraints on the engineering ability to resolve a problem. There may be various costs and benefits, both of the outcomes and of the alternative engineering interventions. There may be government regulations, company policies, or labor agreements regarding the level of protection offered to employees. For example, employees in the National Football League may have lifting tasks similar to employees in nursing homes, but the parameters of the mapping statements and thresholds would certainly vary, even if they could be established. Government legislation and regulation may exist, as in the case of vehicular speed limits. Unions may feel it their duty to negotiate work standards as well as level of protection of their members. Many companies have policy statements regarding the priority given to protection of employees from injury or illness. The same companies will also be very concerned about any mapping of engineering values into quality and productivity outcomes. Forward-looking companies will also employ ergonomics concepts and measures in the enhancement of the motivation of their employees. Thus, the consensus of experts may well have overtones of policy and negotiation in the balancing of the multiple sources of evidence surrounding a particular variable.

A pertinent example lies in the establishment of a guideline regarding the weight of objects that are to be lifted manually. In practice, the designer of an item, such as a parcel, box of hardware, or consumer product, may have no idea regarding the spatial and temporal conditions of lifting or the characteristics of the people who will do the lifting. Observation, over the years, of the local grocery market indicates that there has been a general decrease in the weight of objects, such as drinks and packs of water softening salt. A few decades ago manually handled sacks of coal or corn weighed more than a hundredweight (112 lb) and foundrymen regularly carried very heavy engine blocks. Even today nursing home attendants may have to move patients that are twice their own weight, and the package transportation business has to cater to the demand to move heavy parcels.

The relationship between policy and science in the establishment of mapping statements and upper and lower thresholds is a complex and sometimes contentious affair. The principle components of this process are the scientific evidence and the weighting of emphasis on quality, productivity, protection, and job satisfaction. However, science alone cannot dictate policy.

Human Variability

The key to policy discussions is the scientific evidence of variation in human capability coupled with good domain knowledge of the engineering variables. Engineering variables are much broader than the equivalent range of human capability. For example, human height varies between 3 and 8 ft, whereas the height of shelves in a warehouse may vary from 0 to over 40 ft. Similarly, the weight of stones needed to build a castle may vary between 1 and 1,000 lb, whereas human lifting capability varies between 10 and 500 lb (for a very strong person). Similar comparisons may be drawn for energy, sensory, and operational memory variables. Although there is a general population of humans that includes the very old and the very young, males and females, and healthy and disabled, it is composed of a variety of smaller populations. The population of clerical workers will have very different characteristics from the population of foundry or farm workers. The risk analysis literature differentiates between the general adult population and those exposed to hazardous jobs. The Americans with Disabilities Act (ADA) aims to protect the disabled by providing reasonable accommodation, whereas the World Wrestling Federation thrives on selection of the fittest. Where protection from musculoskeletal injury is concerned, it must be assumed, for want of better evidence, that vulnerability is inversely related to capability. These issues of human variability must be considered in the establishment of decision thresholds. However, those responsible for policy making are rarely comfortable with overtly saying, "We are going to place $X\%$ of our population at an increased risk of injury."

Policy and Percentiles

Traditional anthropometric policy (ergonomics dogma?) considers the 5th and 95th percentiles of stature, segment length, clearance, and functional reach. Even in this relatively simple spatial context the accommodation principles may be very inaccurate because of variability among different body segments and the amalgamation of multiple segments to achieve an accommodation envelope. The pragmatic basis of these percent accommodations lies in the increasing costs and diminishing returns of increasing accommodation from 90% of the population to 95% and to 99% and the practical impossibility of accommodating 100% of a large population. In the area of biomechanics and biomechanical psychophysics suggestions vary between accommodating 99% and 50% of the target population—the 1st and 50th percentiles. In manual labor practice, some accommodation levels may be as low as 5% or 10% of the population—the 90th or 95th percentiles—so only very strong people should apply! The argument for accommodating less of the population on a work dimension by setting the screening and maximum acceptable thresholds high, as in materials handling (biomechanics, psychophysics, or energy,) is that a selection and job-assignment process will separate the capable from the not capable after the hardware and operation design accommodation level has been decided. In other words, those workers whose capabilities are above the threshold are not at risk, and those below the threshold will use individual job matching to eliminate the risk. In many manufacturing organizations this is achieved by seniority arrangements for job choice, temporary medical placement arrangements for workers with a particular injury or vulnerability, and reasonable accommodation for workers who are temporarily or permanently disabled.

Consensus of Experts

In practice, the consensus-of-experts approach to the development of mapping statements and the setting of decision thresholds can work very well (see Fig. 8.4). This is especially true when the expert team is complemented by a group of domain experts who understand both the engineering constraints and the history of human capabilities in the particular work context. In some cases academic, government, and insurance company scientists may contribute to the discussion of these thresholds. In

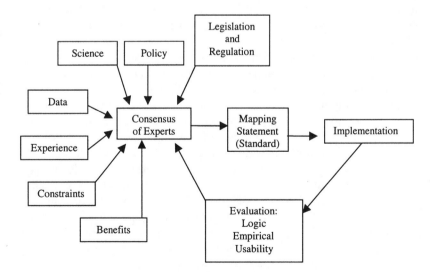

FIG. 8.4. The mapping and threshold development process.

unionized organizations the mapping functions and thresholds may be established by negotiation. This process is not new; there are reports from Biblical times of negotiated labor standards and associated rewards. Another advantage of a consensus approach is that the members of the team, who have contributed to the decision process, will generally support the resulting threshold. National and international organizations such as OSHA, American Society for Testing and Materials (ASTM), ANSI, Society of Automotive Engineers (SAE) and the International Standards Organisation (ISO) base their reputations on this consensus approach. Unfortunately, because of the complexity of the policy component of decision making, even these organizations sometimes have difficulty in reaching consensus. Consequently, the general philosophy to date in the United States has been to leave the specific engineering thresholds to individual companies, with the standards organizations limiting their influence to programmatic guidelines.

Implementation

Once the mapping statement and the associated thresholds have been established, the organization or company has the challenge of implementation and evaluation. The process of implementation of ergonomics measurement and standards will depend on company policy. In large manufacturing companies, there will rarely be enough ergonomists to evaluate every existing and planned job. Consequently, there will have to be a substantial training effort to communicate the ergonomics requirements to the engineers who will be responsible for their implementation. A variety of alternative process models exist, as illustrated in Fig. 8.5.

The alternative processes shown in Fig. 8.5 exist at various stages of the overall product, manufacturing, and production system development and implementation process. Alternative A requires the establishment of acceptable rules (mapping statements and thresholds), a strong management commitment to compliance, and comprehensive engineer training. Alternative B creates an ergonomics bottleneck and unnecessary review of compliant designs; however the task of training engineers may not be as difficult as in Alternative A, and the success of the process will depend on the support of management for ergonomics considerations. Alternative C is probably the most common and preferred approach. This requires basic ergonomics-checklist training of engineers who will be responsible for eliminating extreme discrepancies. Ergonomics specialist analysis and judgement may be brought into play to resolve complex interactions between variables. Modern computer networks have greatly facilitated the widespread deployment of standards within large organizations. Thus, given an agreement on measurement and decision thresholds, the infor-

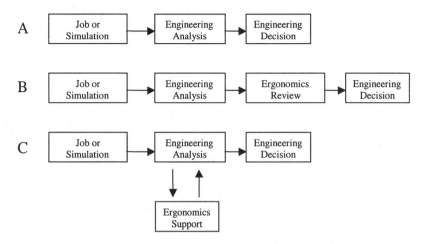

FIG. 8.5.   Implementation process models.

mation and tools can be deployed widely and instantaneously throughout large organizations.

Evaluation

The evaluation of measures and thresholds involves three distinct processes. The first is a review of the logic of the statement and choice of threshold in the context of the target population and the work that has to be performed. The laws of physics and physiology are the basis of such logic. Excessive dissonance will be the cause of conflict and give rise to criticism of the process and those responsible for the process. This is why domain experts as well as ergonomists and other technical specialists need to be involved as early as possible in the analysis system-design process. The second element of evaluation is essentially empirical. There should be an observed relationship between the chosen standards and the outcomes that are the objective of the activity. For example, if the standards involved operator workload in a traffic control situation, and the accident or traffic flow rates did not change, then clearly the prescribed method did not perform as hoped. In the case of musculoskeletal disorders, a reduction of poor postures and high forces should give rise to a reduction of reports of injuries. However, if a substantial cause of injuries was found to be in the modified jobs that had low forces and good postures but high repetition rates, then the scope of the measurement instrument that only addressed posture and force should be brought into question. Unfortunately, because of the lag in the effect of imposition of thresholds on the reduction of work-related musculoskeletal disorder incidence and severity as well as the ever-present variability in response of those affected, the empirical evaluation of these guidelines may take months or years, and even then be faced with noise in the data.

Usability Testing

The final evaluation process of an ergonomics measurement instrument should be its usability. Ergonomics should be ergonomic. The usability of a measurement instrument will be reflected in the interuser and intrauser variability of scores (a quality measure) and the speed of use (a productivity measure). For example, ergonomics analysis tools should be both objective and valid, measuring things that an engineer can change. Checklist and secondary-analysis tools should not require the user to perform complex measurements and carry out calculations under conditions of time stress. This is why the basic measures should be univariate and involve only the resolution necessary. Calculations and other forms of intelligence should be built into the instrument by appropriate table or nomogram design or by the use of computer techniques. The users of a measurement device should be trained in its appropriate usage. For example, the measurement of joint angle is very prone to error, unless appropriate technique is followed. One approach to measuring joint angle is to provide stick figures for the engineer to match, rather than have him or her attempt to measure angles from inappropriate vantage points. A better approach is to not require the engineer to measure joint angles at all. Rather, he or she should measure the deterministic spatial characteristics of the workplace that may have to be designed or redesigned. This usability issue exemplifies a general rule that engineers should measure directly those things that can be changed, and ergonomics checklists to be used by engineers should include the intelligent linkage between engineering (e.g., workplace layout) and human (e.g., posture) variables.

This last point is key to the practice of ergonomics in the field as opposed to the laboratory. In general, the engineer should be provided with an ergonomics measurement and analysis instrument that measures the environment rather than the person.

THE DESIGN OF MEASUREMENT TOOLS

The purpose of ergonomics measurement tools is to extract objective information from the situation in sufficient detail and at a sufficient level of resolution to be useful to the person who will make a decision, implement a change, or both. The issue of what to measure is related to the particular domain. In manufacturing, it will usually be appropriate to measure spatial, force, environmental, informational, and temporal factors. It will generally be easier to measure the things that can be changed, rather than the person, unless of course human changes, such as job assignment or training, are the intervention of choice. In general, it will be easier to measure single variables than complex interactions. However, the tool itself, whether manual or computer based, should incorporate a calculation facility to appropriately assess interactions. For example, a convenient calculation procedure would be to measure horizontal distance and weight and then look up the more meaningful lift moment in a table. Alternatively, if the level of resolution is low, simple integer calculations (as in RULA) may fulfill the purpose of assessing the interaction, reducing the probability of calculation error, and minimizing the time needed to perform the calculation. Similarly, tables can be used to assess the duration of poor postures or the repetition of high forces. Because the tool may be applied to a large number of jobs or products, it will probably be necessary to amalgamate the scores from multiple measurements to provide an overall score for prioritization purposes. Adding together unlike stresses is a questionable process; rather, the counting or profiling of such stresses may be more defensible. In general the final score is likely to be a risk index, such as a green, yellow, red rating, which may indicate the severity of the problem, the urgency of an intervention, or the status of resolution. Some ergonomists prefer a numerical rating to prioritize intervention and color-coding to reflect status—red implies a problem without a solution, yellow indicates a problem with a solution not yet implemented, and green indicates no problem.

The past decade has spawned many manufacturing ergonomics assessment tools, many of which have been validated empirically or have stood the test of time in a real-world environment. The example shown in Fig. 8.6 is illustrative of the kinds of devices that are used.

The job analysis, history, resolution, and evaluation worksheets shown in Fig. 8.6 demonstrate various aspects of measurement in manufacturing ergonomics. First, any investigation must have some classification information that describes the job, the analyst, the time, and a unique reference. The next component will be some description of the reason for the investigation based on a history of outcome problems, such as complaints, product quality, or injuries. All analyses are likely to have links to other investigations and documentation, such as product or medical records, engineering drawings, specifications, and, perhaps, simulations.

The example job analysis worksheet covers various environmental, spatial, force and energy, and information stress factors, each of which are linked to some temporal exposure assessment, such as duration or repetition. The stress factors and temporal exposures are quantitative engineering measures that will indicate both the direction and amount of change

Job Description	Analyst	Date	Reference

		L	M	H	L	M	H
E	**Environmental Factors**					Duration (min/hour)	

		L	M	H	L	H
1.	Light (Lux)	500		50	20	40
2.	Heat (°F)	80		90	20	40
3.	Cold (°F)	60		30	20	40
4.	Noise (dBA)	70		90	20	40
5.	Vibration (m/s^2)	0.5		2	20	40

					Duration (min/hour)	
S	**Spatial / Location / Posture Factors**					
1.	Down (in.)	30		10	20	40
2.	Up (in.)	50		70	20	40
3.	Horizontal Reach (in.)	15		25	20	40
4.	Task Orientation	Front/Above		Behind/Below	20	40
5.	Clearance (in.)	10		1	20	40
6.	Interface (sq.in.)	10		1	20	40

					Exertions / hour	
F	**Force / Work / Energy Factors**					
1.	Lifting/ Carrying (lb)	10		40	30	180
2.	Body Push /Pull (lb)	20		50	30	180
3.	Hand/Arm Push/Pull(lb)	5		20	30	180
4.	Digit Push/Pull (lb)	1		5	30	180
5.	Walking (ft.)	10		20	30	180
6.	Climbing Height (ft.)	2		4	30	180

					Transactions/hour	
I	**Information Factors**					
1.	Target Ratio (%)	75		95	30	180
2.	Working Memory(items)	5		10	30	180
3.	Decision Difficulty(prob)	90/10		60/40	30	180
4.	Pacing	Self		Machine	30	180

J   **Job Profile**

	LL	LM	ML	LH	HL	MM	MH	HM	HH
Total									
Environmental	G	G	G	Y	Y	Y	R	R	R
Spatial	G	G	G	Y	Y	Y	R	R	R
Force	G	G	G	Y	Y	Y	R	R	R
Information	G	G	G	Y	Y	Y	R	R	R
Total									

History and Resolution Summary

Job Description	Analyst	Date	Reference

FIG. 8.6. *(Continued)*

History of Unwanted Outcomes
    Product Defects
    Equipment Damage
    Productivity
    Acute Injuries
    Cumulative Illnesses
    Complaints
    Turnover
    Absenteeism
Other Investigations
    Design Analysis
    Drawings
    Layouts
    Photographs
    Videos
    Medical / Surveillance Records

Priority, Risk Assessment and Status
    Priority based on Analysis
    Priority based on Policy
    Cost of Unwanted Outcomes
    Cost of Alternative Interventions
    Expected Benefits of Interventions
    Decision / Intervention Timeline
Evaluation
    Evaluation Date
    Investment, Piece Cost
    Operational Costs
    Product Quality
    Productivity
    Safety, Health
    Motivation

Intervention Recommendations
Environment
    Lighting
    Thermal Environment
    Acoustic Environment
    Vibration
Product
    Product Architecture
    Component
    Containers and Packaging
    Fasteners
Plant
    Conveyors
    Major Equipment
    Materials Presentation
    Devices / Racks
    Materials Delivery
    Materials Handling Aids
Manufacturing Process
    Workplace Layout
    Jigs and Fixtures
    Tools
Production Organization
    Shift System
    Line Rate
    Job Quotas
    Supervision
    Team Structure
    Job Content
Individual Personnel Arrangements
    Training
    Selection Criteria
    Job Assignment / Choice
    Job Rotation
    Medical Management

Comments

FIG. 8.6.   Job analysis worksheet.

necessary to modify the stress. Both the stress and the temporal exposure factors have two thresholds. The lower threshold represents a level of stress, below which there is unlikely to be an adverse outcome. The upper value is one, above which there is a high probability of an adverse outcome, independent of any interaction with other physical or temporal stressors. The second page provides guidance to the analyst regarding documentation, risk analysis, and intervention. This worksheet is a prototype that could be developed and validated. It is aimed at indicating the basic approach to data capture, amalgamation, reduction, risk assessment, and intervention guidance. The threshold numbers suggested in this worksheet are initial estimates that should be revised in the light of empirical evidence, experience, or policy.

The association between physical stress and temporal exposure does not presume any particular mathematical relationship. A higher level of mathematical association, such as lift moment, posture-minutes, decibel-minutes, or more complex models, such as those found in the NIOSH lift equation, RULA, or the New Production Worksheet, are not addressed here. This is for two reasons. First the assessment of complex interactions on outcomes is based on models of questionable reliability and ease of use in applied contexts. Second, the eventual intervention will inevitably require a decomposition of a complex model for the purposes of engineering intervention. This logic is not meant to imply that there is no value in complex models that more accurately predict outcomes; rather, it is a pragmatic approach to the application of ergonomics measurement in the analysis-decision-design-outcome cycle.

The intervention documentation list also is not claimed to be comprehensive. Rather, it is meant to be illustrative of the range of possible interventions in typical manufacturing situations. An initial objective of manufacturing-facility design is to create a physical environment that is not potentially harmful or otherwise not conducive to human physical or cognitive work. The next opportunity is to design a product that is intrinsically easy to manufacture or assemble. Unfortunately, the complexity of contemporary products often results in undesirable demands on the manufacturing process, such as access to components. The plant design opportunities are generally major, irreversible decisions that may have been made many years ago or will take many years to change. For example, product conveyors, materials transportation, and manufacturing automation all involve substantial planning and investment and are only changeable on long-term cycles based on technology availability and amortization. The smaller manufacturing-process design opportunities are more amenable to short-term substitution or change. Workplace layouts, materials presentation devices, and tooling present the major opportunities for ergonomics interventions. The basic manufacturing process may be appropriate for the production of one item per day, but high production demands impose other constraints and opportunities for intervention. The Fitts lists that compare human and machine capability clearly indicate a trade-off between human versatility and machine repeatability. Unfortunately, contemporary production arrangements have to offset investment and labor costs as well as coincident demands for repeatability and versatility. Thus, the decisions regarding automation, job loading, and job rotation require complex evaluation and trade-offs.

The final intervention opportunity for ergonomists is the choice of who will do a particular job. In an infinitely fluid person-job assignment environment, people with widely different capabilities would be matched with appropriate, widely differing jobs. In reality, companies and unions face situations that are considerably less flexible, particularly with an aging workforce. Consequently the job demands have to be set at levels that will accommodate the capabilities and limitations of a majority of employees who have some security of tenure in that organization. It is infeasible and impractical to set demands so low as to accommodate the full spectrum of limitations of the available population. Also it is unacceptable and infeasible to set the demands so high that the matching of individuals to jobs is very specific. Consequently, measurement in manufacturing ergonomics, coupled with appropriate policy, can optimize the balance between production and protection.

REFERENCES

Chen, J. G., Peacock, J. B., & Schlegel, R. E. (1989). An observational technique for physical work analysis. *International Journal of Industrial Ergonomics*, *3*, 167–176.

Garg, A. (1976). *A metabolic rate prediction model for manual materials handling.* Unpublished doctoral dissertation, University of Michigan.

McAtamney, L., & Corlett, E. N. (1993). RULA: A survey method for the investigation of work related upper limb disorders. *Applied Ergonomics*, *24*(2), 91–99.

Moore, S. J., & Garg, A. (1995). The strain index: A proposed method to analyze jobs for risk of distal upper extremity disorders. *American Industrial Hygiene Association Journal*, *56*, 443–458.

Schaub, K. H., Landau, K., Menges, R., & Grossman, K. (1997). A computer aided tool for ergonomic workplace design and preventive health care. *Human Factors and Ergonomics in Manufacturing*, *7*, 269–304.

Snook, S. H., & Ciriello, V. M. (1991). The design of manual handling tasks: Revised tables of maximum acceptable weights and forces. *Ergonomics*, *34*(9), 1197–1213.

Waters, T. R., Putz-Anderson, V., Garg, A., & Fine, L. J. (1993). Revised NIOSH equation for the design and evaluation of manual lifting tasks. *Ergonomics*, *36*(7), 749–776.

Chapter 9

Mock-Ups, Models, Simulations, and Embedded Testing

Valerie J. Gawron
Veridian Engineering

Thomas W. Dennison
Sikorsky Aircraft

Michael A. Biferno
The Boeing Company

To mitigate risk in system development, it is critical to perform as much testing as possible in the early stages of development. This forward loading of program testing results in the need to turn to methods such as mock-ups, models, and simulations that can provide information about the human system interface before the first article is produced. With the rapid growth of CAD and CAE tools, HFTE practitioners can participate in defining system requirements and testing design concepts much earlier. By conducting human factors testing with CAD/CAE tools early in design, designers can simulate the human actions required to assemble, maintain, and operate equipment and avoid many problems before expensive hardware is built.

HFTE is undergoing a paradigmatic shift away from merely confirming that specific design requirements have been satisfied to exercising the design early on to identify possible problems as part of a multidisciplinary design team. Where traditional testing processes relied on physical development fixtures and development of first articles that could be touched and handled, new product development processes such as concurrent engineering (CE) and integrated product development (IPD), are much broader based and include electronic media, product prototyping, simulation, and customer involvement throughout the process (Biferno, 1993).

This paradigm shift in HFTE will affect the field in dramatic ways for years to come.

This chapter describes many of the human factors tools used in these new design and testing processes. Our intent is to describe specific tools in sufficient detail to enable a professional to judge their utility, as well as to acquaint those new to the field with these tools and how they are used. Our goal is not to provide step-by-step instructions. Rather, we have included the strengths and limitations of each tool and sources and references that will help the person who wants to learn how to apply these tools or learn more about them.

This chapter is organized along the traditional development cycle, that is, in the order the tools and methods would be applied, beginning with inexpensive, low-fidelity examples and progressing to very expensive representations of the product in use in its operational environment. We further distinguish between tools that test the physical operator interface and tools that test the perceptual and cognitive (information) interface, although some tools apply to both categories.

TESTING THE PHYSICAL OPERATOR

An individual evaluating the physical operator interface is concerned that the operator can see, identify, reach, and operate all controls; see and read displays; and fit through access ways without obstruction. Evaluators must make sure that workstations do not cause discomfort, fatigue, or injury. Operators of vehicles must have an adequate field of outside vision. Developers use four types of tools in test and evaluation of the physical operator interface: mock-ups, physical human models, CAE models (also called electronic human models), and simulators.

Guidelines for selecting tools to test the physical operator interface are based on cost, fidelity, and application. Low-cost mock-ups are ideal where the designer does not need high-fidelity tools to test anthropometric accommodation (reach, fit, access, visibility) of smaller interfaces. Larger products, such as vehicle operator stations, can be built with low-cost mock-up materials, but if a CAD model of the interface is available, electronic human modeling is preferable. The highest cost and highest fidelity tools for physical interface testing are full-scale rugged mock-ups, which are difficult to build well but unsurpassed for testing objective as well as subjective physical accommodation of an interface with human test subjects. Where testing may result in harm to human subjects, instrumented full-sized mannequins are used, although they may be expensive. Table 9.1 describes some of the criteria to consider when choosing a physical interface test tool.

TABLE 9.1
Physical Operator Interface Testing Tool Selection Guide

Types of Tools	Selection Criteria							
Test physical operator interface	Validated	Current	Easy to use	Flexible	Widely used	User's manual	PC based	Workstation
Mock-ups	x	x	x	x	x			
Foam core			x		x			
Wood			x		x			
FaroArm	x	x	x	x		x	x	
Stereolithography	x	x		x	x	x		x
Physical human models								
Mannekin	x	x	x	x	x	x		
Full-size mannequins	x	x			x			
Electronic human models								
COMBIMAN	x				x	x	x	x
CREWCHIEF	x				x	x	x	x
Jack	x	x		x	x	x	x	x
Safework	x	x		x	x	x	x	x
RAMSIS	x	x		x	x	x	x	x

Mock-Ups

A mock-up is a three-dimensional model of a product or system that varies in fidelity, complexity, and expense according to the needs of the designer. Although three-dimensional CAD tools can replace mock-ups for some tasks, physical mock-ups are still useful for those applications where budget or training prohibit the use of three-dimensional CAD tools. Also, mock-ups can do some things CAD tools cannot.

Mock-ups enable developers to visualize and evaluate the physical interface when the design is still on paper. Mock-ups are the first three-dimensional look at an interface design and range in scope from an individual control, such as an airplane control yoke, to an entire vehicle control station. They must be full-scale to determine if the expected range of users can reach and actuate controls, whether visibility is acceptable (for example, in an automobile driver's position), and if maintainers have access to equipment.

Foam Core. Mock-ups are very economical for validating spatial relationships because they can be made from inexpensive materials. Foam core is a thin sheet of dense Styrofoam™ (usually ⅛ or ³⁄₁₆ in. thick), covered with white paper on both sides. It is readily available from office supply stores and is very easy to work with. It can be cut with an X-acto™ knife and joined using a hot glue gun. Full-scale, two-dimensional drawings or photographs can be pasted to surfaces for added realism (see Fig. 9.1). Foam core is not, however, a rugged material; it will not support an operator's weight or forceful use. If ruggedness is required, foam core can be built on a structure of wood or some other rigid material (see Fig. 9.2). Cardboard, foam block, or modeler's clay can also be used to construct simple mock-ups. These materials all vary in their cost, skill required, strength, and fidelity. For example, cardboard is relatively flimsy but easy to work with. A mock-up made from modeler's clay can look nearly identical to the finished product, but it is labor intensive and requires skill and experience to develop.

Wood.   Full-scale mock-ups of entire vehicles or control stations are often made of wood. Wood is strong, fairly inexpensive, and can be made to

FIG. 9.1.   Example of cockpit mock-up. Reprinted with permission of United Technologies Sikorsky Aircraft.

FIG. 9.2. Example of mock-up. Reprinted with permission of United Technologies Sikorsky Aircraft.

support the weight of operators, production electronics, and displays for greater realism. However, full-scale wooden mock-ups are time-consuming to make, require considerable skill, and may not be completely realistic (although realism may not be crucial). They are excellent for testing operator reach and visibility envelopes, access, work position comfort, and so on. Mock-ups also serve as a focal point for different disciplines. Conflicts often arise when structure designers, electronics installers, stylists, ergonomists, and others with different agendas vie for the same limited space. A full-scale mock-up is an excellent place to resolve some of these conflicts. This point is expanded by Seminara and Gerrie (1966) using specific historical examples (e.g., Apollo Mobile Lunar Laboratory, Deep Submergence Rescue Vehicle, Earth Orbiting Mission Module).

The FaroArm.   An innovative system exists for confirming or improving the accuracy of a full-scale, three-dimensional mock-up, and for measuring various points in space, including the locations of occupants' extremities while seated in a mock-up. FARO Technologies has developed a fully articulated, six-axis device called the FaroArm™. The system can sense with high resolution the location of the tip of the wand in three-dimensional space, once a zero-point is determined. When used in conjunction with the company's proprietary software, AnthroCAM™, the Faro-Arm™ operator can store the location of a large number of discrete points on any surface of the vehicle by simply touching the tip of the wand to the desired location and pressing a trigger. The locations of measured points can then be compared to corresponding points in a CAD model or on a prototype of the actual product. Vehicle crew station designers can also measure points on human operators seated in the mock-up to determine eye points, reach extent, obstruction and clearance problems, and range of motion or adjustment. The FaroArm™ is used in conjunction with a standard desktop Windows™-based PC. Operation of the FaroArm™ is illustrated in Fig. 9.3.

FIG. 9.3. Operator measuring points in a vehicle using the FaroArm™. Reprinted with permission of FARO Technologies.

Stereolithography. Stereolithography is a method for building solid parts directly from three-dimensional CAD data. It can make detailed mock-ups of handheld input devices, such as cell phones, keyboards, and control grips, complete with cavities for embedding real switches. The advantages are relatively quick turnaround, exact detail, ability to produce complex parts, and ruggedness. The disadvantages are that the cost of machinery and the need for skilled CAD and stereolithography machine operators may put this technology out of reach for some small companies.

The stereolithography device uses three-dimensional solid CAD data to render the object into a series of horizontal slices that defines the object when stacked from bottom to top. The actual part is usually built from a liquid photopolymer resin that solidifies when exposed to a high-radiance light source. Most stereolithography machines use argon or helium-cadmium lasers. A processor moves the laser at a specific rate and in a pattern determined by the shape of the horizontal slices, turning the laser on and off corresponding to where material should and should not be, solidifying the photopolymer, and building the object layer by layer (Jacobs, 1992).

3D Systems of Valencia, California (http://www.3dsystems.com/), has shipped most of the stereolithography systems in use. DTM Corporation (http://www.dtm-corp.com/home.htm) and Cubital (http://www.3dsystems.com/) also produce these systems, although some operate differently. A detailed description of them is beyond the scope of this chapter, but Jacobs (1992) published a good technical review with case studies. A search on the Internet will yield a large number of companies that provide stereolithography rapid-prototyping services.

In summary, mock-ups are a highly useful tool in helping to develop safe, efficient, and effective user interfaces. However, mock-ups are but part of the development process; the human factors test specialist must still factor in the physical and cognitive needs of the user to achieve a wholly representative and functional system.

Physical Human Models

Physical human models are replicas of humans that are used to determine anthropometric fit, physical accommodation, or protection. They can be two- or three-dimensional. Manikin (or Mannequin) is a simple and inexpensive human model developed at the U.S. Air Force Research Laboratory by Dr. Kenneth Kennedy. It is a two-dimensional model of the human form in profile. It is hinged at the neck, shoulders, elbows, hips, knees, and ankles. Designers use Manikin to test anthropometric fit and physical accommodation by placing it flat on a two-dimensional side-view drawing of a workstation (e.g., in the driver's seat) and checking that the design complies with reach requirements, visibility envelopes, and access. It is somewhat imprecise but, because it is both easy to use and inexpensive, it serves as a very good first look at an operator station design. Manikin can represent 5th and 95th percentile stature individuals. It can be made to ½, ¼, or ⅕ scale to correspond to the drawings being used. Complete instructions for easy fabrication are found in MIL-HDBK-759A, including templates for 5th and 95th percentile males that can be traced. They can be made from heavy paper, thin cardboard, or plastic and use swivel grommets for moveable joints.

Use of other full-size mannequins is generally associated with testing impact protection devices (e.g., helmets) and restraint systems (e.g., automobile seat belts) or environments too harsh for human test subjects. These mannequins can be made to faithfully reproduce the characteristics of the human body in size, weight, and density. They can be instrumented with accelerometers, force balances, and strain gauges to measure vibration, g forces, and impacts to predict damage to an actual human. Testers use them in prototypes or on test stands.

Instrumented mannequins are relatively expensive to buy and use and require skill and experience to use. Collection and reduction of data are highly automated and depend on complex statistical models for valid results. Companies or agencies that cannot afford to purchase and use test mannequins can employ independent test laboratories (who have already invested in expensive mannequins for this purpose).

Computer-Based Manikins and Three-Dimensional Human Modeling Systems

Computer-based manikins or digital human models are three-dimensional, computer-graphic representations of the human form. Human modeling systems are interactive software simulation tools that combine three-dimensional human geometry with three-dimensional product geometry, obtained from a CAD system for the purpose of evaluating how well a particular design will accommodate the people who will be interacting with it. Computer-based manikins are controlled by human modeling systems to provide a variety of functions, which include the ability (1) to import and export product data/geometry; (2) to simulate human functions, such as reach and vision; (3) to perform analyses, such as access and interference analyses; and (4) to provide support functions, such as reports, documentation, and user help. Figure 9.4 illustrates how a reach envelope can be used to evaluate control placement. The reach envelope provides a quantitative description of a person's maximum reach, given the seat and restrain assumptions required by the task. The designer then uses this quantitative description to define the requirement that all critical controls be placed within the reach envelope to ensure easy access and op-

FIG. 9.4.   Using a computer-based manikin to assess a reach envelope (top left), physical access (top right), and visual access through the manikin's eyes (bottom).

eration. Verification of a design can be accomplished by merging the reach envelope with a candidate design to ensure controls are within the envelope. As shown in the figure, a similar procedure is used to assess visual and physical access.

The size and shape of digital human models are based on anthropometric data from one or more standard databases, such as the 1988 survey of U.S. Army personnel (Gordon et al., 1989). The movements of digital human models are based on biomechanical models of human motion so that the behaviors required for tasks, such as data entry, installation of line replaceable units, or egressing through hatches, can be modeled and tested. Human modeling systems are primarily used for accessibility and interference analyses to ensure that a substantial percentage, if not all, of the likely users can see and reach all parts of the product's human interface. As the human modeling field continues to mature, however, human strength assessment is becoming increasingly important. Human modeling systems provide a means to determine user posture and then reference the underlying strength data to perform accurate assessments.

The COMputerized BIomechanical MAN-Model (COMBIMAN) was developed by the Air Force Research Laboratory to support evaluation of the physical accommodations in an airplane cockpit. These evaluations include analyses of fit, visual field, strength for operating controls, and arm and leg reach (MATRIS, 2001). COMBIMAN includes anthropometric survey data for U.S. Air Force (USAF) male pilots, female pilots, USAF women, U.S. Army male pilots, U.S. Army women, and U.S. Navy male pilots (MATRIS, 2001). The output includes a table of strength percentiles, visual field plots, and a three-dimensional model. A manual is available (Korna et al., 1988). COMBIMAN currently runs on an IBM workstation in conjunction with computer-aided three-dimensional interactive application (CATIA) and on a personal computer (PC) in conjunction with AutoCAD.

Crew Chief is a three-dimensional model of a maintainer. It has been used to evaluate physical accessibility, visual access, and strength required to perform maintenance tasks (MATRIS, 1992). The output can be used to identify design-induced maintenance problems. A user's manual is available (Korna et al., 1988). Crew Chief runs either on a mainframe in a stand-alone mode or on workstations as a subprogram to CADAM V.20, CADAM V.21, CADS 4001/4X.5B, and CADD Station for Unix (MATRIS, 2001). Crew Chief includes an animation capability (Boyle, Easterly, & Ianni, 1990).

The Computer Graphics Research Laboratory at the University of Pennsylvania developed Jack. Originally designed for use by NASA to both design and evaluate the space station, Jack has been used throughout the aerospace industry and has been used to evaluate reach and vision envelopes. Jack provides an articulated figure, enabling multilimb positioning, view assessment, and establishment of strength requirements (Badler, 1991). Jack is strong in areas of manikin movement control and in supporting the visualization of operator interactions with product geometry. Jack has been integrated into several systems, including the Man-Machine Integration Design and Analysis System (MIDAS). The rights to market and support Jack have been purchased by Engineering Animation, Inc.

(EAI), and subsequently Unigraphics Solutions has purchased EAI. Current information about the functions and features of Jack can be found at http://www.eai.com/products/jack/.

Safework was originally developed by Genicom Consultants of Canada, which is now a subsidiary of Dassault. Safework is a human modeling system that is strong in its application of anthropometric data to mannequin size and shape. Safework also runs as a process under CATIA, which greatly simplifies geometry access and preparation to do a human modeling analysis. Current information about the functions and features of Safework can be found at http://www.safework.com/.

Rechnergestütztes Anthropologisch-Mathematisches System zur Insassen-Simulation (RAMSIS) is a product developed by Tecmath. It is supported by a consortium of German auto manufacturers and derives some of its technology from Anthropos, a human model also developed in Germany. It is strong in passenger automotive applications, such as comfort assessment, evaluation of passenger ingress and egress, and the acquisition of whole-body surface anthropometry using three-dimensional scanning devices. Current information about the functions and features of RAMSIS can be found at http://www.tecmath.com/.

TESTING THE INFORMATION INTERFACE

The information interface is composed of the medium by which information is presented to the user, the form the information takes (e.g., light, sound, shape), the organization of the information (e.g., binary signals, printed language, spoken words, icons, video), and the logic and the structure that determines the basic appearance of the interface. Designers and testers need to confirm early on that the presentation of information, and the devices used for the input of information, complements the operator's cognitive and perceptual capabilities. The tools described in the sections to follow can help the tester address these issues throughout the design process. They are described in increasing order of complexity, fidelity, and expense (see Table 9.2).

Rapid Prototyping

Rapid prototyping is used primarily in the design of information interfaces. Although it can refer to the generation of solid mock-ups from three-dimensional CAD data, we will discuss rapid prototyping of menu-based, hierarchical, and multifunctional interfaces that accompany information-intensive systems. Some designers build rapid prototypes of aircraft avionics (i.e., integrated cockpit displays), business-related software

TABLE 9.2
Information Interface Testing Tool Selection Guide

Types of Tools	Validated	Current	Easy to use	Flexible	Widely used	User's manual	PC based	Workstation
Rapid prototyping								
Designers Workbench	x	x		x	x	x		x
Altia	x	x		x		x		x
VAPS	x	x		x	x	x		x
Simulators	x	x			x			
DNDS	x	x						

packages that run on desktop computers, and even simple automated teller machines (ATMs). Rapid prototyping enables these designers to construct quickly the structure, layout, and hierarchy of the interface, operate it, change it, allow potential users to evaluate it, and perform experiments comparing performance of various configurations. This technique can also be used as a training device to teach operators to navigate menu hierarchies. Most rapid prototyping packages run on PCs or more sophisticated workstations. Some examples of rapid prototyping packages are described in the following sections.

VAPS/CCG. VAPS/CCG (Virtual Applications Prototyping System/C-Code Generator) runs on Windows, on Silicon Graphics, and Sun workstations, using versions of Unix. It also runs on embedded systems. VAPS enables developers to build prototypes of simple displays or entire operator stations, such as a car instrument panel or an airplane cockpit. It has a graphics editor that can generate high-resolution pictures of the object. The designer draws using a mouse or digitizer. Graphics can also be imported from three-dimensional CAD models for greater realism. VAPS provides a logic editor to determine the behavior of the elements of the graphics in response to inputs from a control or aspects of the simulated world in which the prototype is running. Real controls or simulated controls (i.e., mouse, touch screen, and voice) can be used (Virtual Prototypes, Inc., 1988). For example, developers can evaluate a fighter aircraft multifunction display design with multiple display modes, using a joystick for the flight controller as an actual control stick from an aircraft. The attitude indicator, altimeter, airspeed indicator, and so on, move in response to control inputs. The compass rose rotates during turns, and menu pages and specific options on those pages can be selected using virtual bezel buttons. These prototypes can be built relatively quickly. In fact, using VAPS in the late 1980s, an entire fighter cockpit prototype was built in just over 2 weeks. In the early 1980s, a comparable operator-in-the-loop simulation took 1 to 2 years to build ("Crew Systems Lab," 1987). Developers can run VAPS using data generated internally (within VAPS) as a desktop prototype, they can integrate VAPS into a full mission simulation environment, or they can operate using real external data. Another major advantage is the ability to rehost the code generated by VAPS into the actual product with few restrictions, which can result in savings of cost and time. VAPS has an Ada Code Generator and a C Code Generator that produce the code, which is rehosted (Virtual Prototypes, Inc., 1992).

Virtual Prototypes now markets a family of products including FlightSim/HeliSim, which is a basic, configurable, aircraft flight and handling qualities model that enables developers to tailor the flight characteristics of their simulation to the product being designed. Scenario Toolkit And Generation Environment (STAGE) is used to create interactive simulated environments in which VAPS/CCG models can run. Objects in STAGE (e.g., threats on a battlefield) can be programmed to behave according to rules selected by the modeler. Virtual Prototypes' products have an open architecture that enables them to interface with other simulations, including distributed interactive simulations. The developer, Virtual Prototypes, Inc., of Montreal, Quebec, Canada (514-483-4712; www.virtualprototypes.ca), offers 3- and 5-day training courses and recommends that participants be proficient in C.

Designer's Workbench. Designer's Workbench, a product of Centric Software (www.centricsoftware.com), is an interactive graphics rapid-prototyping system that runs on Silicon Graphics workstations. Designer's Workbench (DWB) can produce complex three-dimensional graphics and features a display editor and a runtime environment. Three-dimensional models are translated into Open GL/C for multiple platform or embedded system use. DWB has an application that creates synthetic environments (called EasyTerrain), in which prototypes of models built in DWB can be run. EasyTerrain can import Digital Feature Analysis Data (DFAD). An example of synthetic terrain is shown in Fig. 9.5.

Altia. Altia Design, like VAPS, enables developers to build dynamic, interactive prototypes of graphical interfaces without programming. Altia Design (www.altia.com) is composed of several applications. The Graphics Editor provides the capability to draw objects or import high-resolution, bitmapped, photo-quality graphics. Altia also provides a library of graphical objects. The Animation, Stimulus, and Control Editors enable designers to

FIG. 9.5. Synthetic environment generated with Designer's Workbench
EasyTerrain.

impart behaviors to graphical elements. The Properties Editor specifies
what object properties can be modified (i.e., color, scaling). The Connection
Editor connects graphics, and their inherent behaviors and properties, to
an external simulation. Altia runs in Windows and Unix environments.

Juran (1998) recommends employing graphics professionals or industrial designers for graphics that will be used for customer demonstrations.
In these cases, realism and fidelity are extremely important, and although
rapid prototyping systems are becoming more powerful, they are also becoming more complex. Many companies that develop products with complex user interface designs are now employing dedicated graphical user
interface designers to model human factors engineers' ideas. Despite the
cost of employing these people, the life-cycle savings realized by rapid
prototyping outweighs the cost of the poor or dangerous designs that can
result from skipping this step.

Simulators

Often, displays and controls that appear to be well designed as separate
units, and have passed muster in successive evaluations through the development process, fail to play together well. For example, rapid prototyping
may indicate that individual displays present their information well and
are fully acceptable to users. It may take full simulation of the system in its
environment, however, to reveal that the same displays present difficulties
when integrated into a synchronous system. A design can be shown in a
new light when the operator has system dynamics to control and operational time pressures to manage. Problems can become especially acute if
the operator's information-gathering and integration strategy changes at

different phases of the operation. Information may not be allocated to the proper controls and displays, or in the optimal form, after all. Only a simulator or the actual product prototype can uncover these problems because they put together all aspects of use of the product, including the temporal factor.

There is no shortage of information and research on what makes a simulator more effective. This section is a summary of some of that research. Presentation of visual information is a critical aspect of simulator effectiveness. Most vehicle simulators use a wide field-of-view (FOV) display to present an interactive picture of the world through which the simulated vehicle is traveling. Several studies cited in Padmos and Milders (1992) recommend FOV sizes depending on the application. Dixon, Martin, and Krueger (1989) recommend a 160° × 60° FOV for bomber simulators and a 300° × 150° FOV for fighter and helicopter simulators. The minimum acceptable FOV for driving simulators is 50° × 40°, but more realistic applications call for up to 180° (Haug, 1990). Ship simulators should afford a 280° × 45° FOV (van Breda, Donsalaar, & Schuffel, 1983). Display quality itself also impacts simulator effectiveness. Testing contrast ratios from 2.3:1 to 24:1, Kennedy, Berbaum, Collyer, May, and Dunlap (1988) found that the greatest improvements in subjects' performance in a flight simulator was accounted for by increases in contrast from 5.4:1 to 10.4:1.

In the past, some flight simulator visual systems were provided by a small, movable camera, mounted above a terrain board. This board resembled the miniature world that a model railroad hobbyist would create, with little trees, buildings, and so on; control inputs in the cockpit drove the camera over the terrain, and the camera fed back a visual representation of the world. Modern simulator displays are almost exclusively computer-generated graphics displays. Images are formed by joining large numbers of polygons to form a moving mosaic. As the vehicle moves through the world, objects change shape and size to indicate perspective and distance, and things are rotated as the point of view changes. Performance improvements can be obtained by increasing the number of objects in the environment (Rogers & Graham, 1979; cited in Padmos & Milders, 1992). Kleiss and Hubbard (1993), however, found that, although large numbers of objects improved altitude awareness in a flight simulator, object realism did not. Terrain realism improved performance, but the authors suggest that, if memory and processing limitations force tradeoffs, put available resources into the number of objects. This finding was confirmed by Lintern, Sheppard, Parker, Tates, and Dolan (1989), who found that level of visible detail, manipulated by day versus dusk conditions, had no effect on subjects' performance.

The apparent location of the source of the visual scene also influences simulator effectiveness. In the real world, the brain sees a picture presented from the eyes' location in space. In a simulator, the user perceives a world from the location of the computer's eye. This can cause problems in simulators with head-mounted virtual displays. Imagine driving a car with only a periscope, mounted 10 ft above the roof, for a visual reference. Collimated objects create an illusion that they are distant by focusing them artificially at infinity. Iavecchia, Iavecchia, and Roscoe (1988) maintain that this causes consistent overestimation of depth (i.e., things seem farther away than they really are). They recommend magnifying projected visual images by 1.25 times to compensate for this illusion. Padmos and Milders (1992) cite many relevant studies and describe a number of physical variables that affect simulator quality. Anyone procuring a simulator, designing studies to be performed in simulators, or desiring to know more details about simulator/user interface issues would benefit from this review.

One of the longest running controversies in simulation concerns the efficacy of motion. It is a more important issue in the realm of training simulators but is worth mentioning in a discussion of development simulators because it is believed to provide greater fidelity. Fidelity of motion is important in training simulators because higher fidelity can involve more of the senses and better prepare the student for the real world. Although most of the research on simulator motion centers on training, both training and testing are more effective with greater motion realism. Because the brain naturally interprets reality by integrating not just visual but also kinesthetic, proprioceptive, and vestibular cues, the concept that motion increases fidelity has face validity. The best argument against using motion is cost. Motion is very expensive, and the benefits do not always justify the cost. Another argument against motion (for flight simulation) is that aviators are trained not to trust their vestibular and kinesthetic perceptions if they contradict what the instruments present to the eyes. Does motion train aviators to rely falsely on vestibular and kinesthetic cues? The counter argument is that experiencing vertigo in a simulator teaches the aviator not to trust the seat of the pants.

Koonce (1974, 1979) found significantly higher error rates for pilots flying a simulator with no motion than with motion. When the same maneuvers were flown in an actual aircraft, there was a significant rise in error rates for the no-motion group but not the motion group, compared with performance in the simulator. Performance is, therefore, more predictable when the pilot goes from a simulator with motion to an actual aircraft. Koonce attributes the superior performance in the motion-based simulator to the possibility that, in visual flight, motion provides an alerting cue when the aircraft is straying out of constraints, and the pilot can make corrections more quickly than without the aid of the vestibular and kinesthetic cues. Caro (1979) makes the distinction between maneuver cues (pilot-induced motion) and disturbance cues (externally induced motion, e.g., turbulence). He believes one of the deciding factors for motion and the type of motion is the type of aircraft being simulated. If it is a very stable aircraft, motion may not be worth the cost. He points out the lack of empirical data to aid in making the decision systematically. For evaluators, the answer will undoubtedly come down to cost versus need for realism. There are probably fewer situations in test and evaluation simulation where motion is absolutely necessary, with the possible exception of evaluating flight control systems. In that case, designers can make incremental adjustments that pilots can evaluate subjectively. This is beneficial, but not essential, for flight control system design, however.

Simulator sickness is a problem associated with motion. Motion simulators typically have a different update rate for the visual system (30 Hz) and the motion system (15 Hz). Frank, Casali, and Wierwille (1988) assert that this can decouple visual perception of the world from kinesthetic and vestibular sensations, leading to discomfort or sickness. The thinking among some simulator designers is that the brain processes kinesthetic and vestibular cues faster than visual cues; therefore, motion should lead visual presentation. FAA Advisory Circular 120-40B (1993) states that motion may lead visual cues in simulators but not by more than 150 ms for level C and D simulators and 300 ms for level A and B simulators. Frank et al. (1988) present research suggesting that, although no lag is preferable, performance is better and that wellness is more likely to be maintained when the visual system leads the motion system.

Simulators are beneficial because they relate most closely to physical and cognitive stimulation of the real world, other than the product under development itself. This benefit does not come without a price. Simulators are expensive (although cost varies with fidelity). They require extensive computing power, operators skilled in hardware and software integration, and, generally, a dedicated maintenance crew. In fact, some require a dedicated building. However, dollars saved in design missteps, rework, and solution of problems that would be correctable if discovered in the prototype phase usually justifies the cost when developing large-budget items, such as civil or military aircraft.

Distributed Network of Dissimilar Simulations (DNDS)

DNDS is the current state of the art in development and training simulation. In a DNDS exercise, individual simulators in different locations simulate the same environment. For example, a pilot flying an attack helicopter simulator sees a tank that is actually being controlled by an individual in a tank simulator somewhere else. The U.S. Army Simulation Training and Instrumentation Command has conducted both research and battlefield training using this technology, with promising results (Farranti & Langhorst, 1994). It would also be appropriate for evaluating large-scale tactics. The technology is extremely expensive, but it is not nearly as expensive as what it has the potential to replace (e.g., battlefield operational testing and large military training exercises). Moreover, simulation participants are safe from the real-world risks of crashes, explosions, and other mishaps.

TESTING PERFORMANCE

Up to this point, this chapter has described test tools that are generally used when the product has already taken some tangible form, however primitive. The decisions that designers make with the help of these tools usually consist of rearranging displays; improving legibility, visibility, or comfort; changing the location of a control; altering the structure of a menu to improve accessibility of key information; and so on. This section describes a different class of tools that are used to test the entire system (i.e., both human and machine) performance. These tools enable designers to make some of the fundamental decisions about the product, often before they have in mind any clear picture of its physical form. A representative sample of these simulations is listed in Table 9.3. Some tools are

TABLE 9.3
Performance Testing Tool Selection Guide

Types of Tools	Selection Criteria							
Test performance	Validated	Current	Easy to use	Flexible	Widely used	User's manual	PC based	Workstation
AIRT		x				x		x
ATHEANA	x	x	x	x		x	x	
CASHE	x	x	x	x		x	x	
CREWCUT	x	x				x	x	
Data Store	x					x		
HOS	x	x	x	x	x	x		
KOALAS	x	x				x		x
MAPPs	x							
MIDAS	x	x	x	x		x		x
OASIS		x	x	x		x		
OCM	x	x				x	x	
SAINT	x	x	x	x	x	x	x	
TEPPs	x					x		
THERP	x					x		
UIMS	x	x						
WITNESS	x	x	x	x	x	x		

described in great detail whereas others are described only in historical context.

Developers use digital simulations for many reasons. Seventeen of these have been listed in AGARD-AR-356 (1998):

1. examine the impact of technology on operational effectiveness
2. predict the types of human error that may be associated with a system design and the frequency of these errors
3. take into account the effects of stressors
4. determine the level of human performance required to meet system performance requirements
5. identify areas where an investment in training will have significant human performance benefits and help assess cost-benefits of training system options
6. provide criteria for assessing total system (human and system) performance
7. aid in function allocation
8. predict whether the operator will be able to resume manual control after automation failure
9. determine the number of operators required for a particular system
10. aid operator selection
11. predicting crew workload
12. characterize the flow of information required to perform specific tasks
13. model communication
14. predict the benefits, if any, of modifications
15. predict the effects on performance of control variables
16. ensure that equipment layout is optimized
17. guide development of procedures (pp. 3–4)

Digital simulations for human factors testing have been around since the 1960s. They began as databases and evolved through subjective assessments, treating the human as a system, language development, error prediction, assessing human–computer interactions, and developing tool sets.

Databases

The earliest simulations were based on collection of empirical data. One of the first of these was the Data Store developed by Munger, Smith, and Payne (1962). It contained human performance data organized by type of control and display used. The data included probability of successfully completing the task and time to complete the task.

Because data were not available for all tasks, the next step in the history of digital simulation was to add subjective assessments to fill in the gaps in objective data. A good example of such a simulation is the Technique for Human Error Rate Prediction (THERP). THERP was developed by Swain (1964) to assess human error in military systems. It begins with a decomposition of tasks into individual actions, such as flipping a switch or reading a dial. For each action, a probability tree is built for subsequent actions. According to Meister (1984), THERP has been used in 11 agencies and countries.

Subjective Assessments

It was not long after digital simulation appeared that there were attempts to build simulations totally derived from subjective assessments. For example, The Establishment of Personnel Performance Standards (TEPPS) was developed in the late 1960s to predict the probability of task completion and the time to complete the task (Meister, 1984). It required (1) the development of a Graphic State Sequence Model (GSSM) for the system to be evaluated, (2) application of expert judgments of task completion probabilities and times, (3) conversion to a Mathematical State Sequence Model (MSSM) so that the probability equations can be applied, and (4) exercising the MSSM to obtain system-level task completion probabilities and times. The method was abandoned after a single validation test (Smith, Westland, & Blanchard, 1969).

Human As System

In the 1970s the quest began to model the human like any other engineering system. One example is the Optimum Control Model (OCM), which assumes that human nervous and muscular systems react like an optimal state estimator. In the OCM, the human observes target-tracking error and uses a closed-loop transfer function to nullify the observed error. The human's response is not perfect, and the error is due to remnant, the portion of the human's input not related to the observed target error (Hess, 1980).

Language Development

The demand for digital simulations grew as the complexity of the systems to be simulated grew. This demand created a need for a simulation language to support the development of operator models. The U.S. Air Force met the need with Systems Analysis of Integrated Networks of Tasks (SAINT), a Fortran-based simulation language. It was developed to model complex human-machine systems (Wortman, Seifert, & Duket, 1976). It provides both discrete and continuous simulation. SAINT has been used during safety analysis of nuclear power plants, interface evaluation of remotely piloted vehicles, and testing the operator-machine interface for a process-control console in the steel industry (Wortman, Duket, & Seifert, 1977). It has also been used as a software verification and validation tool (Hoyland & Evers, 1983). SAINT has been widely used throughout the world (Sestito, 1987). A user's manual is available (Duket, Wortman, Seifert, Hann, & Chubb, 1978).

MicroSAINT was designed to run on microcomputers or PCs. Like its predecessor, SAINT, it is a discrete-event network simulation. It has been used almost solely for task-based system evaluations, such as firing howitzers (Bolin, Nicholson, & Smootz, 1989), loading U.S. Navy ships (Holcomb, Tijerina, & Treaster, 1991), and performing surgery (*Industrial Engineering*, 1992). Attempts have been made to use MicroSAINT to model aircraft carrier landings. The effort was abandoned because of lack of funds (Stanny, 1991). It has also been used with target acquisition and met with mixed results (See & Vidulich, 1997). MicroSAINT requires detailed task information, including sequencing, resource consumption, and performance times (Bogner, Kibbe, Laine, & Hewitt, 1990). It has been used to develop several stand-alone models, including a human performance model that estimates work load (Bogner et al., 1990).

Both SAINT and MicroSAINT have received recent criticism:

> However, these models have not been shown to be able to accommodate . . . knowledge-based descriptions of performance. . . . Typically, a task network model will require the user to provide estimates for the times and accuracies of behaviors at the finest level of model representation, whereas a knowledge-based model affords the promise of obtaining basic performance data from general perceptual, motor, and cognitive models. (Zachary et al., 1998, p. 6)

An earlier NATO report (Beevis et al., 1994) was even more negative:

> SAINT is not very easy to use. The program "interface" requires knowledgeable users, and they find the documentation difficult to use. Some published versions of the code have included errors, which users have had to identify and correct. The commercial versions such as MicroSAINT are easier to use, but are limited in their capability to run large networks, and their use of dynamic state variables. They do not have all the features of the full SAINT program. They do not include an interface to a programming language, but use a reduced set of FORTRAN, so they cannot represent complex logic. (p. 1)

Error Prediction

The 1980s saw an increased concern in predicting human error as a result of several industrial accidents. One of the first error models was the Maintenance Performance Prediction System (MAPPs), a discrete-event simulation model that uses Monte Carlo techniques to predict the number of maintenance errors as a function of the number of attempts to complete an action (Seigel, Bartter, Wolf, & Knee, 1984). A second example is the Human Operator System (HOS) IV that predicts the number of maintenance errors. HOS IV is recommended for evaluation of new or modified systems. It requires detailed descriptions of the operator-system interface, the operator–machine interactions, and the environment in which the system will be used (Bogner et al., 1990). HOS IV is part of the Manpower and Personnel Integration (MANPRINT) Methods Program (Dick, Bittner, Harris, & Wherry, 1989). HOS micromodels have been incorporated into MicroSAINT (Plott, Dahl, Laughery, & Dick, 1989). HOS has been applied beyond maintenance to piloting. For example, Glenn and Doane (1981) reported a +.91 correlation between predicted and actual pilot eye-scans during simulated landings of the NASA Terminal Configured Vehicle.

Digital simulations to predict human error are still being developed today. For example, A Technique for Human Event Analysis (ATHEANA) is a human reliability model being developed by the NRC to identify errors of commission and dependencies associated with the design of nuclear power plants (U.S. Nuclear Regulatory Commission, 1998).

Assessing Human–Computer Interactions

The 1980s could be characterized as a period of rapid changes in human-computer interfaces. One of the major challenges to human factors engineers was to evaluate these interfaces. A number of digital simulations were developed to help in this evaluation. One, the User Interface Management System (UIMS) was originally developed to describe human–computer interactions (Johannsen, 1992). Its use has been expanded to include evaluation of an online-advisory system, fault management in a power plant, and the ability of professional pilots to detect errors in an automated aircraft guidance system.

As UIMS was evolving to broaden its applicability from simple human-computer interfaces to the more general problems of automation design, other models were also being developed to address automation challenges to human operators. One of the best known is Automation Impacts Research Testbed (AIRT), an advanced system modeling technology to aid system designers and evaluators in understanding the impact of automation alternatives and distributed operations on operator performance (Young & Patterson, 1991). AIRT was developed for evaluating the effect of automation on air defense system command and control operators (Corker, Cramer, Henry, & Smoot, 1990). The system had three components: (1) human performance process (HPP) models (Young, 1993), (2) rapid operator-machine interface prototyping tools, and (3) scenarios in which to exercise the models and interfaces developed. The HPP models emulated human behavior through simulation of specific human information processes and attributes. Within AIRT, HPP models served as artificially intelligent team members. AIRT's ability to substitute an actual human in-the-loop operator for a modeled operator, during identical simulation trials, enabled differing aggregate models to be compared with actual human performance. This ability to perform side-by-side comparisons of operators and models created a unique vehicle for human performance research, enabling developers to compare side-by-side (and different) theories and techniques. AIRT development continued as the Operability Assessment System for Integrated Simultaneous Engineering (OASIS).

OASIS was a collection of tools to support designers in building soft prototypes (i.e., a mock-up of the system where the system functionality is emulated in software) of competing multicrew station designs. The soft prototypes were designed to be used to investigate operability issues, such as fielding the system under test in conditions for which it was not designed (BBN, 1993). The software functional specification (Deutsch, Adams, Abrett, Cramer, & Feehrer, 1993) stated the need for four components: (1) simulation tools, (2) knowledge acquisition tools, (3) runtime environment, and (4) a user interface. For the simulation tools, an object-oriented program was required. The knowledge acquisition tools included text and graphics editors. The runtime environment enabled the user to define and control scenarios; event-driven signaling to HPP submodels was included. The model was applied to single and multitasking behavior of en route air traffic control. The user interface included a graphics user interface, table-editing facilities, status displays, and animations to support simulation monitoring and development.

Digital simulations have also been developed for automated assembly lines. One example is WITNESS, a discrete-event simulation program, developed to support manufacturing operations (Gilman & Billingham, 1989). WITNESS enables the user to visualize the world from the perspective of an individual manufacturing component. Basic elements include physical elements (parts, machines, conveyors, buffers, labor, vehicles, and tracks), control elements (attributes, variables, distributions, functions, files, and part files), and reporting elements (time series and histograms). The benefits of WITNESS include (1) improved performance of manufacturing operations, (2) reduced capital cost, (3) identification of production contributions of individual components, (4) reduction of risk in decision making, (5) an understanding of operations before they change, (6) improvement in discipline in decision making, and (7) exploration of design ideas (Hollicks, 1994).

A key component of evaluating automation is understanding human decision making in automated systems. At one time, models were developed specifically for this purpose. The Knowledgeable Observation, Analysis Linked Advisory System (KOALAS) models four phases of decision making: (1) sensing the environment, (2) interpreting and assessing the situation, (3) generating options, and (4) selecting an option. The sensor data correlator uses algorithms to detect common events among sensors. This information is presented on the Crew Systems Interface and is used to generate a Hypothetical World Model that is evaluated by the Evidence Manager. The Evidence Manager uses advice and data to feed back to the sensor data correlator, which, in turn, updates the Hypothetical World Model. KOALAS has two major uses: (1) developmental and operational test and evaluation (DT&E and OT&E) test planning and (2) decision process evaluation. KOALAS has been used to support F-14D multisensor integration, F/A-18 electronic warfare integration, and AN/SLQ-32 evaluation of embedded training and to combine radar tracking and navigation data for the Mariner's Eye (MATRIS, 1994).

Tool Sets

KOALAS is one of many tool sets that have been developed to support digital simulations. One of the first was Computer Aided Systems Human Engineering (CASHE). It has four components: (1) soft-copy human factors documents, including MIL-STD-1472D and *Data Compendium* (Boff & Lincoln, 1988); (2) tools to personalize the database; (3) an integrated Perception and Performance Prototyper; and (4) data manipulation tools (http://dticam.dtic.mil/hsi/index/hsi7.html).

MIDAS is an integrated set of tools for crew station designers to apply human factors principles and human performance models to the conceptual design of helicopters. In the developer's own words:

> MIDAS is intended to be used at the early stages of conceptual design as an environment wherein designers can use computational representations of the crew station and operator, instead of hardware simulators and man-in-the-loop studies, to discover first-order problems and ask "what if" questions regarding the projected operator tasks, equipment and environment. (MATRIS, 2001, p. 1)

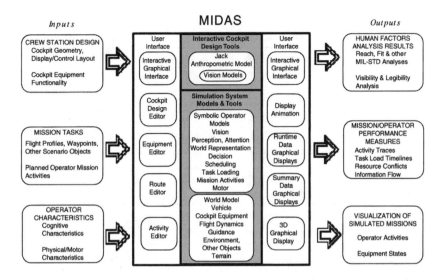

FIG. 9.6.   MIDAS architecture.

A functional view of MIDAS is given in Fig. 9.6, from Corker and Smith (1993).

TESTING MANPOWER, PERSONNEL, AND TRAINING REQUIREMENTS

Some of the largest costs associated with a manned system are the labor of operators and the amount and type of their training. To predict some of these costs, several tools have been developed. These are listed in Table 9.4 and are described in the following sections.

Design Evaluation for Personnel, Training, and Human Factors (DEPTH)

DEPTH was designed to support an IPD process. It has been used during system design, modification, and testing. DEPTH has four major components: (1) a graphic human model (Jack); (2) human performance databases defining human abilities; (3) a CAD program; and (4) a logistics support analysis (LSA) process, integrated with the Computer-Aided Acquisition and Logistics Support (CALS) system (Glor & Boyle, 1993). The databases include anthropometric and ergonomic data from CREWCHIEF (MATRIS, 1994). DEPTH version 2.0 also includes HOS V (MATRIS, 2001).

TABLE 9.4
Manpower, Personnel, and Training Testing Tool Selection Guide

Types of Tools	Selection Criteria						
Test manpower, personnel, & training	*Validated*	*Current*	*Easy to use*	*Flexible*	*Widely used*	*User's manual*	*PC based*
DEPTH	x	x	x	x		x	x
HARDMAN	x	x	x	x	x	x	x
MANPRINT	x	x	x	x	x	x	x

Hardware Versus Manpower (HARDMAN)

HARDMAN refers to a family of models: (1) the HARDMAN Comparability Methodology (HCM), (2) HARDMAN II, and (3) HARDMAN III. Each of these is described in this section.

HCM has been used to predict manpower, personnel, and training requirements based on requirements for similar systems (U.S. Army, 2001). It is extremely data-collection intensive and requires mission analysis, functional requirements analysis, equipment comparability analysis, reliability and maintainability analysis, and task identification. However, there is an excellent guide (Mannle, Guptil, & Risser, 1985) that provides the analyst with descriptions of each step in an analysis, the rules to be applied at each step, and the interface points to other steps and to data. The guide also provides managers with planning and management procedures.

Further, this guide includes descriptions for identifying data requirements, identifying data sources, selecting among those sources, collecting data, evaluating those data, and maintaining/storing the data. Requirements analysis includes identifying manpower; training resources, including both equipment and courseware; and personnel requirements based on pipeline requirements and flow rates. For manpower, the analyses include determining Military Operational Specialties (MOSs), grades, work load, force structure, and work capacity. Data evaluation includes both impact and tradeoff analyses. Impact analysis is used to determine resource availability, criticality, and impacts at the force level. Tradeoff analyses are used to identify tradeoff areas, alternatives, and results.

HCM has been thoroughly evaluated (Zimmerman, Butler, Gray, Rosenberg, & Risser, 1984). The results indicated that HCM (1) met the majority of user requirements; (2) conformed to other manpower, personnel, and training models; (3) provided reasonable results on 85% of test issues; (4) had internal reliability; and (5) was 80% to 90% accurate. One shortcoming was the sensitivity of the manpower estimates to life-cycle costs.

HARDMAN II, also called Man Integrated Systems Technology (MIST), was developed to support front-end analysis of Army weapon systems. Herlihy, Secton, Oneal, and Guptill (1985) provided the following list of advantages of MIST: 1) "cost effective because it reduces the time to collect, input, and analyze data; 2) provides trade-off capability; 3) stores information in a common database; 4) eliminates mathematical errors; and 5) provides a structured approach to conducting a front-end analysis" (pp. xiii, xiv).

HARDMAN II has ten components. The first component is the Introduction and Version History/Exit MIST (INT). It contains the entry/exit screens and enables the user to record changes to MIST files and see a record of past changes. The second component, System Requirements Analysis (SRA), has the structure, formats, and worksheets required for front-end analysis. The third component, Military Occupational Specialty (MOS) Selection Aid (MSA), provides cognitive and training resource requirements on which to base a MOS selection decision (Herlihy et al., 1985). The fourth component, Manpower Requirements Determination (MRD), provides estimates of the number of man hours required at each MOS to maintain a system for a period of time specified by the user. The fifth component, the Personnel Requirements Determination (PRD), uses three flow factors, Trainees, Transients, Holdees, and Students (TTHS), promotion rates, and attrition rates, to estimate steady-state personnel requirements for each MOS by pay grade (Herlihy et al., 1985).

The sixth component, the Training Cost and Resources Determination (TCR), enables the user to examine the effects of training resources (e.g., optimum class size, type of instruction) on training costs (i.e., total cost, cost per graduate, instructor requirements, and training time; Herlihy et al., 1985). The seventh component, Training Media Selection (TMS), enables the user to optimize training efficiency or training cost (Herlihy et al., 1985). The eighth component, the Manpower, Personnel, and Training Summary Report (MPT), lists manpower requirements by MOS level, number of personnel, training days, number of instructors, and training cost (Herlihy et al., 1985). The ninth component, the MIST Administrator Area, is password protected and is used by the administrator to manage and maintain the MIST software (Herlihy et al., 1985). The final component, Datatrieve Reports, enables the user to generate reports listing functional requirements, workload data, training analysis results, and personnel data (Herlihy et al., 1985).

HARDMAN III can be used to predict the manpower, personnel, training, and human engineering requirements for systems under development (Kaplan, 1991). It requires extensive data collection (usage rates, reliability data, and career data) and mission decomposition (mission, functions, and subfunctions). The output is work load predictions and manpower and training requirements. HARDMAN III has seven linked software modules that run on PCs.

Module 1, the System Performance and RAM Criterion Development Aid (SPARC), identifies performance (time and accuracy) and RAM requirements (Bogner et al., 1990). Module 2, Manpower Constraints Estimation Aid (M-CON), identifies manpower resource constraints (MATRIS, 2001). Specifically, M-CON calculates four measures of manpower constraints: maximum operator crew size, maximum maintenance hours, total operator manpower requirements, and total maintenance manpower requirements (Herring & O'Brien, 1988). The third module, the Personnel Constraints Aid (P-CON) is used to predict the MOSs required to operate a system (Bogner et al., 1990). Specifically, P-CON calculates the numbers and percentages of personnel that will be available with the characteristics needed to use and maintain a system (Herring & O'Brien, 1988). The fourth module, the Training Constraint Aid (T-CON), is used to predict training times for operators and maintainers as a function of system characteristics (Bogner et al., 1990).

The fifth module, the Manpower-Based System Evaluation Aid (MAN-SEVAL), predicts maintenance and operator manpower requirements for competing system designs (Bogner et al., 1990). MAN-SEVAL includes a library of reliability, availability, and maintainability (RAM) data for previous Army systems. The model uses component failure rates, time to perform maintenance, and the mission scenario to estimate work load, accessibility, and performance time (Laughery, Dahl, Kaplan, Archer, & Fontenelle, 1988). The sixth module, the Personnel-Based System Evaluation Aid (PER-SEVAL), has been used in test and evaluation to determine operator aptitudes required to meet required system performance (Bogner et al., 1990). PER-SEVAL has three parts: (1) models to predict performance as a function of training, (2) models to predict the effects of environmental stresses (e.g., heat), and (3) models to aggregate individual performance estimates to team performance estimates. The seventh and last module, Manpower Capabilities (MANCAP), is used to evaluate operator and maintainer manpower required for a particular system (MATRIS, 2001).

HARDMAN III has been applied to Product Improvement Programs (PIPs) (Bogner et al., 1990) and to the evaluation of numerous air defense systems, including the Patriot Missile system. It has been updated and is now called the Improved Performance Research Integration Tool (IMPRINT) and is part of MANPRINT.

MANPRINT

MANPRINT "is the Army's systematic and comprehensive program for improving the effectiveness of system performance at minimum costs for personnel, maintenance and repairs throughout their entire life cycle"

(U.S. Army, 2001). Executive guidance requires that MANPRINT requirements be used within the Army during source selection. Specifically, "Effective this date, all required and appropriate MANPRINT requirements and opportunities will be evaluated and considered in the best value trade-off analyses associated with source selection for acquisition of all Army systems."

MANPRINT includes a collection of tools ranging from documents to software that runs on personal or mainframe computers. The documents include a guidebook for using MANPRINT (updated September 1998) that contains objectives and key concepts. A handbook also has been developed to help system acquisition personnel apply MANPRINT (Army Publications & Printing Command, 1994). The handbook has chapters on the acquisition program, tailored MANPRINT support, MANPRINT management, activities throughout the life-cycle phases, and the MANPRINT assessment. There is also a MANPRINT directory as well as a quarterly bulletin and proceedings of a yearly symposium.

MANPRINT software tools include the Automated Information Systems (AIS) MANPRINT Management Tool for estimating manpower, personnel, and training resources; the Army Manpower Cost System (AMCOS) for predicting the life-cycle cost of a new system by year for each MOS; IMPRINT for modeling soldier performance throughout the life cycle of a system; the Early Comparability Analysis (ECA) standardized questions to assess task difficulty and performance; WinCrew for predicting workload as a function of task design; Operator Workload Knowledge-Based Expert System Tool (OWLKNEST) for selecting appropriate techniques for assessing operator workload; HCM; and Parameter Assessment List—MANPRINT Automated Tool Edition (PAL-MATE) to assess soldier survivability. A more complete description of these tools can be found at http://www.manprint.army.mil/manprint.

MANPRINT has been used throughout the entire cycle of system procurement. MANPRINT provisions have been made to support development of a request for procurement (Chaikin & McCommons, 1986). MANPRINT has also been used to forecast system effectiveness and availability (Allen Corporation, 1988). Use of MANPRINT early in the acquisition of the Airborne Target Handover System/Avionics Integration resulted in reduced task times and simplified procedures and maintenance (Metzler & Lewis, 1989). It was an integral part of the integrated logistics support/reliability, availability, and maintainability (ILS/RAM) evaluations of the Army's Light Family of Helicopters (LHX) (Robinson, Lindquist, March, & Pence, 1988). The MANPRINT database management system was especially helpful in the LHX program (Jones, Trexler, Barber, & Guthrie, 1988). MANPRINT has been used during test and evaluation of the Personnel Electronic Record and the Worldwide Port System. It was used after operational testing of the Aquila remotely piloted vehicle. Traditional operational test revealed major deficiencies as a result of manpower, personnel, training, and human factors problems (Stewart, Smootz, & Nicholson, 1989). MANPRINT has also been used to identify problems with existing systems, for example, the AN/TRC-170 Digital Troposcatter Radio (Krohn & Bowser, 1989). The evaluation included structured interviews, on-site observations, and performance measurement.

EVALUATION SUPPORT TOOLS

Evaluation support tools are available to the human factors engineer to support selection of the appropriate digital simulation, work load metrics, evaluation of software usability, and test planning and execution. These tools are rated in Table 9.5.

Human Operator Modeling Expert Review (HOMER)

HOMER is an expert system developed to help users select the appropriate digital simulation for their application. The selection is made based on responses to users responses to 19 questions (see Table 9.6).

The Software Usability Evaluator (SUE)

SUE was developed to implement a software version of Air Force Operational Test and Evaluation Center (AFOTEC) Pamphlet 800-2, volume 4, *Software Usability Evaluation Guide* (Charlton, 1994). SUE contains 171 attribute statements with which a software evaluator agrees or disagrees. The statements were derived from MIL-STD-1472D (1989) and Smith and Mosier (1986). A help table within SUE lists the attributes and the

TABLE 9.5
Evaluation Support Tool Selection Guide

Types of Tools	*Selection Criteria*							
Support evaluation	*Validated*	*Current*	*Easy to use*	*Flexible*	*Widely used*	*User's manual*	*PC based*	*Workstation*
HOMER	x	x	x	x		x		
OWLKNEST	x	x	x	x		x	x	
SUE	x	x	x	x		x	x	
Test PAES	x	x	x	x	x	x	x	

TABLE 9.6
List of HOMER Questions

Questions	*Choices*
My primary interest is	Crew component
	Team interactions
	Display format and dynamics
	Control design and dynamics
	Automation
	Procedures
	Workspace geometry/layout
	Communications
	Environmental stressors
The design phases(s) I will analyze are	Operation analysis/research
	Conceptual design
	Feasibility: dem/val
	System development
	Test and evaluation
The equipment/system I will analyze is	Off the shelf
	Mod of existing system
	A completely new system
The crew I plan to analyze is	A single operator
	Two or more operators
Max time available for completing analysis is	Days
	Weeks
	Months
The funds available for software purchase are	$0–5000
	$500–50,000
	>$50,000
I am NOT willing to use a	IBM-type PC (with Windows)
	PC or Sun (with UNIX)
	Silicon Graphics
	Macintosh
	Any computer
Available personnel skills include	Subject matter experts
	Human factors experts
	Computer programmers
	Modeling/systems analyst
Available data include	Timelines
	Tasks network
	Parameters
	Analysis of similar systems
	Model of relevant dynamics
The model should represent work-load peaks by	Mission duration
	Errors
It is important that the model sup-ports	A vehicle control model
	Crew station layout
	State transitions
	System/automation logic
	Physical sim of the work space
	View of the external scene

(Continued)

TABLE 9.6
(*Continued*)

Questions	Choices
The model must run in	Real time; scenario based
	Faster (Monte Carlo sims)
For decisions the model must	Emulate decision processes and generate decisions
	Generate decisions by following user-spec rules
	Introduce user-spec decisions as user-spec points
For errors, the model must	Generate reasonable errors at likely points
	Insert user-specified error at likely points
	Insert user-specified errors at user-spec points
Model outputs must include	Response times
	Accuracy estimates
	Crew work load estimates
	Task list
	Task networks
	Procedure list
	Timeline
	Function/task allocation
	Biomechanical measures
	Fit, reach, visual envelopes
	Training requirements
	Selection requirements
	Estimate of sys effectiveness
	Maintainability
	Data flow analysis
The output must be in the form of	Real, absolute values
	Figures of merit
The model must be capable of generating	Mission, task, crew summary
	Segment-by-segment summary
	Second by second events
The model must	Generate dynamic visualization (animation)
The model must estimate the impact in system performance of	Human characteristics
	Equipment characteristics
	Environmental factors
	Stressors

Note. From *A Designer's Guide to Human Performance Modeling* (pp. 24–25), by AGARD AR-356, 1998, Ceduy, France: Advisory Group for Aerospace Research and Development. Copyright 1998.

source of each. Some of the statements require subjective responses (e.g., it is always easy to know where you are in the menu hierarchy), whereas others are objective observations (e.g., the system contains a cursor). These 171 statements are categorized into usability characteristics: consistency, descriptiveness, error abatement, and responsiveness. This aggregation is optional because some users feel that important diagnostic information is lost when the results are aggregated. The 171 statements can be reduced for evaluation of a particular statement by setting qualifiers describing the human-computer system to be evaluated. For example, if the system does not have a mouse, attributes of the mouse will not be presented to the evaluator. Each of the attributes is accompanied by notes to clarify its meaning. Prior to evaluation, demographic information is requested from the evaluator: name, identification number, duty position, and years of experience. During evaluation, the evaluator is presented with an attribute statement and asked to give one of two ratings: agree or disagree. If the evaluator agrees with the statement, the next statement is presented for rating. If the evaluator disagrees with the statement, the evaluator is asked to provide a comment on the problem and then to estimate the operational impact of the problem.

After the evaluation is completed, the analyst can request one of two summary reports. The first is a table of attribute statements by respondent by comment. The second is a table of attribute statements by impact rating by frequency of occurrence of that rating. Either report can be output into an ASCII file (Taylor & Weisgerber, 1997). A user's manual is available (Charlton, 1994).

Test Planning, Analysis, and Evaluation System (Test PAES)

Test PAES has four components: (1) a Structured Test and Evaluation Process (STEP), (2) a set of databases containing both reference material and test-specific information, (3) a set of Structured Test Procedures (STPs), and (4) custom software to integrate these tools in a seamless system using commercial off-the-shelf (COTS) hardware and custom software.

STEP. The STEP is a series of steps that are required during crew station evaluation. The steps start with identifying test issues and end with documenting lessons learned. The complete set of steps is listed in Table 9.7. The STEP is implemented in Test PAES through a series of menus that call up needed information and applications as the user navigates through the process.

Databases. The Test PAES databases include reference-material databases and test-specific databases. The former include (1) a card catalog of technical reports, journal articles, government regulations, and forms; (2) an encyclopedic dictionary with human factors engineering, operational, statistical, and scientific terms; (3) a lessons-learned database with entries from the Joint Unified Lessons Learned System and the AFFTC; (4) a measures database describing the strengths and limitations of over 80 performance, work load, and situation-awareness measures; and (5) an STP

TABLE 9.7
Test PAES Steps

Step Number			Step Title
1			Plan Test
	1.1		Identify test issues
	1.2		Define test objectives
	1.3		Review previous tests and mishap data
	1.4		Identify test conditions
	1.5		Determine safety constraints
	1.6		Design flight or ground test
		1.6.1	Define test points
		1.6.2	Select measures and thresholds
		1.6.3	Prepare data list
		1.6.4	Develop test profiles
	1.7		Finalize test plan
		1.7.1	Develop questionnaires
		1.7.2	Prepare briefing materials
		1.7.3	Prepare test plan
2			Conduct Flight or Ground Test
	2.1		Brief and train test crew
	2.2		Collect data
		2.2.1	Flight test
		2.2.2	Ground test
	2.3		Process data
		2.3.1	Format data
		2.3.2	Time synchronize data
		2.3.3	Verify data
		2.3.4	Assemble archive
		2.3.5	Generate event log
	2.4		Collect debriefing data
	2.5		Analyze data
		2.5.1	Calculate and verify measures
		2.5.2	Test for artifacts in data
		2.5.3	Compare measures with thresholds
3			Evaluate Results
	3.1		Consider statistical power of results
	3.2		Project from test to operational conditions
	3.3		Determine crewstation deficiencies
4			Prepare Test Reports and Briefings
5			Follow-Up
	5.1		Critique STPs
	5.2		Document lessons learned

database containing all the STPs currently available. The test-specific databases include (1) an audit trail for documenting test planning and data analysis decisions; (2) an event log to identify significant events that occurred in each flight; (3) a flight card planning database to maintain copies of flight cards for each crew station evaluation; (4) a note pad for identifying problems, listing things to do, and annotating data; (5) a questionnaire database to store copies of questionnaires to be used in the current evaluation; (6) a test-point database to identify the status of evaluation test points; and (7) a time-sampled database to store data to be used in the current crew station evaluation.

STPs. Test PAES includes a library of STPs. The STPs recommend methods to evaluate crew station interfaces (e.g., side controllers), specific systems (e.g., autopilots), and specific tasks (e.g., precision instrument approach). The STPs include (1) background information on the crew station component being evaluated, (2) previously used test objectives, (3) a list of relevant standards including thresholds for further evaluation and for definite crewstation deficiency, (4) objectives addressed, (5) resource requirements, (6) measurement guidance, (7) procedures, (8) measurement conditions and tolerances, (9) training and briefing materials, (10) data analysis plan, (11) reporting guidelines, (12) references, (13) a summary of lessons learned while performing these evaluations, and (14) a glossary of terms.

Computer System. The Test PAES computer system provides tools for test planning, data collection, debriefing, data processing, data analysis, result evaluation, and reporting and improving the system (i.e., follow-up). Data handling is a key component. Test PAES enables the user to manipulate time-synchronized multimedia data (see Fig. 9.7). It is a highly integrated crew station evaluation system comprising both COTS hardware and software and custom software.

In summary, the main functional goal of Test PAES is to support the user in efficiently executing the STEP. The Test PAES computer system assists in accomplishing this by providing the tools necessary to assist in the process, the data necessary to drive the process, and help and guidance in successfully completing the process. Test PAES is currently being operationally tested in a number of Air Force, Department of Transportation, and Navy agencies as well as several contractor facilities.

EMBEDDED TESTING

In human factors test and evaluation, we strive for valid and reliable measurements taken from representative users in test environments that are as close to the actual mission conditions as possible. Typically, the goal is to

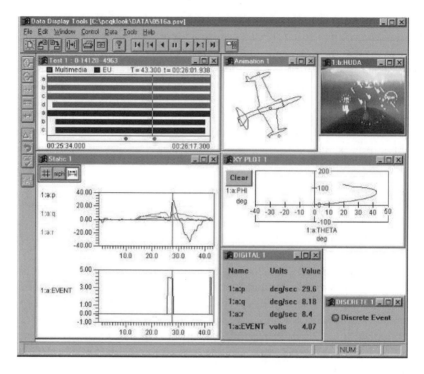

FIG. 9.7.   Test PAES data display capability.

predict product performance or acceptance under actual operating conditions. The more realistic the test conditions, and the more the test subjects match the users of the product, the more likely the accuracy of the predicted product performance. Embedded testing is one approach to product testing that comes close to achieving these ideal goals.

Embedded testing during normal product operation is conducted in some industries to improve performance-prediction accuracy. One of the most comprehensive applications of this approach is found in an FAA sponsored program, where commercial airline pilot performance is measured and documented on an advanced flight recorder as part of standard airline operations. The Flight Operations Quality Assurance (FOQA) program provides a means to manage risk and detect trends in flight crew performance before they turn into costly incidents (Penny & Giles Aerospace, 1996). In this situation we have the real operators in the real operational environment being measured on parameters that are directly related to safety (e.g., takeoff and landing performance). When an airline's pilots show unacceptable performance in a class of tasks, the airline and manufacturer can look into the need for improved systems, training, work/rest schedules, or other remedies. Embedded testing offers the operator and the manufacturer an opportunity to identify and understand performance problems before they cause injury or economic loss.

There are other benefits of embedded testing. Product designers can have access to high quality data on user behavior and states to improve their product's design. Embedded testing can provide the basis for continuous product improvement and improved training and user support functions. The methods used for embedded testing include capturing of user responses as they interact with digital systems, such as Internet software, cable television interfaces, and telephone interactive menus. Another form of embedded testing is remote sensing, where the user does not have to contact the measurement system with any voluntary behavior. Video cameras can be used to remotely sense eye movements and gross-motor activity to determine user reaction to the product on video display monitors. Voice parameters, navigation behavior, and even thermographic features are also candidate measures for embedded testing.

WHERE DO WE NEED TO GO?

One of the major barriers to be overcome in using CAE tools in test and evaluation is the development of an accurate, complete, anthropometric database of operators (Roebuck, 1991). Another barrier is the lack of data standards for system acquisition (Boyle, 1991). How should test and evaluation data be presented? How should human resource requirements be stored with the system data? This barrier is being addressed, at least partially, by the CALS system (Alluisi, 1991). A third barrier is the lack of integration among the complete suite of tools needed to support human factors test and evaluation (Glor, 1991; Thein, 1991). Other areas for improvement include (1) automatic task composition, (2) detailed hand models, (3) an enhanced vision model, (4) analysis of multiperson tasks, (5) expanded task-analysis criteria, and (6) interaction with training systems (Boyle et al., 1990). Other authors have argued that HFTE should also consider the cultural impact of the system—specifically, the humanization of technology (Martin, Kivinen, Rijnsdorp, Rodd, & Rouse, 1991). These authors have identified the following measurement issues (1) viability: Are the benefits of system use sufficiently greater than its costs? (2) acceptance: Do organizations/individuals use the system? (3) validation: Does the system solve the problem? (4) evaluation: Does the system meet requirements? (5) demonstration: How do observers react to the system? (6) verification: Is the system put together as planned? and (7) testing: Does the system run, compute, and so on?

It is clear that human factors test and evaluation, early in the product development process, will continue to grow and expand because there are sound business reasons to conduct testing to reduce risk. It also seems that those methods that are easy to acquire, learn, and interpret will be likely candidates for continued application. What is not so clear is how valid the methodology will be or how much of the test and evaluation methodology will be based on sound theory and data. In the race for faster, better, cheaper, the human factors community generally has better to offer, but they may not get to practice their craft if they are not able to do their job faster and cheaper using mock-ups, electronic and physical human models, simulators, and embedded testing.

REFERENCES

AGARD AR-356. (1998). *A designer's guide to human performance modeling.* Cedux, France: Advisory Group for Aerospace Research and Development.

Allen Corporation. (1988). *Handbook for quantitative analysis of MANPRINT considerations in Army systems* (ARD-TR-86-1). Falls Church, VA: Author.

Alluisi, E. (1991). Computer-aided Acquisition and Logistic Support—Human System Components (CALS-HSC). In E. Boyle, J. Ianni, J. Easterly, S. Harper, & M. Korna (Eds.), *Human-centered technology for maintainability: Workshop proceedings* (AL-TP-1991-0010). Wright-Patterson Air Force Base, OH: Armstrong Laboratory.

Arditi, A., Azueta, S., Larimer, J., Prevost, M., Lubin, J., & Bergen, J. (1992). *Visualization and modeling of factors influencing visibility in computer-aided crewstation design* (SAE paper 921135). Warrendale, PA: Society of Automotive Engineers.

Army Publications and Printing Command. (1994, October). *Manpower and Personnel Integration (MANPRINT) in the System Acquisition Process* (AR 602-2 [Online]). Available: http://books.usapa.belvoir.army.mil/cgi-bin/bookmgr/BOOKS/R602_2/COVER.

Badler, N. I. (1991). Human factors simulation research at the University of Pennsylvania. In E. Boyle, J. Ianni, J. Easterly, S. Harper, & M. Korna (Eds.), *Human-centered technology for maintainability: Workshop proceedings* (AL-TP-1991-0010). Wright-Patterson Air Force Base, OH: Armstrong Laboratory.

Bay Medical Center improves quality with MicroSAINT. (1992). *Industrial Engineering, 24,* 23–24.

Beevis, D., Bost, R., Döring, B., Nordø, E., Oberman, F., Papin, J. P., Schuffel, I. R, & Streets, D. (1994, November). *Analysis techniques for man-machine system design.* (AC/243 (Panel 8) TR/7 Vol. 2, Issue 2). Brussels, Belgium: NATO Headquarters.

Biferno, M. A. (1993, August). Human activity simulation: A breakthrough technology. *SAE Aerospace Engineering, 4,* 4.

Boff, K. R., & Lincoln, J. E. (Eds.). (1988). *Engineering data compendium: Human perception and performance.* Wright-Patterson Air Force Base, OH: Armstrong Laboratory.

Bogner, M. S., Kibbe, M., Laine, R., & Hewitt, G. (1990). *Directory of design support methods.* Department of Defense Human Factors Engineering Technical Group, Washington, DC: Pentagon.

Bolin, S. F., Nicholson, N. R., & Smootz, E. R. (1989). Crew drill models for operational testing and evaluation. *Proceedings of the MORIMOC II Conference* (pp. 151–156). Alexandria, VA: Military Operations Research Society.

Boyle, E. (1991). Human-centered technology: Ends and means. In E. Boyle, J. Ianni, J. Easterly, S. Harper, & M. Korna (Eds.), *Human-centered technology for maintainability: Workshop *proceedings* (AL-TP-1991-0010) (pp. 38–56). Wright-Patterson Air Force Base, OH: Armstrong Laboratory.

Boyle, E., Easterly, J., & Ianni, J. (1990). Human-centered design for concurrent engineering. *High Performance Systems, 11*(4), 58–60.

Caro, P. W. (1979). The relationship between flight simulator motion and training requirements. *Human Factors, 21*(4), 493–501.

Chaikin, G., & McCommons, R. B. (1986). *Human factors engineering material for manpower and personnel integration (MANPRINT) provisions of the request for proposal (RFP)* (Technical Memorandum 13-86). Aberdeen Proving Ground, MD: U.S. Army Human Engineering Laboratory.

Charlton, S. G. (1994, June). *Software usability evaluation guide.* (AFOTEC Pamphlet 99-102, Vol. 4). Kirtland AFB, NM: Air Force Operational Test and Evaluation Center.

Corker, K. M., Cramer, M. L., Henry, E. H., & Smoot, D. E. (1990). *Methodology for evaluation of automation impacts on tactical command and control (C2) systems: Implementation* (AFHRL-TR-90-91). Logistics and Human Factors Division, Wright-Patterson Air Force Base, OH: Air Force Human Resources Laboratory.

Corker, K. M., & Smith, B. R. (1993). *An architecture and model for cognitive engineering simulation analysis: Application to advanced aviation automation.* Paper presented at the AIAA Computing in Aerospace 9 Conference, San Diego, CA.

Crew systems lab offers rapid feedback on cockpit concepts. (1987, November 23). *Aviation Week & Space Technology, 27*(21), 98–99.

Deutsch, S. E., Adams, M. J., Abrett, G. A., Cramer, M. L., & Feehrer, C. E. (1993). *Research, development, training, and evaluation (RDT&E) support: Operator model architecture (OMAR) software functional specification* (AL/HP-TP-1993-0027). Wright-Patterson Air Force Base, OH: Human Resources Directorate.

Dick, A. O., Bittner, A. C., Jr., Harris, R., & Wherry, R. J., Jr. (1989). *Design of a MANPRINT tool for predicting personnel and training characteristics implied by system design* (ARI Research Note 89-04). Alexandria, VA: U.S. Army Research Institute for the Behavioral and Social Sciences.

Dixon, K. W., Martin, E. L., & Krueger, G. M. (1989). The effect of stationary and head-driven field of view sizes on pop-up weapons delivery. *Proceedings of the 11th Interservice/Industry Training Systems Conference* (pp. 137–140). Arlington, VA: American Defense Preparedness Association.

Duket, S. D., Wortman, D. B., Seifert, D. J., Hann, R. L., & Chubb, J. P. (1978). *Documentation for the SAINT simulation program* (AMRL-TR-77-63). Wright-Patterson Air Force Base, OH: Air Force Aerospace Medical Research Laboratory.

Farranti, M., & Langhorst, R. (1994). *Comanche PMO & first team dissimilar simulation networking at the May 1993 AUSA Show.* Paper presented at American Helicopter Society Annual Conference, Washington, DC.

Federal Aviation Administration. (1993). *Airplane Simulator Qualification* (Advisory Circular 120-40B, Appendix 1). Washington, DC: U.S. Department of Transportation.

Frank, L. H., Casali, J. G., & Wierwille, W. W. (1988). Effects of visual display and motion system delays on operator performance and uneasiness in a driving simulator. *Human Factors, 30*(2), 201–217.

Gilman, A. R., & Billingham, C. (1989). A tutorial on SEE WHY and WITNESS. In *1989 Winter Simulation Conference Proceedings* (Cat. No. 89CH2778-9, pp. 192–200). New York: SCS.

Glenn, F. A., & Doane, S. M. (1981). *A human operator simulator model of the NASA terminal configured vehicle (TCF)* (NASA-15983). Washington, DC: NASA.

Glor, P. J. (1991). Human-centered design evaluation for enhanced system supportability. In E. Boyle, J. Ianni, J. Easterly, S. Harper, & M. Korna (Eds.), *Human-centered technology for maintainability: Workshop proceedings* (AL-TP-1991-0010) (pp. 253–260). Wright-Patterson Air Force Base, OH: Armstrong Laboratory.

Glor, P. J., & Boyle, E. S. (1993). Design evaluation for personnel, training, and human factors (DEPTH). *Proceedings of Annual Reliability and Maintainability Symposium*, 18–25.

Gordon, C. C., Churchill, T., Clauser, C. E., Bradtmiller, B., McConville, J. T., Tebbetts, I., & Walker, R. A. (1989). *Anthropometric survey of U.S. Army personnel: Methods and summary statistics* (U.S. Army, Natick TR-89/044). Natick, MA: Natick Soldier Center.

Haug, E. J. (1990). *Feasibility study and conceptual design of a national advanced driving simulator.* Iowa City: University of Iowa, College of Engineering, Center for Simulation and Design Optimization of Mechanical Systems.

Herlihy, D. H., Secton, J. A., Oneal, J., & Guptill, R. V. (1985). *Man integrated systems technology user's guide* (R-4734). Alexandria, VA: U.S. Army Research Institute for the Behavioral and Social Sciences.

Herring, R. D., & O'Brien, L. H. (1988). MANPRINT aids to assess soldier quality and quantity. *Proceedings of the 30th Annual Conference of the Military Testing Association*, 3–8.

Hess, R. A. (1980). A structural model of the adaptive human pilot. *Guidance and Control Conference Boulder, Colorado, August 6–8, 1979, collection of technical papers* (A79-45351 19-12, pp. 573–583). New York: American Institute of Aeronautics and Astronautics.

Holcomb, F. D., Tijerina, L., & Treaster, D. (1991). *MicroSAINT models of the close-in weapon system (CIWS): Sensitivity studies and conformal mapping of performance data (CISWENS). Exploring the utility MicroSAINT models: CIWS loading operation models under normal and MOPP IV conditions. Final report* (NTIS AD-A250 098/1). Bethesda, MD: Naval Medical Research and Development Command.

Hollocks, B. W. (1994). Improving manufacturing operations with WITNESS. *Computer Simulation, 6,* 18–21.

Hoyland, C. M., & Evers, K. H. (1983). Dynamic modeling of real-time software using SAINT. *Proceedings of the IEEE 1983 National Aerospace and Electronics Conference (NAECON)*, 907–911.

Iavecchia, J. H., Iavecchia, H. P., & Roscoe, S. N. (1988). Eye accommodation to head-up virtual images. *Human Factors, 30*(6), 689–702.

Jacobs, P. F. (1992). *Rapid prototyping & manufacturing: Fundamentals of stereolithography.* Dearborn, MI: Society of Automotive Engineers.

Johannsen, G. (1992). Towards a new quality of automation in complex man-machine systems. *Automatica, 28*(2), 355–373.

Jones, R. E., Trexler, R. C., Barber, J. L., & Guthrie, J. L. (1988). *Development of LHX MANPRINT issues* (ARI Research Note 88-88). Alexandria, VA: U.S. Army Research Institute for the Behavioral and Social Sciences.

Juran, M. T. (1998). *Simulation graphics.* Paper presented at Industrial Truck and Bus Meeting and Exposition, Indianapolis, IN.

Kaplan, J. D. (1991). Synthesizing the effects of manpower, personnel, training, and human engineering. In E. Boyle, J. Ianni, J. Easterly, S. Harper, & M. Korna (Eds.), *Human-centered technology for maintainability: Workshop proceedings* (AL-TP-1991-0010) (pp. 273–283). Wright-Patterson Air Force Base, OH: Armstrong Laboratory.

Kennedy, R. S., Berbaum, K. S., Collyer, S., May, J. G., & Dunlap, W. P. (1988). Spatial requirements for visual simulation of aircraft at real world distances. *Human Factors, 30*(1), 153–161.

Kleiss, J. A., & Hubbard, D. C. (1993). Effects of three types of flight simulator visual scene detail on detection of altitude change. *Human Factors, 35*(4), 653–671.

Koonce, J. M. (1974). *Effects of ground based aircraft simulator motion conditions on predictions of pilot proficiency* (Technical Report ARL-74-5/AFOSR-74-3 [AD783256/257]). Savoy, IL: University of Illinois Research Laboratory.

Koonce, J. M. (1979). Predictive validity of flight simulators as a function of simulator motion. *Human Factors, 21*(2), 215–223.

Korna, M., Krauskopf, P., Haddox, D., Hardyal, S., Jones, M., Polzinetti, J., & McDaniel, J. (1988). *User's guide for CREW CHIEF: A computer-graphics simulation of an aircraft maintenance technician (version 1-CD20*, AAMRL-TR-88-034). Wright-Patterson AFB, OH: Armstrong Aerospace Medical Research Laboratory.

Korna, M., Krauskopf, P., Quinn, J., Berlin, R., Rotney, J., Stump, W., Gibbons, L., & McDaniel, J. (1988). *User's guide for COMBIMAN programs (computerized biomechanical manmodel) version 8.* Wright-Patterson AFB, OH: Aerospace Medical Research Laboratory.

Krohn, G. S., & Bowser, S. E. (1989). *MANPRINT evaluation: AN/TRC-170 digital troposcatter radio system* (Research Report 1524). Alexandria, VA: U.S. Army Research Institute for the Behavioral and Social Sciences.

Laughery, K. R., Dahl, S., Kaplan, J., Archer, R., & Fontenelle, G. (1988). A manpower determination aid based upon system performance requirements. *Proceedings of the Human Factors Society*, 1060–1064.

Lintern, G., Sheppard, D. J., Parker, D. L., Tates, K. E., & Dolan, M. D. (1989). Simulator design and instructional features for air-to-ground attack: A transfer study. *Human Factors, 31*(1), 87–99.

Mannle, T. E., Guptil, R. V., & Risser, D. T. (1985). *HARDMAN comparability analysis methodology guide.* Alexandria, VA: U.S. Army Research Institute.

Martin, T., Kivinen, J., Rijnsdorp, Rodd, M. G., & Rouse, W. B. (1991). Appropriate automation—Integrating technical, human, organizations, and economic and cultural factors. *Automatica, 27*, 901–917.

MATRIS. (2001). *Directory of design support methods.* San Diego, CA: DTIC-AM. http://dticam. dtic.mil/hsi/index/hsi11.html.

Meister, D. (1984). Human reliability. In F. Muckler (Ed.), *Human factors review.* Santa Monica, CA: Human Factors Society.

Metzler, T. R., & Lewis, H. V. (1989). *Making MANPRINT count in the acquisition process* (ARI Research Note 89-37). Alexandria, VA: U.S. Army Research Institute for the Behavioral and Social Sciences.

MIL-HDBK-759A. (1981). *Human factors engineering design for Army materiel.* Redstone Arsenal, AL: U.S. Army Missile Command.

MIL-STD-1472D. (1989). *Human engineering design criteria for military systems, equipment and facilities.* Redstone Arsenal, AL: U.S. Army Missile Command.

Munger, S. J., Smith, R. W., & Payne, D. (1962). *An index of electronic equipment operability: Data store* (AIR-C43-1/62 RP [1]). Pittsburgh, PA: American Institute of Research.

Padmos, P., & Milders, M. V. (1992). Quality criteria for simulator images: A literature review. *Human Factors, 34*(6), 727–748.

Penny & Giles Aerospace. (1996, August). *Unlocking the value of your flight data.* Santa Monica, CA: Author.

Plott, C., Dahl, S., Laughery, R., & Dick, A. O. (1989). *Integrating MicroSAINT and HOS.* Alexandria, VA: U.S. Army Research Institute.

Robinson, R. E., Lindquist, J. W., March, M. B., & Pence, E. C. (1988). *MANPRINT in LHX: Organizational modeling project* (ARI Research Note 88-92). Alexandria, VA: U.S. Army Research Institute for the Behavioral and Social Sciences.

Roebuck, J. A., Jr. (1991). Overcoming barriers to computer human modeling in concurrent engineering. In E. Boyle, J. Ianni, J. Easterly, S. Harper, & M. Korna (Eds.), *Human-centered technology for maintainability: Workshop proceedings* (AL-TP-1991-0010) (pp. 8–30). Wright-Patterson Air Force Base, OH: Armstrong Laboratory.

Rogers, B., & Graham, M. (1979). Motion parallax as an independent cue for depth perception. *Perception, 8*, 125–134.

See, J. E., & Vidulich, M. A. (1997). *Computer modeling of operator workload during target acquisition: An assessment of predictive validity* (AL/CF-TR-1997-0018). Wright-Patterson Air Force Base, OH: Human Engineering.

Seminara, J. L., & Gerrie, J. K. (1966). Effective mockup utilization by the industrial design-human factors team. *Human Factors, 8,* 347–359.

Sestito, S. (1987). *A worked example of an application of the SAINT simulation program* (ARL-SYS-TM-93). Melbourne, Australia: Defence Science and Technology Organisation.

Siegel, A. I., Bartter, W. D., Wolfe, J. J., & Knee, H. E. (1984). *Maintenance personnel performance simulation (in APPs) model* (ORNL/TM-9041). Oak Ridge, TN: Oak Ridge National Laboratory.

Smith, R. L., Westland, R. A., & Blanchard, R. E. (1969). *Technique for establishing personnel performance standards (TEPPs), results of Navy user test, vol. III* (PTB-70-5). Washington, DC: Personnel Research Division, Bureau of Navy Personnel.

Smith, S., & Mosier, J. N. (1986). *Guidelines for designing user interface software* (ESD-TR-86-278). Bedford, MA: MITRE Corporation.

Stanny, R. R. (1991). *Modeling for human performance assessment (NTIS AD-A249 646/1).* Pensacola, FL: Naval Aerospace Medical Research Laboratory.

Stewart, J. E., Smootz, E. R., & Nicholson, N. R. (1989). *MANPRINT support of Aquila, the Army's remotely piloted vehicle: Lessons learned* (Research Report 1525). Alexandria, VA: U.S. Army Research Institute.

Swain, A. D. (1964). *THERP* (SC-R-64-1338). Albuquerque, NM: Sandia National Laboratories.

Taylor B. H., & Weisgerber, S. A. (1997). SUE: A usability evaluation tool for operational software. *Proceedings of the Human Factors Society 41st Annual Meeting,* 1107–1110.

Thein, B. (1991). Human performance modeling: An integrated approach. In E. Boyle, J. Ianni, J. Easterly, S. Harper, & M. Korna (Eds.), *Human-centered technology for maintainability: Workshop proceedings* (AL-TP-1991-0010) (pp. 148–159). Wright-Patterson Air Force Base, OH: Armstrong Laboratory.

U.S. Army. (2001). *HARDMAN.* http://www.manprint.army.mil/manprint/references/handbookaquis/appendix/appendixC.htm.

U.S. Nuclear Regulatory Commission. (1998). *Technical basis and implementation guidelines for a technique for human event analysis* (NUREG-1624). Washington, DC: Author.

van Breda, L., Donselaar, H. V., & Schuffel, H. (1983). *Voorstudie ten Behoeve van de Navigatiebgug der 90-er Jaren* (Report No. 4769 6.4/R52). Rotterdam: Maritime Research Institute for the Netherlands.

Virtual Prototypes, Inc. (1988). *VAPS product overview.* Montreal, Canada: Virtual Prototypes, Inc.

Virtual Prototypes, Inc. (1991). *VAPS 2.0 new features.* Montreal, Canada: Virtual Prototypes, Inc.

Virtual Prototypes, Inc. (1992). *Aerospace and defense solutions.* Montreal, Canada: Virtual Prototypes, Inc.

Wortman, D. B., Duket, S. D., & Seifert, D. J. (1977). *Modeling and analysis using SAINT: A combined discrete/continuous network simulation language* (AMRL-TR-77-78). Wright-Patterson Air Force Base, OH: Air Force Aerospace Medical Research Laboratory.

Wortman, D. B., Seifert, D. J., & Duket, S. D. (1976). *New developments in SAINT: The SAINT III simulation program* (AMRL-TR-75-117). Wright-Patterson Air Force Base, OH: Air Force Aerospace Medical Research Laboratory.

Young, M. J. (1993). *Human performance models as semi-autonomous agents.* Presented at the 4th Annual Conference on AI, Simulation, and Planning in High Autonomy Systems, Denver, CO.

Young, M. J., & Patterson, R. W. (1991). Human performance process model research. In E.
    Boyle, J. Ianni, J. Easterly, S. Harper, & M. Korna (Eds.), *Human-centered technology for
    maintainability: Workshop proceedings* (AL-TP-1991-0010) (pp. 57–71). Wright-Patterson
    Air Force Base, OH: Armstrong Laboratory.

Zachary, W., Campbell, G., Laughery, R., & Glenn, F. (1998). *Application of human modeling
    technology to the design, operation, and evaluation of complex systems* (CHI Systems Technical
    Memo 980727.9705). Arlington, VA: Office of Naval Research.

Zimmerman, W., Butler, R., Gray, V., Rosenberg, L., & Risser, D. T. (1984). *Evaluation of the
    HARDMAN (hardware versus manpower) comparability methodology* (Technical Report 646).
    Alexandria, VA: U.S. Army Research Institute for the Behavioral and Social Sciences.

Chapter 10

Questionnaire Techniques for Test and Evaluation

Samuel G. Charlton
Waikato University
and Transport Engineering Research New Zealand Ltd.

Most of us working in the area of human factors testing have spent a considerable amount of time developing, reviewing, and analyzing questionnaires. These questionnaire duties are typically met with an air of resignation or ambivalence by human factors testers. Many of us, trained with the objective data of physics held out as the model of scientific method, cannot help but regard the subjective questionnaire data as less than scientific. This aversion persists, even with the knowledge that some degree of subjectivity pervades even the most objective scientific methodologies. We can often warm ourselves to these duties by turning to elaborate multidimensional scaling and psychometric scale development techniques to regain a sense of scientific merit and self-esteem. In the time- and dollar-sensitive environment of test and evaluation, a significant expenditure on questionnaire development typically cannot be justified. A great deal of test data are collected through the use of questionnaires, most of them designed for a single system or test, a disposable resource with a finite shelf life.

This chapter is about those disposable questionnaires. It makes no attempt to cover the intricacies of psychometry and scale validation; these subjects could (and do) fill books in their own right. This chapter will introduce, or remind, the reader of the salient issues involved in developing and using simple questionnaires in test and evaluation. These simple questionnaires are an important test resource, only rarely employed to their full potential. Questionnaires can be used to quantify difficult-to-quantify aspects of human factors design, such as user preferences and satisfaction with economy and a high degree of precision. Further, humans are excellent data integrators; asking operators about some aspects of system performance can be much more cost-effective and useful than other forms of measurement. In short, the use of quick-and-dirty questionnaires in test and evaluation is an inescapable situation, but we could do far better with them than we typically do. Although this chapter will not make anyone an expert, it will point out some of the common pitfalls and identify some simple techniques to improve the quality of questionnaires in test and evaluation.

WHEN TO USE QUESTIONNAIRES

The most common pitfall associated with questionnaires is to use them in situations where they are simply not needed to answer the test issues at hand. Because questionnaires are frequently perceived as an easy way to collect data, they are often casually prepared and used to evaluate a system, procedure, or interface. Poorly prepared questionnaires, used in situations where other data sources are readily available, can be worse than collecting no data at all; they can convey false information about the issues under test.

The most appropriate use for test and evaluation questionnaires is as a source of supplementary or explanatory data, not as a criterion measure of success. In some cases, however, objective measures of the success of a human factors design effort or the effectiveness of a system are not possible or practical. In these cases, two types of questionnaires can be used as test criteria. The first type is a standardized questionnaire or rating scale, in which a particular result or score has been recognized as a limit, or red line, by the scientific or test community (e.g., a SWAT score of 40 for aviation). Not to put too fine a point on it, but these are rare as hen's teeth, and even the example above has yet to achieve any general consensus or acceptance. This leaves the tester with the second type of questionnaire, a system-unique questionnaire used to measure performance or design attributes particular to the system under test. Questionnaires of this second type, when used as test criteria, are almost always the subject of some controversy. The source of the controversy can concern the criterion value used as the threshold between good and bad results, the rating scale used, or even the phrasing of the questions. Because nearly everyone has a passing familiarity with questionnaires, nearly everyone has an opinion about how they should be designed.

Whether or not the test and evaluation questionnaires will be used as a criterion measure or simply as a supporting or informational measure, there are several rules of thumb for developing reliable test questionnaires. An examination of questionnaires used in the operational test and evaluation of U.S. Air Force systems (Charlton, 1993) identified five principles for the development of test and evaluation questionnaires:

1. A statistically representative sample size must be employed. Based on a desired sampling error of 10% at 80% confidence, and for a generic population size of 1,000, the recommended sample size is 40 or more ratings or subjects per evaluation area.

2. A parametric rating scale/descriptor set should be used. That is, a balanced, equal-interval scale with normative values should be used so that the data produced will approximate interval as opposed to ordinal data. (Note: This step applies only to questionnaires using rating scale questions.)

3. Questions should be based on narrowly focused evaluation areas. To provide good agreement and reliable data, the questions should reflect specific, well-defined tasks or attributes, not broad areas of system performance.

4. A well-defined threshold of acceptance must be identified in advance. The criterion for a positive evaluation should be described in terms of the minimally acceptable distribution of questionnaire responses (e.g., a criterion based on a median score of 5 and 80% of the ratings greater than or equal to 4 on 6-point effectiveness scale) prior to data collection.

5. Questionnaires should be associated with objective performance measures where feasible. Questionnaire data should not be used as a principal or single method of evaluation without first exhausting other efforts to obtain objective measures and requirements.

The recommendations described above were developed to ensure that questionnaires would provide the kind of data required to support conclusions made by the test team and documented in test reports. For example, a sample size of 40 is recommended based on a sampling error of ± 10%, a statistic most of us are used to seeing in conjunction with political polls. This means that when we observe 80% of the subjects rating a system characteristic *effective* or better, we can estimate that between 70% and 90% of the population at large would rate the system effective or better. (See the topic of statistical considerations later in this chapter.) Similarly, the recommendation to use approved parametric rating scales is based on experience with presenting questionnaire results to higher headquarters' staff. A poorly designed questionnaire scale can undermine the whole test effort by calling into question the validity of test results based on disagreement on the appropriateness and meaning of the rating scale and descriptor set used. Finally, the recommendation to base questions on well-defined and narrowly focused aspects of system performance is based on our experience that these types of questions produce better agreement in the subjects' responses and therefore make data analysis and interpretation of the questionnaire results easier.

CREATING A QUESTIONNAIRE

Once a decision has been made to use a questionnaire as a test measure, whether as a criterion measure or as a source of supporting information, there are several well-defined steps involved in their construction.

Step 1

The first step is to select a questionnaire type. To make a selection you will first need to make a list of all the test areas for which you may use questionnaires and think through the way you would analyze the data to support test reporting. This will enable you to select the right type of questionnaire to provide you with the data you need for the test report. There are a number of questionnaire types suitable for human factors testing. These include open-ended questions, multiple choice questions, and rating scales. Some common examples of these questionnaire types are presented in Fig. 10.1, along with brief descriptions of the situations where they can be used to best advantage. Of these techniques, the rating scale has been used most frequently for collecting data from relatively large samples of representative operators or users. Moreover, rating scales can produce easily quantifiable data that are readily integrated with other sources of data. In contrast, for relatively small samples of subject matter experts, open-ended questions or structured interviews would be more appropriate because of the wealth of unanticipated information these participants can provide.

Step 2

The second step in questionnaire design is the selection of the response scale and descriptor set. Although the wording of the questions can be thought of as the stimulus for the respondent's answer, the response scale and descriptor set will determine the form of the answer. This is a critical consideration in questionnaire construction, so we will describe it in some detail. The response scale defines the distribution of responses by providing the number and type of allowable answers to a question. The balance, polarity, and number of values are all important considerations in selecting a response scale for a questionnaire. A response scale that has an equal

A. **Rating Scale Questionnaire**: Used for most T&E situations.
Advantages: Structured, specific questions, can be answered quickly.
Data are very reliable and amenable to summary statistics.
Disadvantages: Structure leaves little room for unanticipated answers.
Questions must be well thought out to ensure adequate coverage.

Example: Rate your overall reaction to the rations supplied to you
during the field exercise.

| ├—————————┼—————————┼—————————┼—————————┤ |
| Strongly   Dislike   Neutral   Like   Strongly |
| dislike   like |

B. **Semantic Differential**: Used for measurement of values, attitudes, or
complex relationships. Simultaneous ratings across multiple dimensions.
Advantages: Provides data on the relative similarity of various concepts or
attributes. Easy to prepare and administer.
Disadvantages: Limited utility for T&E. Correct analysis and interpretation
requires experience.

Example: Place an X in each of the following rows to describe your
experiences with the TRAKS System.

Accurate   __ __ __ __ __ __ __   Inaccurate

Slow   __ __ __ __ __ __ __   Fast

Unreliable   __ __ __ __ __ __ __   Reliable

Easy to use __ __ __ __ __ __ __   Difficult to use

Ineffective   __ __ __ __ __ __ __   Effective

C. **Multiple Choice Questionnaire**: Used for screening respondents,
collection of demographic data. Respondent selects one category.
Advantages: Answers are easy to summarize and may be very reliable.
Disadvantages: Cannot ask complex, detailed questions. Question may
force the respondent to make a choice even though they don't know the
answer or feel there is no difference between the answers.

Examples:
What is the amount of training you received for TRAKS?
___ 1 day overview ____ 2 week ops course ____ 6 week course

Which did you prefer using, the FINN radio or the PVU message system?
_____ FINN _____ PVU _____ No preference

D. **Open-ended Questionnaire**: Open-ended questions are most useful when
there are too many possible responses to be listed or foreseen.
Advantages: May obtain answers that were not anticipated by the
questionnaire author. Open-ended questions are easy to write.
Disadvantages: Open-ended questions require more time to answer.
Answers are difficult to summarize and may overload data analysts.

Example: What is the most important improvement needed for the
KC-135 navigation system? _____

_____

FIG. 10.1. Common test and evaluation questionnaire types.

number of positive and negative alternatives is called balanced. Historically, balanced scales have been preferred by researchers because they tend to produce distributions that are more nearly normal. Unbalanced scales are typically selected only when there is reason to suspect that a tendency to select extreme responses will produce an uneven distribution. In such cases, additional response alternatives may be provided in the range where clustering is expected to spread the distribution of responses more evenly along the continuum.

The presence or absence of a neutral midpoint does not inherently affect a scale's balance, but it may affect the response distribution. Denying a neutral midpoint tends to increase the variability about the theoretical center and thus reduces the discriminability near the center (Charlton, 1993). In addition, some respondents resent being forced to select a choice that departs from true neutrality. This occasionally results in the omission of responses to some questions. In a few situations, where there is reason to believe that respondents will be unwilling to provide anything but a neutral response, the midpoint can be dropped from the response scale. The consequences of forcing the respondents to make a choice must be carefully weighed against the potential benefit of obtaining nonneutral responses.

The number of alternatives along the response scale is determined on the basis of the degree of discrimination required. Increasing the number of response alternatives can decrease the number of nonresponses and uncertain responses. Increasing the number of response alternatives, however, tends to increase the questionnaire administration time. Perhaps the best basis for selecting the number of alternatives is to consider how easy each response is to differentiate from the others. Research shows that clear discriminability can be obtained with up to seven alternatives (Babbit & Nystrom, 1989). More than seven alternatives increases the response variability and reduces the likelihood that clear discriminability will be obtained. For example, giving the respondents 10 choices when they can only tell the difference between 6 of the values will increase the variability of their responses and thus lower the reliability of the questionnaire. For most test and evaluation questionnaires, a balanced, bipolar scale with 5 to 7 points is best.

Response alternatives that accompany response scales are called descriptors or semantic anchors. Descriptors must be chosen for consistency, discriminability, and comprehensibility to be effective. The words selected for descriptor sets should be clearly understood by respondents and should have no ambiguity of meaning. Many studies have been conducted to determine the probable favorability and discriminability of verbal alternatives (Babbit & Nystrom, 1989). These studies have determined scalar values and standard deviations for words and phrases that can be used to order response alternatives for custom applications.

Scales may be anchored at the ends, at the center, or at any number of points along the continuum. Some evidence suggests that greater scale reliability is achieved with complete verbal anchoring. Inappropriate anchors present the respondent with an ambiguous question, which must be reinterpreted before an answer can be selected. This in turn increases the variability of respondents' answers and presents the analyst with data that were influenced by unanticipated and irrelevant factors.

In general, it is good practice to construct response sets with parallel wording for balanced scales. For example, if the phrase *substantially agree* were to be used, the phrase *substantially disagree* could also be used. It should be noted that parallel wording may not always provide equally distant pro and con response alternatives, although they will generally be perceived as symmetrical opposites. Several recommended sets of response alternatives are presented in Table 10.1. These example sets are based on normative response distributions such that they are equidistant and will yield nearly linear (equal interval) response distributions.

Step 3

The third step in questionnaire construction is wording the questions. There are seven rules to follow in wording your questions:

1. Vocabulary. It is important to speak to the level of the individuals who will be answering the questionnaire. Avoid using jargon, acronyms, or overly technical terms that may be misunderstood by the respondents.

2. Negatives. Use positive phrases wherever possible. Negative phrases, such as "Rate the degree to which the system possesses no voids or gaps," are frequently misread or misunderstood by the respondents; the single word "no" can be eliminated to make the question more easily understood. Finally, double negatives should never be used.

3. Double-barreled questions. Double-barreled questions pose two questions simultaneously, such as "Rate the responsiveness and reliability of the system." If the responsiveness is good, but the reliability is poor, the respondent will have difficulty answering the question. In this example, two separate questions should be asked, one for responsiveness and one for reliability.

4. Leading/loaded questions. Leading questions presuppose some event or state. Questions such as "Rate the lack of responsiveness of the system" presume that the system is unresponsive. Loaded questions, like leading questions, presume some state but also carry with them a charged emotional content, as in the following example: "Rate your lack of ability with respect to the duties you carried out today." Leading questions may

TABLE 10.1
Recommended Scale Anchors for Test and Evaluation

Completely effective	Very effective	Extremely useful
Very effective	Effective	Of considerable use
Effective	Borderline	Of use
Borderline	Ineffective	Not very useful
Ineffective	Very Ineffective	Of no use
Very ineffective		
Completely ineffective		
Totally adequate	Very adequate	Always
Very adequate	Slightly adequate	Frequently
Barely adequate	Borderline	Now and then
Borderline	Slightly inadequate	Seldom
Barely inadequate	Very inadequate	Never
Very inadequate		
Totally inadequate		
Completely acceptable	Largely acceptable	Very important
Reasonably acceptable	Barely acceptable	Important
Barely acceptable	Borderline	Borderline
Borderline	Barely unacceptable	Not important
Barely unacceptable	Largely unacceptable	Very unimportant
Moderately unacceptable		
Completely unacceptable		
Undoubtedly best	Undoubtedly best	Strongly like
Conspicuously better	Moderately better	Like
Moderately better	Borderline	Neutral
Alike	Noticeably worse	Dislike
Moderately worse	Undoubtedly worst	Strongly dislike
Conspicuously worse		Excellent
Undoubtedly worst		Good
		Only fair
		Poor
		Terrible

produce biased data for that question; loaded questions may have a carry-over effect for the entire questionnaire.

5. Emotionality. Related to the issue of loaded questions described above, questions containing emotional, sensitive, or pejorative words have the potential for invalidating the data for the entire questionnaire. Questions perceived as emotional or sensitive frequently involve the personal qualities, capabilities, and knowledge of the person completing the questionnaire. Because it is unreasonable to expect people to objectively evaluate their own performance, questionnaire items should be directed to the adequacy of the system, rather than the user.

6. Brevity. Keep your questions short; a single sentence is best. The more words it takes to ask a question, the more complicated it is to understand and the greater the opportunity for misunderstanding.

7. Relevance. Always keep in mind the relationship between the questions you are asking and the purpose of the test. Once embarked on a questionnaire creation effort, it is all too easy to add more nice-to-know questions. If the questions are irrelevant, however, all you are doing is imposing an unnecessary burden on the respondents and the data analysts.

Step 4

The fourth step is the assembly of the questionnaire elements. Although individual questions may have balanced response scales, good descriptor sets, and appropriate wording, they still must be combined into a complete package. There are several characteristics of good questionnaires, including clear instructions, consistent format, brevity, and appropriate materials. All of these areas require attention to ensure the quality of the questionnaire data. The questionnaire cover sheet is a key ingredient in obtaining the respondents' cooperation by informing them about the purpose of the questionnaire and the procedures for completing it. A good cover sheet contains a method of tracking the data (space for control codes, operator ID or position, and date/time group), information regarding the use of the data, assurance of confidentiality (if needed), and instructions for completing the questionnaire. An example cover sheet for a human factors questionnaire is presented in Fig. 10.2.

For most types of questionnaires, the questions should flow from the most general topics to the most specific, or from the most frequent/common events to the rare or unusual. This is done to minimize carry-over effects, instances where early questions influence or bias the respondent's answers to later questions.

Questionnaires should be as brief and to the point as possible. One of the dangers of overly long questionnaires is the tendency for respondents to choose neutral or extreme responses out of expediency, fatigue, or simple carelessness. These problems may be detected by scrutinizing the performance or response patterns of individual subjects. Carelessness may be detected in unusually short amounts of time taken to complete the questionnaire, as well as a tendency to give answers that deviate consistently from the norm. If you find that it takes more than 15 minutes to answer all the questions, you should consider dividing the questions among two or more separate questionnaires to be administered at different times during the test. Another option is to use a hierarchical questionnaire where filter questions are used to eliminate questions that are irrelevant to the tasks or issues immediately at hand. Hierarchical questionnaires can contain a

---

User Questionnaire

Directions: Your personal responses to this questionnaire are very important in helping to evaluate the performance of System X. Your answers will help to identify improvements in hardware, software, and procedures. Your name and individual answers will be kept strictly confidential. There are 17 questions for which you will be asked to provide a response on a 7-point scale.

For each question, please select and circle the term which best describes your answer as in the example shown below. If after providing a rating there are any problem areas you wish to identify, select ALL of the areas that apply by circling the number. If you select "Other/Comments", please write your comment on the lines immediately following the question and continue on the back of the page if necessary. Your comments are encouraged and will be valuable to the success of this questionnaire.

Example: Rate the overall quality of the Denver Broncos' performance this season.

| Completely Acceptable | Largely Acceptable | Somewhat Acceptable | Borderline | Somewhat Unacceptable | Largely Unacceptable | Completely Unacceptable |

Problem area(s)
1. Offensive line.
2. Defensive line.
3. Special teams.
4. Quarterback.
5. Coaching.
6. Other/Comments. I'm switching to rugby !!

Remember, this is your opportunity to provide your assessment of the performance of the _____ system. If you have any questions just ask a test team member.

Please write down the name of your position title and organization or office in the space provided below:

Position title: _____   Organization: _____

- -

Test team use only:
    QUESTIONNAIRE #: _____   DATE/TIME _____

FIG. 10.2.   Questionnaire cover sheet.

large pool of potential questions but can speed administration times markedly by focusing on only those areas where the respondents have something meaningful to say. An example of one set of questions from a test of a multimedia presentation interface is shown in Fig. 10.3. Hierarchical questionnaires will take longer to prepare, and care must be taken that essential questions appear at the top level (i.e., are not filtered out), but once prepared their brevity and efficiency is appreciated by the respondents.

The most common medium for questionnaires is paper and pencil. The paper-and-pencil method is quite flexible and easily prepared, but there

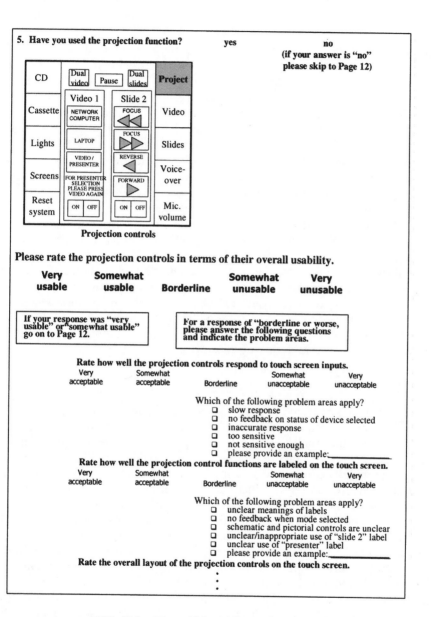

5. Have you used the projection function?   yes   no

(if your answer is "no" please skip to Page 12)

CD	Dual video	Pause	Dual slides	Project

Projection controls

Please rate the projection controls in terms of their overall usability.

Very usable   **Somewhat usable**   **Borderline**   **Somewhat unusable**   **Very unusable**

If your response was "very usable" or "somewhat usable" go on to Page 12.

For a response of "borderline or worse, please answer the following questions and indicate the problem areas.

Rate how well the projection controls respond to touch screen inputs.

Very acceptable   Somewhat acceptable   Borderline   Somewhat unacceptable   Very unacceptable

Which of the following problem areas apply?
- ❑ slow response
- ❑ no feedback on status of device selected
- ❑ inaccurate response
- ❑ too sensitive
- ❑ not sensitive enough
- ❑ please provide an example:_____

Rate how well the projection control functions are labeled on the touch screen.

Very acceptable   Somewhat acceptable   Borderline   Somewhat unacceptable   Very unacceptable

Which of the following problem areas apply?
- ❑ unclear meanings of labels
- ❑ no feedback when mode selected
- ❑ schematic and pictorial controls are unclear
- ❑ unclear/inappropriate use of "slide 2" label
- ❑ unclear use of "presenter" label
- ❑ please provide an example:_____

Rate the overall layout of the projection controls on the touch screen.

FIG. 10.3.   Hierarchial questionnaire format.

are important considerations in its use. First, consider whether adequate time and manpower will be available to manually code and enter the data into a database or statistical analysis package. If you decide to use paper questionnaires, be sure to leave adequate room to respond, particularly in the comments area. The amount of space you provide will determine the amount of detail you receive in the answers. Conversely, don't leave an excessive amount of space because some respondents feel compelled to fill all the available space. If you plan to use separate answer sheets, be sure that they have the same number of questions and alternatives as the questionnaire. Readability of the questionnaire is another important consideration, particularly in situations where illumination levels are low and distractions are high (cockpit or nighttime situations). Consider also the requirements for the size and stock of paper where handling considerations and field use make compact or semirigid forms necessary.

Computer-based questionnaires can be an attractive alternative to paper and pencil questionnaires for a variety of reasons. The major advantage to questionnaires presented on a laptop or other computer is that the data will not require later key entry by a data clerk or analyst. For complex questionnaire formats (e.g., hierarchically organized questionnaires), the computer can present the questions in a clear and easy-to-follow sequence. Some obvious disadvantages are the hardware and logistics requirements of acquiring and transporting a laptop computer and potential difficulties with screen glare and viewing distances.

Step 5

The fifth and final step in the construction of a questionnaire is the review or pretest. It must, of course, be reviewed for grammatical and typographical errors, but it must also be reviewed for content and clarity. A quality check of the questionnaire is imperative to avoid hours of wasted time and energy in collecting useless or incomplete data. Further, a poor-quality questionnaire will tend to increase resistance on the part of the respondents such that you may not be welcomed back for a second visit after you have wasted their time by giving them a poor-quality questionnaire. Quality checks can be performed either by tabletop review with subject-matter experts or by actually administering the questionnaire to a small number of system users as a pretest. Whichever method is used, the quality check is directed at identifying questions that are ambiguously worded and questions that are inappropriate or not applicable to the operator's experience and duties. The following three inspection elements are recommended in the review and approval of test and evaluation questionnaires:

1. Question relevance. The first feature of the questionnaire to be examined is the correspondence between the questionnaire and the meas-

ures identified in the test plan. Question relevance is as important a consideration as correspondence to the wording and intent of the test measures. Irrelevant questions or questions that do not address the test issues or measures are identified at this point and excluded from subsequent review steps.

2. Question wording. The next area to be examined is the wording of the individual questions. The questions are examined to ensure that there are no double-barreled, leading/loaded, or poorly worded questions, and that the questions match the descriptors used on the response scale. Spelling, typographical, and grammatical errors are also identified at this time.

3. Questionnaire format. The final area examined is the overall questionnaire format, the clarity of instructions, adequacy of any comment fields, and the overall length of the questionnaire. This is typically accomplished by attempting to complete the questionnaire from the perspective of the average respondent.

WHEN TO ADMINISTER A QUESTIONNAIRE

The most common time to administer a questionnaire is at the end of the day or duty shift. This also happens to be one of the worst times to try and administer a questionnaire. Most test participants are tired, eager to go home, and will spend the minimum amount of time they possibly can to complete a questionnaire. A better time to administer a questionnaire is at the start of a duty cycle or during a break in the respondent's shift. The best of all possible conditions is to designate a period of time during the shift when the respondents complete the questionnaires as part of the day's activities such that it doesn't cost them any of their off-duty or break time. Most important, schedule questionnaire administration to occur when the questionnaire topic is still fresh in the respondent's memory.

As mentioned previously, written instructions are an absolute necessity and should be provided on the cover page and throughout the questionnaire as appropriate. An additional technique that frequently increases compliance and the quality of the answers is an oral introduction prior to the first administration of the questionnaire. Much of the information presented orally will be the same as the information provided on the cover sheet, but the respondents now identify the questionnaire with a person and will be more likely to give thoughtful consideration in their answers and will know who to contact if they have a concern or don't understand a question.

Another key to successful use of questionnaires is to be involved and available during the time or times the questionnaires are being filled out. A questionnaire that is simply distributed or left on a desk in the break room will be perceived as unimportant and will be given little effort or even ignored altogether. Your involvement, however, should not be so overwhelming as to be a nuisance or to bias the respondents' answers. The number and type of participants in the test should guide the amount and type of interaction you have with the respondents during questionnaire administration. Where there are only a very few respondents, selected because of their expertise rather than their representativeness as actual users, you may wish to discuss with them the issues involved in the test, encourage comments, and treat them as subject-matter experts, rather than using the questionnaire to provide a simple snapshot of the day's test activities. In cases where multiple questionnaire administrators are used, written instructions and answers to frequently asked questions should be provided to the administrators to ensure uniformity in how the administrators interact with the respondents.

One final issue in the subject of questionnaire administration is how often to administer questionnaires to test participants. A good rule of thumb is to administer the questionnaire only as often as is absolutely necessary. Repeatedly administering a questionnaire to the same subjects does not increase the sample size for data analysis purposes; sample size calculations are based on numbers of respondents, not volume of responses. Respondents who are repeatedly presented with the same questionnaire will quickly stop putting their time and energy into answering the same questions. The data analyst is also confronted with the problem of deciding how to reduce and present the large volume of completed questionnaires. If data are needed for each of five test scenarios, then questionnaires should be prepared for each of the scenarios and administered once during each scenario. General questions on topics that are not scenario-dependent should only be administered once or twice. One very successful practice is to administer the questionnaire to each test participant twice: once at the very beginning of the test and again at the end. The first administration informs the respondents of the kinds of test issues you are interested in, and the second time provides the analyst with good data based on their experiences with the system during the test. Only the data from the end of test are used in the analysis; data from the initial administration are discarded (although the respondents are not typically informed of this fact). The initial administration at the start of the test is also useful in finding questions that need revision or rewording.

STATISTICAL CONSIDERATIONS

Questionnaire data differ from the type of data that experimental psychologists and statisticians usually encounter. Obviously, questionnaire data are not a ratio measurement scale like meters or feet. Instead, questionnaire data are often ordinal measures, for example, *very acceptable* is better than *effective*. Under the best cases, where response scales and descriptor sets are based on normative data (as in the scales and descriptors recommended earlier), questionnaire data will approximate an interval scale. At their worst, questionnaires based on scales and descriptors with unknown properties represent qualitative, or categorical scales; nothing can be said about the order of merit or degree of difference between various ratings, just as right is not better than left and apples are not better than oranges.

With well-developed questionnaires and carefully selected response scales and descriptor sets, an analyst can be assured of a relatively greater degree of power in his or her analysis. Questionnaire data, however, still should not be subjected to statistical analyses without careful consideration of the way in which the data are distributed. If statistical comparisons between groups, scenarios, or procedures are required for a particular test design, questionnaire data are typically analyzed using nonparametric tests, such as the Mann-Whitney, chi-square, squared ranks, or median tests. In cases where you find bimodal distributions, even nonparametric tests can be distorted by the extreme variability of the data. For these situations, descriptive analyses are better suited to identify the respondent characteristics associated with each cluster of ratings.

This brings us to the topic of sampling error. The calculation of sampling error provides the analyst with an indication of the degree to which his or her sample of questionnaire responses represents the population about which he or she is attempting to draw conclusions. The subject of sampling error is usually raised in the context of determining appropriate sample sizes, before any data are collected. Further, sampling error is often confused with the calculation of confidence intervals. Although related, confidence intervals and sampling error are distinct entities and calculated quite differently. For example, a confidence interval of 90% is interpreted as a range of scores or data into which the population parameter of interest (e.g., the mean) would fall for 90 out of 100 randomly drawn samples. In contrast, a sampling error of plus or minus 10% is interpreted as meaning that our sample mean lies within 10% of the true population mean and can be calculated at any level of confidence, for example, 10% sampling error at 90% confidence, 80% confidence, and so on. The calculation of sampling errors is based on the size of the sample relative to the size of the population of interest as follows:

$$sE = Z\sqrt{\frac{P(1-P)}{n}} \times \left[\sqrt{\frac{N-n}{N-1}} \right]$$

Where: sE = Sampling error
Z = Z score for desired confidence level

P = Percent of sample in one category
n = Sample size
N = Population size

A more frequently used formula involves the calculation of the sample size required to support a sampling error of a desired amount. This formula is as follows:

$$n = \frac{NZ^2}{\dfrac{sE^2(N-1)}{P(1-P)} + Z^2}$$

Where: sE = Sampling error
Z = Z score for desired confidence level
P = Percent of sample in one category
n = Sample size
N = Population size

Rather than repeating these calculations for each new questionnaire, human factors practitioners frequently consult tables containing a range of sampling errors, confidence intervals, and population errors. Table 10.2 presents the required sample sizes for some of the most frequently used levels of confidence intervals and sampling error.

In a similar vein, reliability and validity are two issues that often arise in the context of questionnaire development. In general terms, a questionnaire has validity if the component items measure the variable that it was intended to measure and not some other variable. Reliability refers to the extent to which the same results can be obtained with the same questionnaire when repeatedly applied to the same group of raters. Establishing the validity and reliability of a questionnaire require statistical and analytical practices that typically exceed the resources and needs of most test-specific human factors questionnaires and are certainly beyond the scope of this chapter.

The approach to designing valid and reliable questionnaires adopted by this handbook is to develop the questionnaires in accordance with widely accepted practices in questionnaire construction. These practices and principles include the wording of items, length of questions and questionnaires, and general ability of the respondents to understand the requirements of the test situation. Wording of the stem of the question should not favor any of the response alternatives to the question. Moreover, the wording of any one item should not influence the response to any other item. As described earlier, potential problems can frequently be avoided by pretesting the questionnaire. Other questionnaire develop-

TABLE 10.2
Questionnaire Sample Sizes As a Function
of Confidence Intervals and Sampling Error Rates

Confidence Interval	80%			90%			95%		
Sampling Error (+ or −)	5%	10%	15%	5%	10%	15%	5%	10%	15%
Population Size					Sample Size				
50	39	23	14	42	29	19	44	33	23
100	63	30	16	73	40	23	80	49	30
250	100	36	17	130	53	27	152	70	37
500	125	38	18	175	59	28	217	81	39
1,000	143	40	18	212	63	29	278	88	41
5,000	161	41	18	255	66	30	357	94	42
15,000	164	41	18	264	67	30	375	95	43
50,000	166	42	18	268	67	30	381	96	43
150,000	166	42	18	268	67	30	383	96	43

ment practices that apply to the issues of validity and reliability include the representativeness of the respondents to the population of system users and the relevance of questions to the system aspect under assessment.

QUESTIONNAIRE ANALYSIS

Analysis of questionnaire responses is not inherently difficult. For most types of rating scales, the simple method of assigning values of 1 through 5 (in the case of 5-point scales), or their linear transform, should be used. This approach simplifies the scoring process, allowing it to be performed with minimal error by relatively untrained supporting personnel. In addition, a simple scoring system allows for ready visual checking, computer checking, or both for scoring errors as well as manual or computer-based summarization of the results.

The primary means of analyzing the questionnaire data will be to compute percentages of ratings for each question. These percentages provide the basis for comparison and will provide insight into any problems experienced during the test. For example, the data in Table 10.3 were collected with a 7-point rating scale ranging from *Totally Unacceptable* to *Totally Acceptable*. The table represents a simple and expedient method of presenting questionnaire results. As can quickly be seen from the distribution of response percentages in the table, ratings of workload, workspace, and facility layout are generally high, controls received mixed ratings, dis-

TABLE 10.3
Example Questionnaire Percentile Distribution

Percent of Respondents Rating:

Rating Areas	Totally Unacceptable	Very Unacceptable	Somewhat Unacceptable	Borderline	Somewhat Acceptable	Very Acceptable	Totally Acceptable
Workload	0	0	0	0	78	22	0
Facility layout	0	0	0	0	44	56	0
Workspace	0	0	0	0	22	67	0
Controls	0	0	0	22	22	33	22
Displays	0	0	11	56	11	22	0
Noise	0	22	11	33	33	0	0
Temperature	11	11	11	0	22	33	11

plays received borderline ratings, and there was wide disagreement on the subject of noise levels and temperature.

Also useful in the presentation of questionnaire data are measures of central tendency, a way of summarizing the typical rating for each question. As stated earlier, questionnaire data represent an ordinal (or at best, interval) measurement scale. Because of this, use of the sample mean or average is inappropriate. Instead, a mode or median statistic should be used to summarize questionnaire ratings. The mode is the most frequently selected rating category. In the example data, the mode for workspace, facility layout, and controls is 2 (*Very Acceptable*), 1 (*Somewhat Acceptable*) for workload, 0 (*Borderline*) for displays, and split evenly between 0 and 1 (0.5) for noise levels. The ratings for temperature represent a special kind of pattern known as a bimodal distribution, in which there are two distinct groups of responses. When bimodal distributions are encountered the questionnaires should be examined to see if different crew positions, duties, or conditions (night vs. day shifts) led to the difference in ratings.

The definition of a median is the 50th percentile, the point at which 50% of the responses lie above and 50% lie below. In our example, the median for facility layout, workspace, and controls is 2, for workload it is 1, for displays and noise levels it is 0. A median value of 1 can be calculated for temperature, but it would hide the significant number of negative ratings. It is important to note that calculation of the median for this question also hides the fact that the responses fall into two distinct groups (a bimodal distribution), as described in the previous paragraph. When using a median to summarize questionnaire data, bimodal distributions, such as the one observed for the responses to the question on temperature, need to be explicitly described and investigated. Thus, in our example, the two groups or distributions of ratings were found to be the result of different room temperatures associated with different shifts. The solution was to report the ratings of temperature separately for each shift. Most statistics programs will calculate the median for you from raw data, so it is important to always examine the response distributions and look for bimodal distributions or extreme ratings, rather than simply rely on the median values reported by a statistics program.

Another way to view these types of test data is to describe the amount of agreement in the questionnaire responses by examining the variability of the ratings. This is especially important for results like those obtained for the questions on noise levels and temperature. One way to capture both the variability and central tendency data in a single format is to use box plots, bar charts, or histograms. The advantage of bar charts and histograms is that they quickly show the overall pattern of responses including bimodal distributions and the amount of agreement overall. Although graphical presentations are typically more appealing to the eye, they can take up more page space than simply reporting the median values. In general, for briefings, small numbers of questions, or for detailed reports, histograms, present the data very nicely. Where page space is limited or where the number of questions is quite large, tables, box plots, or simple narrative descriptions may be required to present questionnaire data.

When you obtain a wide range of responses to a question, this indicates that the subjects disagreed in their ratings of the system's performance. In these cases the analyst should examine the raw data or a frequency distribution of the data to determine the reason for the disagreement (or if there were different groups of people or different test conditions that produced the differing opinions on the question, as in the case of the bimodal distribution of ratings of temperature described in the example). Once again, where significant disagreement is observed in the results, the analyst may need to report the results separately for different test scenarios, duty shifts, or personnel categories.

After you have summarized the data by one of the techniques described previously, the next step is to identify the specific problem areas and relate them to the rest of the test data. The specific problem areas can be identified by reading the comments written for the questions (or by examining the problem areas checked on a hierarchical questionnaire). Comments and problem areas identified by the respondents should be categorized and tabulated for those questions receiving negative ratings. Additional information about specific problems can sometimes be collected after the fact by interviewing the respondents 1 or 2 days after questionnaire administration. Delays greater than 2 days will usually result in the respondents being unable to accurately recall the reasons for their negative ratings for a specific question. Once the problems have been identified they should be related back to system performance and user requirements. In the case of the example data presented in Table 10.3, the specific problem associated with noise levels was excessive noise.

EVALUATION

Finally, a word about comparison of questionnaire data to test criteria. As alluded to earlier, there are two elements to a good questionnaire test criterion: a measure of central tendency and a measure of dispersion or cutoff point. The measure of central tendency, such as a median, is used to identify the direction (positive or negative) and magnitude of the ratings, whereas a measure of dispersion, such as the variance or the rating corresponding to the 80th percentile, captures the amount of agreement in the ratings. For example, a questionnaire criterion that states the median rating must be *very effective* or better, with 80% of the ratings falling at *effective* or better, ensures that not only will a majority of favorable responses be required for a positive evaluation but also that there will be substantial agreement among the respondents. This 80-50 rule for questionnaire data can be quite successful, provided the rating scale is a good one (i.e., bipolar and equal-interval) and there is an adequate sample of respondents available.

Although this type of criterion will work well in most situations, there are some cases where it does not, specifically, evaluations where the responses to many different questions must be considered jointly to arrive at a single pass/fail decision. In these cases, some combinatorial rule must be developed to arrive at a single conclusion. One simple method is the weakest-link rule, which states that all of the questions must meet a minimum criterion, such as the 80-50 rule, to receive a positive overall evaluation. This approach is preferable to one that relies on an overall average across different questions because a few very negative attributes can be masked by many positive (and perhaps less important) attributes. Another solution is to use a weighting system where results from individual question areas are multiplied by some index of importance or criticality prior to calculation of an overall average. The weights used to indicate degree of importance can either be chosen a priori according to an analysis of system requirements or can be selected by the respondents themselves at the time they take the questionnaire. For example, in a standardized metric used to assess software usability (Charlton, 1994), respondents were asked to rate both the degree to which each usability attribute is present and the importance or mission impact of each attribute. The overall usability of the system was then assessed by summing the mission impact scores for each question across all subjects rating the attribute negatively. The result was a ranking of usability problems that considers both the number of negative ratings received and the relative importance of each problem identified. A similar technique has been employed to evaluate the effectiveness of training systems by obtaining and combining ratings of both the trainability and the importance of each of the tasks or concepts to be taught.

In summary, like them or not, questionnaires are a mainstay of human factors test and evaluation. As a source of supporting test information, they are unquestionably a cost-effective and highly flexible test methodology. Using questionnaire data as the principal evaluation item in a test, however, must be accompanied by a well-constructed and accepted criterion measure of success. Employing the principles and practices of good questionnaire design, administration, and analysis will help to ensure the success of your questionnaire ventures.

REFERENCES

Babbit, B. A., & Nystrom, C. O. (1989, June). *Questionnaire construction manual* (Research Product 89-20). Fort Hood, TX: U.S. Army Research Institute for the Behavioral and Social Sciences.

Charlton, S. G. (1993). *Operational test and evaluation questionnaire handbook* (AFOTEC Technical Paper 7.2). Kirtland AFB, NM: Air Force Operational Test and Evaluation Center.

Charlton, S. G. (1994). *Software usability evaluation guide* (AFOTEC Pamphlet 99-102, Vol. 4). Kirtland AFB, NM: Air Force Operational Test and Evaluation Center.

Chapter 11

Testing the Workplace Environment

Thomas G. O'Brien
O'Brien & Associates

Although the last 50 years have seen many new and innovative methods for measuring and evaluating human-system interfaces, much of HFTE still requires examination of the physical environment in which humans must perform. Armed with a tool belt of calipers, tape measures, and light and sound level meters, the tester may have to consider the more mundane issues of the workplace environment. In an earlier chapter, we addressed anthropometic testing and its role in evaluating the form and fit aspects of systems designs. In this chapter, we present methods and measures to evaluate the environmental aspects of the workplace.

Evaluating workspace and anthropometry often entails measurement of static dimensional characteristics of human-occupied systems, such as automobiles, aircraft cockpits, and computer workstations, to name a few. Measuring environmental phenomena (i.e., noise, lighting, and others), as dynamic characteristics, requires the HFTE specialist to conduct measurements while the system is up and running. Indeed, it is often the case that when an environmental problem is suspected, the only way to capture those problems at their worst is during realistic operations.

Human performance depends in large part on the quality of the system's operating environment. Years of research have yielded much information about the limits within which people can perform safely and without undue performance degradation; exceed those limits and human performance degrades. The only way to ensure that those limits are not excessive is to measure those phenomena in question: noise, lighting, shock and vibration, atmospheric composition, temperature, humidity and ventilation, and visibility. Measures are then compared with established criteria to assess whether parameters are within tolerable limits. Even if a system fits its operational crews along its anthropometric dimensions, what good would it be if the environment in which the user must perform is dangerous, causes physical and psychological distress, or contributes otherwise to performance degradation? To help assure that environmental factors have been adequately controlled in the design of a system, and to identify any such problems that might have slipped past the initial prototype phase of development, it becomes necessary to perform environmental tests.

This chapter describes a number of the more common environmental tests. The design standards and criteria referred to herein were extracted from several sources, including MIL-STD-1472 (1999) and DOT/FAA/CT-96/1 (1996). In fact, any reference to one of these sources will also mean that the same design criteria can be found in the other. Test procedures were drawn in part from the U.S. Army *Test Operations Procedures* (U.S. Army Test & Evaluation Command, 1983) and from personal experience. The reader should become familiar with the test procedures and should practice data collection methods before attempting to draw conclusions from test results.

LIGHTING

For the practitioner, both ambient illumination (i.e., the light surrounding an operator) and glare (i.e., light reflected from surfaces) are of primary concern. A bright light shone into the operator's eyes as he or she attempts to observe a dim object (for example, a computer display next to a large window at a busy airport security terminal) and reflected glare from work surfaces are common causes of reduced performance in visual tasks. Glare can also cause visual discomfort leading to headaches and other physical maladies, even with relatively short exposure. Measuring illumination and glare is not always a simple matter of setting up the right test equipment and recording measurements, especially in a chaotic operational environment. (Nor is it always economically feasible to do so). When measurement is not possible, the practitioner should at least observe when and under what circumstances an operator may be having trouble in his or her task because of poor illumination or glare. Later, if the magnitude of the problem warrants, it may be possible to investigate the matter further under controlled conditions. (Or, if it is an economic issue, the apparent severity of the problem may warrant the cost of measurement equipment.)

When measurement is possible, it is important to understand what to measure. Illumination is the amount of light (luminance flux) falling on a surface measured in lumens/m^2 = lux = 0.093 ft-c (footcandles). Illumination decreases with the square of the distance from a point source. On the other hand, luminance is the amount of light per unit area emitted or reflected from a surface. Luminance is measured in candela per square meter (cd/m^2), footlamberts (ft-L), or millilamberts (mL). 01.0 cd/m^2 = 0.31 mL = 0.29 ft-L. The luminance of a surface does not vary with the distance of the observer from the surface being viewed. Instruments for measuring both these phenomena are available through various sources, for example, Texas Instruments.

Sufficient interior and exterior lighting is required for the safe, efficient, and effective operation of vehicles and other systems. This section presents test methods and measures for the assessment of workspace lighting. Test procedures are primarily intended for measuring light levels of enclosed spaces but can be extended to external work sites to the degree that requirements for external lighting are consistent with those defined for internal illumination. Various criteria for task illumination requirements are presented in Table 11.1. Although these criteria have changed little over the years, the reader may want to verify the latest by referring to MIL-STD-1472.

Lighting Issues

Lighting issues include more than just ambient room illumination or the available light falling upon a work surface. Other environmental issues associated with lighting include direct glare, reflected glare, transilluminated display brightness, CRT (cathode ray tube) brightness, and brightness ratio.

Direct Glare. Glare-control methods assume the operator is using unaided vision. Eyeglasses reflect glare into the eyes if a bright light behind the viewer is between 30° above and 45° below the line of sight, or if it is within 20° left or right of the line of sight.

Reflected Glare. Reflected glare from work surfaces often leads to a reduction in task efficiency where visual performance is critical. Specific criteria for acceptable range of reflectance values for different surfaces can be found in MIL-STD-1472. When measurement is not possible, look for the following:

- Specular reflectance (i.e., from a shiny surface) greater than 3 times the average luminance of the surrounding area
- Highly reflective work surfaces

TABLE 11.1
Recommended Illumination Levels for Common Tasks

Work Area or Type of Task	LUX* (FOOTCANDLES)	
	Recommended	Minimum
Assembly, general:		
coarse	540 (50)	325 (30)
medium	810 (75)	540 (50)
fine	1,075 (100)	810 (75)
precise	230 (300)	2,155 (200)
Bench work:		
rough	540 (50)	325 (30)
medium	810 (75)	540 (50)
fine	1,615 (150)	1,075 (100)
Business machine operation		
(calculator, digital, input, etc.)	1,075 (100)	540 (50)
Console surface	540 (50)	325 (30)
Emergency lighting	no maximum	30 (3)
Hallways	215 (20)	110 (10)
Inspection tasks, general:		
rough	540 (50)	325 (30)
medium	1,075 (100)	540 (50)
fine	2,155 (200)	1,075 (100)
extra fine	3,230 (300)	2,155 (200)
Machine operation	540 (50)	325 (30)
Office work, general	755 (70)	540 (50)
Reading:		
large print	325 (30)	110 (10)
newsprint	540 (50)	325 (30)
handwritten reports, in pencil	755 (70)	540 (50)
small type	755 (70)	540 (50)
prolonged reading	755 (70)	540 (50)
Stairways	215 (20)	110 (10)
Transcribing and tabulation	1,075 (100)	540 (50)

Note. Some tasks requiring visual inspection may require up to 10,000 lux (1,000 ft.-c). As a guide in determining illumination requirements, the use of a steel scale with 1/64 in. divisions requires 1950 lux (180 ft.-c) of light for optimum visibility. From MIL-STD-1472, 1978. Adapted with permission.

*As measured at the task object or 760 mm (30 in.) above the floor.

- Angle of incidence of a direct light source equal to the operator's viewing angle
- Use of highly polished surfaces within 60° of the operator's normal visual field
- Light source placed behind the maintainer

Transilluminated Display Brightness. The brightness of transilluminated displays should be 10% greater than the surrounding surface but
never more than 300% of the surrounding surface. The brightness contrast of the figure-ground relationship should be at least 50%.

CRT Brightness. The brightness of the surface around a CRT should
be from 10% to 100% of that of the CRT. None of the surfaces surrounding the CRT should exceed it in brightness, with the exception of warning
lights.

Brightness Ratio. This is one of the more common problems in illumination that a practitioner will face. The brightness ratios between the
lightest and darkest areas or between a task area and its surroundings
should be no greater than specified in Table 11.2.

Environmental Lighting Test Procedures

A spot brightness meter and a photometer will be needed to measure light
levels. (Note: When purchasing or renting test equipment, we sometimes
find sufficient instructions on how to use the instruments but it would be a
good idea to review the following procedures anyway.) The test specialist
should identify all areas on workstations within a test item where lighting

TABLE 11.2
Recommended Brightness Ratios

	Environmental Classification		
Comparison	*A*	*B*	*C*
Between lighter surfaces and darker surfaces	5:1	5:1	5:1
Between tasks and adjacent darker surroundings	3:1	3:1	5:1
Between tasks and adjacent lighter surroundings	1:3	1:3	1:5
Between tasks and more remote darker surfaces	10:1	20:1	*
Between tasks and more remote lighter surfaces	1:10	1:20	*
Between luminaries and adjacent surfaces	20:1	*	*
Between the immediate work area and the rest of the environment	40:1	*	*

Note. A = interior areas where reflectances of entire space can be controlled for optimum visual conditions; B = areas where reflectances of nearby work can be controlled, but
there is only limited control over remote surroundings; C = areas (indoor and outdoor)
where it is completely impractical to control reflectances and difficult to alter environmental
conditions. From DOT, FAA/CT-9611, by FAA Technical Center, 1996, NJ: FAA Technical
Center. Adapted with permission.
*Brightness ratio control not practical.

could be a problem. Alternate light sources should also be identified for assessment, that is, panel lighting, map light, dome light, general area lighting, and others. Potential glare problems on areas where the test participant is subjected to high levels of direct or reflected light should also be identified.

To measure illumination in areas where you may suspect a problem with lighting, a photometer is used. Measurements should be made at maximum and minimum illumination for cases where the light intensity is controllable. The ambient illumination should be measured and reported for all lighting tests. For workbenches and consoles several readings should be made in about 1-ft increments in a grid pattern over the surface to be evaluated. Record the illumination level on a data sheet reflecting the same grid pattern. Measurements should include illumination levels in low and daylight conditions of ambient light. The brightness of displays is measured using the spot brightness meter. Several areas should be measured over each display surface to identify hot spots or areas of non-uniform luminance.

Reflectance is measured using a spot brightness probe and photometer. It should be measured on all control panel surfaces, workstation surfaces, and other surfaces where reflected light may cause discomfort or interfere with visibility. Reflectance is the amount of light reflected from a surface and is directly dependent on the amount of light falling on the surface. Thus, measurements of both illuminance and luminance must be made at each measurement point. The test specialist should establish measurement points in a grid pattern along the surface. Grids should be of 1- and 2-ft increments. Measurements should be made under all potential lighting conditions. Luminance measurements should be taken using the spot brightness probe held perpendicular to the surface with the distance from the surface dependent on the field of view of the probe. Illuminance measurements should be taken using the photometer probe mounted flat on the surface. Reflectance is then calculated as follows:

$$\text{Reflectance} = \text{Luminance/Illuminance} \times 100.$$

Display lighting should be assessed for lighting uniformity, lighting balance, trim range, and sunlight readability. Lighting and uniformity and balance should be assessed by measuring the brightness (in candela per meter square (cd/m^2)) at six equal distant points around each display surface. Use the spot brightness meter for this measurement. Trim range should be assessed by measuring the maximum brightness of each display to ensure visibility in the brightest expected ambient environment. Assess the smoothness and evenness of a lighting variable control (rheostat) by decreasing the illumination from brightest to dimmest control settings.

Assess the minimum brightness to ensure that the display will be visible under all expected use conditions. Assess the readability of displays in direct sunlight by observing the brightness contrast and hue contrast of displays under all expected use conditions.

Data Reduction and Analysis

As mentioned earlier, illumination data include illumination values of light falling on a surface or area (in lux) and the brightness levels of illuminated displays (in candela per square meter). Data analysis should identify areas where either sufficient light is not available or where too much light (glare) is present. Present the data in the form of line drawings of panels, consoles, or work areas with brightness data indicated for locations of light measurements. Present specific data points, such as CRT brightness, by means of tables. Analyze lighting uniformity by calculating the mean brightness and standard deviations of points measured within each display. Determine the ratio of the standard deviation divided by the mean and compare this figure with the appropriate criteria. Lighting balance should be analyzed by calculating the mean brightness of each display and the brightness ratio of each pair of displays within a functional group. Compare these brightness ratios with the appropriate criteria.

Regarding subjectivity, it may not always be possible to measure environmental lighting, as was indicated at the beginning of this section. Here, notes should be made indicating potential problems with illumination or glare. If possible, conduct a user survey and ask whether poor illumination or reflected light is a problem. Operator personnel are usually quick to express an opinion and may rate a problem sufficiently high on a rating scale to indicate to the practitioner that further investigation (i.e., measurement) is needed.

NOISE

Noise is defined as "that auditory stimulus or stimuli bearing no informational relationship to the presence or completion of the immediate task" (Burrows, 1960). Primarily, noise measurements are made to determine whether noise levels generated by equipment, facilities, or systems present hazards to people. From a purely human-performance standpoint, noise is measured to determine whether levels exceed limits that have been known to cause performance degradation. Measures are also made to judge whether individual hearing protection or special attenuation devices are required for certain tasks, or whether extraordinary steps should be taken in systems design to protect against the deleterious effects of high levels of noise (Meister, 1986). The most common sources of noise include those from airports, motor vehicles and aircraft, weapons, power tools, and highways. But generally, whatever the source, excessive noise should be evaluated with respect to safety and performance degradation so that steps can be taken toward abatement.

Noise Testing Issues

There are two types of noise with which the human factors test specialist is concerned: steady state and impulse. Steady-state noise is a periodic or random variation in atmospheric pressure at audible frequencies; its duration exceeds 1 s, and it may be intermittent or continuous. Impulse noise is a short burst of acoustic energy consisting of either a single impulse or a series of impulses. The pressure-time history of a single impulse includes a rapid rise to a peak pressure, followed by somewhat slower decay of the pressure envelope to ambient pressure, both occurring within 1 s. A series of impulses may last longer than 1 s.

Steady-state and impulse noise criteria are provided as limits for peak pressure levels and B duration in MIL-STD-1474 (1984) but are also provided in MIL-STD-1472 and DOT/FAA/CT-96/1. Tables 11.3 and 11.4 provide respective steady-state noise categories and noise limits for personnel-occupied areas. Additional acoustic noise criteria for offices and shop areas can be found in Beranek (1957).

TABLE 11.3
Steady-State Noise Limits for Occupied Areas

Octave Band	Category			
Center Frequency	A	B	C	D
63 Hz	130 dB	121 dB	111 dB	106 dB
125	119	111	101	96
250	110	103	94	89
500	106	102	88	83
1,000	105	100	85	80
2,000	112	100	84	79
4,000	110	100	84	79
8,000	110	100	86	81
dB (A) Criteria	108	100	90	< 85

Note. For Categories A through D, in those cases where the mission profile for the system or equipment being tested exceeds 8 hours of operation in each 24 hours, the limits specified shall be reduced sufficiently to allow for an exposure for longer than 8 hours, as approved by the developer in conjunction with the Surgeon General, Occupational Safety and Health Administration (OSHA) or other medical authority.

TABLE 11.4
Steady-State Noise Categories and Requirements

System Requirement	Category
1. No direct person-to-person voice communication required, maximum design limit. Hearing protection required.	A
2. System requirement for electrically aided communication via sound attenuating helmet or headset. Noise levels are hazardous to unprotected ears.	B
3. No frequent, direct person-to-person voice communication required. Occasional shouted communication may be possible at a distance of one foot. Hearing protection required.	C
4. No frequent, direct person-to-person voice communication required. Occasional shouted communication may be possible at a distance of 2 ft. Levels that exceed Category D require hearing protection.	D
5. Occasional telephone or radio use or occasional communication at distances up to 1.5 m (5 ft) required.	E**
6. Frequent telephone or radio use or frequent direct communication at distances up to 1.5 m (5 ft) required.	F**

**For design of mobile or transportable systems.
Note.   Categories A, B, C, and D are based primarily on hearing conservation priorities, whereas the remaining categories are based primarily on communications requirements. From MIL-STD-1472, 1978. Adapted with permission.

Environmental Noise Test Procedures

Generally, sound measurement and recording instrumentation should conform to current American National Standards Institute (ANSI) specifications. Measurement of steady-state noise requires the following equipment:

1. Microphones. Microphones shielded against wind effects should be used, having a flat response at grazing incidence of 90° or having an essentially flat response at normal incidence (0°). A random incidence corrector should also be used with a 1-in. microphone. Frequency response should be flat, between 20 Hz and 18 kHz.

2. Sound level meters. A portable, battery operated sound level meter conforming to the requirements for Type I as specified by current ANSI standards.

3. Octave band filter set. This filter set should be used, which conforms to Type E, class II, as specified by current ANSI standards.

4. Magnetic tape recorder (optional). A magnetic tape recorder having a flat frequency response from 20 Hz to 18 kHz ($\pm$ 2 dB) should be used for recording sounds for later analysis. A computer-based sound recorder may also be used.

Instrumentation required to test impulse noise follows:

1. Microphones/transducers. These should have a flat dynamic response of ± 2 dB over the frequency range of 20 Hz to 70 kHz. Microphones having the appropriate dynamic range and rise-time characteristics should be used for measurements up to 171 dB; transducers (blast gauges) should be used for measurements above 171 dB.
2. FM (frequency modulated) magnetic tape recorder. The FM recorder should have a frequency response up to 80 kHz, ± 0.5 dB.
3. Digital oscilloscope. A digital oscilloscope or other suitable equipment should be used to digitize peak intensity and duration of impulse noise.

The following preparations should be conducted prior to steady-state and impulse noise tests (U.S. Army Test & Evaluation Command, 1983):

1. Sketch the location and general orientation of the test item and microphone with respect to the test site.
2. Select the appropriate instrumentation from the previous list of instruments, making sure that instrumentation has a valid calibration certification.
3. Prepare an acoustical test data form (see Fig. 11.1).
4. Check to make certain all personnel exposed to steady-state noise levels above 85 dB(A), and impulse noise exceeding 140 dB(A), are wearing hearing protection.
5. Position sound recording devices and transducers at the center of the expected head positions of operators. Recommended position is 65 in. (165 cm) above the ground plane for standing personnel and 31.5 in. (80 cm) above the seat for seated personnel.
6. Make sure the ambient steady-state noise is at least 10 dB below the noise being measured (25 dB for impulse noise).
7. Do not conduct tests when wind velocities exceed 12 mph (54 m/s), when relative humidity exceeds 85%, or during active precipitation (wind and humidity may affect the sound levels).
8. Check instruments for alignment after each series of tests.
9. Make sound recordings with the test item configured and operated as expected during normal operation.

To test steady-state noise levels associated with stationary equipment (generators, pumps, and other power equipment), microphones should be placed approximately at the center of the probable head positions of equipment operators and at positions 5 ft (1.5 m) from each side and each

VEHICULAR ACOUSTICAL TEST DATA FORM | TIME: | DATE:

TEST ITEM: | TEST CONDUCTOR: | TEST ITEM OPERATOR:

REG MODEL NO.: | SERIAL NO.: | ODOMETER: | HOUR METER: | TEST ITEM CONDITION:

AMBIENT TEMPERATURE: | RELATIVE HUMIDITY: | SURFACE: (Paved/Stone/Dirt) | TERRAIN: (Level/Hills)

BAROMETRIC PRESSURE: | SKY COVER: | OPERATIONAL CONDITIONS: (Stationary Operation/Hiway Driving/Drive-By) | MEASUREMENT LOCATION: (Interior/Exterior)

WIND DIRECTION: | WIND VELOCITY: | MICROPHONE: | SOUND LEVEL METER:

MICROPHONE LOCATION: | TAPE RECORDER: | TAPE NO.: | OCTAVE ANALYZER:

GEAR	RPM	APPROACH SPEED (MPH)	dB(A)	dB(B)	dB(C)	ALL PASS	63	125	250	500	1,000	2,000	4,000	8,000	REMARKS
										DRIVER'S EAR POSITION					
1															
2															
3															
4															
										PASSENGER'S EAR POSITION					
1															
2															
3															
4															

INTERIOR OCTAVE-BAND SOUND LEVELS (Hz)

Maximum allowable limits for unprotected hearing

FIG. 11.1. Sound level data form for vehicular acoustics tests.

end of the test item, 65 in. (165 cm) above the ground. Operate the equipment in its normal operating mode. Record dB(A), dB(C), and octave-band pressure levels at all microphone locations. The test should be conducted under the equipment's highest noise-producing operating conditions. When the noise level exceeds 85 dB(A) at a distance of 5 ft (1.5 m), determine the distances and directions from the noise source at which the noise level is equal to 85 dB(A). Take as many readings as necessary to accurately plot a 85 dB(A) contour curve as shown in Fig. 11.2.

To test steady-state noise associated with moving equipment (e.g., vehicles, aircraft, other noise-producing vehicles) interior noise microphones should be placed 6 in. (15 cm) to the right of the driver's right ear and at the center of the probable head location of assistant drivers or passengers. Operate the vehicle at two-thirds maximum rated engine speed in each of its forward gears over a paved test course, with any auxiliary equipment that might add to the overall noise levels (e.g., heaters, blowers, air conditioners) in operation. Referring back to Fig. 11.1, record dB(A), dB(C), and octave-band pressure levels at each microphone location for each gear range with the windows or hatches first open, then closed.

To test the exterior noise associated with moving equipment (drive by) place a microphone 50 ft (15.2 m) from, and perpendicular to, the center-

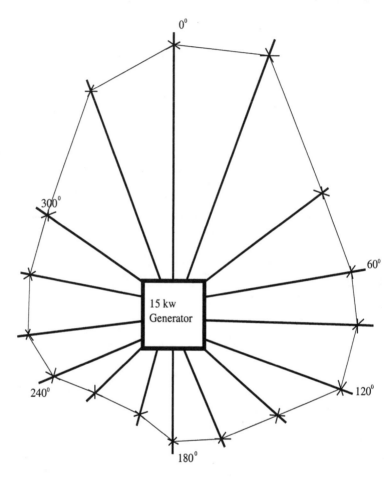

Sound level limits for exterior noise		
Examples, Types of Equipment	Distance from Centerline	Sound Level Limit (dB(A))
Motor vehicle	15.2 m (50 ft)	83
Generator (60 kW)	7 m (23 ft)	80
Air compressors	7 m (23 ft)	76

FIG. 11.2. Steady-state noise plot showing 85 dB(A) contour curve for a 15 kW generator.

line of the vehicle path and 4 ft (1.2 m) above the ground plane. Place a marker 50 ft (15.2 m) in front of the microphone centerline. Set the sound level meter for fast response on the A-weighted network. Drive the vehicle toward the marker at two-thirds maximum rated engine speed, in a gear that will allow the vehicle, when fully accelerated, to reach the maximum engine speed between 10 and 50 ft (3 and 15 m) beyond the microphone centerline, without exceeding 35 mph (56 km/h) before reaching the endline. Record the maximum dB(A) values as the vehicle is driven past the microphone. The applicable reading is the highest sound level obtained for the run. Three measurements should be made on each side of the vehicle, unless it becomes obvious after the first run that one side is definitely higher in sound level, in which case record the sound level for that side only. Report the sound level as the average of the 2 highest readings that are within 2 dB of each other.

Recording impulse noise (i.e., noise with a trough-to-peak wave duration of less than 1 s) requires somewhat different procedures. When the expected pressure levels are in excess of 171 dB, use suitable pressure transducers (e.g., blast gauges); when expected levels are below 171 dB, use fast-response microphones. Choose one of two techniques for recording impulse noise: Photograph the trace obtained on a cathode-ray oscilloscope connected to the transducer system, or record the impulse noise with an FM tape recorder. Measure the peak pressure level and B duration from data recorded during each of at least three separate tests. Use the arithmetic means of the peak pressure levels and B-durations from these (three or more) tests (if consistent) to define the impulse noise when the range of peak pressure levels does not exceed 3 dB. If the range of the peak pressure levels exceeds 3 dB, conduct additional tests until the number of measurements equals or exceeds the range in decibels.

To make a 140-dB noise contour curve, place a transducer 5 ft (1.5 m) above the ground as close to the test item as is considered safe and on each 30° radial line centered at the test item. Place a second series of transducers, a third series of transducers twice the distance as the second, and, finally, a fourth series of transducers twice the distance as the third. From this configuration of transducers, a 140-dB noise contour curve around the test item can be predicted by interpolation.

Data Reduction and Analysis

Steady-State Noise. Tabulate all direct measurement data, or, for data that has been recorded on magnetic tape, analyze the data in the laboratory for each specified requirement. If a requirement is not specified, analyze the data for dB(A), dB(C), and octave-band sound levels in each of eight octave bands and extract a portion of each test segment from the magnetic tape, computer disc, or other recording media, and plot a spectral analysis. Then, present the data taken for 85 dB(A) contour curves. Assess the noise condition for the following, as applicable: minimum distance people may approach without hearing protection; type of hearing protection required; type of communication possible; distance of probable communication; speech intelligibility; maximum detectable distance; and primary sources of noise (e.g., exhaust, tracks).

Impulse Noise. The oscilloscopic trace photographs or the magnetic tape records should be analyzed to determine peak pressure level, A duration and B duration. Referring to Fig. 11.3, tabulate the data to determine safety conditions. When making comparison noise tests, only peak pressure levels are required and reported by round number. Present the data taken for 140-dB contour curves. Assess the noise condition for the minimum distance people can be from the noise source before they must wear hearing protection and note the type of hearing protection required. Fig. 11.4 shows the peak pressure level curves based on peak impulse noise levels. A noise curve (W, X, Y, or Z) is selected from Fig. 11.4 and applied to Table 11.5 to determine the type of hearing protection required.

IMPULSE NOISE TEST DATA FORM		TIME:	DATE:
TEST CONDUCTOR:		TEST ITEM:	
POSITION NO. 1		POSITION NO. 2	
Round No. Peak Pressure Level (dB) A-Duration (ms) B-Duration (ms)		Round No. Peak Pressure Level (dB) A-Duration (ms) B-Duration (ms)	
1		1	
2		2	
3		3	
4		4	
5		5	
REMARKS:			

FIG. 11.3. Impulse noise level data form.

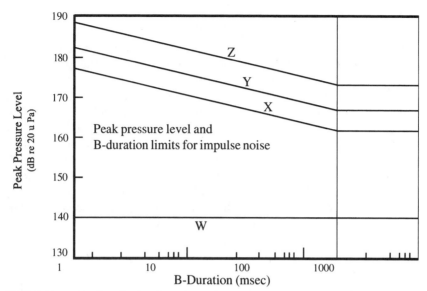

NOTE: Formulas for plotting impulse noise curves are:

$$PPL_Y = 167 + \frac{2 \log_{10}(200/T)}{\log 2}, \text{ for } T < 200; \quad PPL_Y = 167, \text{ for } T > 200$$

where: PPL_Y = peak pressure level of curve Y, in dB,

   T = B-duration, in msec

Also, curve Z is 6.5 dB above curve Y and curve X is 5 dB below curve Y.

FIG. 11.4.   Peak pressure level curves (use Table 11.5 to select appropriate hearing protection).

TABLE 11.5
Impulse Noise Protection Criteria

Maximum Expected Number of Exposures in a Single Day*	Impulse Noise Limit		
	No Protection	Ear Plugs or Muffs	Both Plugs & Muffs
1,000	W	X	Y
100	W	Y	Z
5	W	Z	Z**

*A single exposure consists of either (a) a single pulse for nonrepetitive systems (systems producing not more than 1 impulse per second, such as rifle fire), or (b) a burst for repetitive systems, such as automatic weapons fire.

**Higher levels than Curve Z not permitted because of the possibility of other nonauditory physiological injury.

Because people's sensitivity to noise varies somewhat, it is important to measure noise. When measurement is not possible, the practitioner should make notes of areas in which noise levels may exceed criteria. For steady-state noise, it remains a subjective opinion of the practitioner, but barring any impairment in his or her hearing, the author has found that noisy environments usually exceed some level of criteria and will recommend further investigation (measurement). It helps to ask the operators (such as we might do in a survey) whether they think noise interferes with their job and to rate the severity. Impulse noise typically means very loud sounds, such as explosives, canon, or rifle fire, and the common sense recommendation, even without the benefit of measurement, would be that hearing protection is required. However, impulse noise from machinery, for example, may require measurement to determine just how much hearing protection should be provided to workers, exposure durations, and others.

SPEECH INTELLIGIBILITY

Generally, speech falls in the frequency range of 500 to 5,000 Hz. When extraneous noise interferes with speech sounds at the same frequency, it may become necessary to assess this masking effect. To do this, the tester may want to measure the clarity with which a speaker can be understood over a voice communications system, speaking in an auditorium, holding a simple conversation in a vehicle, or other similar applications. The method for conducting such tests is referred to as intelligibility testing.

In speech intelligibility testing, it is essential that the content of the message not be an uncontrolled variable. Consequently, words or messages are used over repeated test trials that are of equal difficulty to comprehend. The lists chosen for speech intelligibility tests have nearly the same phonetic composition. The speech sounds used are representative of a range of words and are balanced across lists for approximate equality in difficulty.

There are three lists of words/sentences from which the specialist can choose to assess speech intelligibility: the Phonetically Balanced Word Lists; the Sentence Lists; and the Modified Rhyme Test (MRT) Lists. We have selected the MRT as an example of a reliable procedure for testing speech intelligibility. For a more in-depth discussion about the other methods, and for calculating the articulation index, see Van Cott and Kinkade (1972).

The MRT has three distinct advantages: It requires little practice, errors cannot result from misspelling the words when pronounced, and instrumentation and procedures make it a convenient and easy-to-apply tool for identifying speech communications problems. The only disadvantage is that it assesses only the intelligibility of the front or end consonants of a word and does not test the intermediate vowel sound.

Criterion scores for speech intelligibility tests depend on the system and application being tested. Some system specifications identify intelligibility criteria; at other times, it is left to the human factors test specialist to judge whether scores sufficiently meet or exceed design requirements or whether a serious enough problem exists to warrant further testing. Things to be considered include the vocabulary of the talker and listener; whether numbers, single words, short phrases, or complete sentences are to be used on the system; and what percentage of each kind of item must be heard correctly for the system to operate (Van Cott & Kinkade, 1972). For the MRT, a score of 90% or better is generally considered adequate for most communications systems.

Speech Intelligibility Test Procedures

Use the appropriate communication system, facility, or environment appropriate for the application under test. Prepare a talker's sheet and listener's sheet for the MRT, samples of which are shown in Tables 11.6 and 11.7. A suite of instruments should be available to measure temperature, humidity, and air speed as well as ambient acoustical noise.

Test participants should be screened for abnormal hearing and speech disorders. For the MRT, talkers and listeners should be thoroughly versed on talking and listening procedures and should practice the method several times before the test. Listeners and talkers should be rotated between test samples so that a test sample, listener, or talker does not unduly bias the test. Also, care should be taken by the test specialist to ensure that boredom and fatigue do not create bias in the data. The test specialist

TABLE 11.6
Modified Rhyme Test: Talker's Sheet

1. kick	11. sale	21. pat	31. pin	41. came
2. feat	12. peak	22. did	32. seed	42. boil
3. pup	13. king	23. kit	33. ray	43. fib
4. book	14. sag	24. tin	34. test	44. cud
5. lip	15. sip	25. tier	35. pave	45. keel
6. rate	16. told	26. bent	36. bath	46. lark
7. bang	17. bun	27. sun	37. pop	47. heave
8. fill	18. lake	28. shed	38. pig	48. den
9. mass	19. gun	29. pot	39. tan	49. saw
10. tale	20. bust	30. duck	40. cape	50. beat

Note. U.S. Army Test and Evaluation Command by U.S. Army Test and Evaluation Command, 1983, Aberdeen Proving Ground, MD: Author. Adapted with permission.

TABLE 11.7
Modified Rhyme Test: Listener's Sheet

1 lick pick stick
  pick tick wick
	17 bun bus buff
	   bug but buck
		33 may pay way
		   day say ray
			49 saw raw jaw
			   law paw thaw

2 feat heat neat
  seat meat beat
	18 late lace lay
	   lane lake lame
		34 best vest rest
		   nest test west
			50 beam beat bead
			   bean beak beach

3 pup puff pus
  puck pun pub
	19 bun nun sun
	   run fun gun
		35 pane page pale
		   pave pace pay	

4 hook took look
  book cook shook
	20 bust dust rust
	   just gust must
		36 bat bath bad
		   ban back bash	

5 hip lip rip
  sip dip tip
	21 pack pan pass
	   path pat pad
		37 pop shop mop
		   cop hop top
			Scorer: _____

6 rate race rave
  ray raze rake
	22 did dig dip
	   din dill dim
		38 dig rig big
		   pig wig fig
			Score: _____

7 gang hang sang
  bang fang rang
	23 bit hit kit
	   wit fit sit
		39 tam tang tan
		   tack tab tap	

8 will till hill
  fill kill bill
	24 tin win din
	   sin pin fin
		40 cane cake cape
		   tame same name	

9 mat mass mad
  map math man
	25 teal tear tease
	   team teak teach
		41 came game fame
		   tame same name	

10 gale pale tale
   male sale bale
	26 tent sent bent
	   rent went dent
		42 oil soil toil
		   foil coil boil	

11 sake sale safe
   sane save same
	27 sun sub sum
	   sud sup sung
		43 fit fizz fin
		   fill fig fib	

12 peace peat peak
   peas peal peach
	28 wed red bed
	   shed fed led
		44 cuss cut cup
		   cup cub cud	

13 kick kid kit
   kill kin king
	29 hot tot not
	   pot lot got
		45 peel heel eel
		   feel keel reel	

14 sack sag sass
   sap sat sad
	30 dun duck dub
	   dug dud dung
		46 bark hark park
		   dark lark mark	

15 sip sit sin
   sick sill sing
	31 pin pip pig
	   pit pill pick
		47 heal heap heat
		   heave heath hear	

16 sold cold told
   gold hold fold
	32 seep seem seethe
	   seen seed seek
		48 then men ten
		   hen den pen	

Note. From U.S. Army Test and Evaluation Command, by U.S. Army Test and Evaluation Command, 1983, Aberdeen Proving Ground, MD: Author. Adapted with permission.

should provide a schedule including the order of stations to be used for transmitting and receiving, the talkers and listeners to be used, and the word lists to be transmitted.

After procedural messages have been exchanged, the designated talker will transmit the designated words in a clear, strong voice, spacing the words to allow a 4-s interval between monosyllabic words. Listeners at the receiving stations will transcribe all test words. Listeners should be required to add any appropriate comments relative to words that may be missed or about which they are uncertain. Sufficient log entries should be made at all stations to fully identify each test message. Transmissions may be recorded at the receiving stations for comparison with recordings made at the transmitting stations to assess the quality of the communications equipment being tested. Each test message, consisting of a complete word list manually transcribed by the listeners, will be scored to determine the speech intelligibility percentage for that message. Sufficient test messages (about 100) should be transmitted to give a high degree of confidence in the results. These messages should be equally distributed among the listeners, who should number about 20 but in no case less than 5 (U.S. Army Test & Evaluation Command, 1983).

Data Analysis and Reduction

Data should include a list of test equipment used, a description of the facility in which the test was conducted, a list of test participants, and transcripts of transmitted and received messages with the date, time, talker, listener, test setup (test equipment, frequencies and modes used, and other pertinent information), and speech intelligibility score annotated. Data should also include the location and description of the test site, including elevation, height, type, and orientation of antennas, if radio communications are involved. It should include signal strength measurements, description and measurement of ambient acoustical noise, comments of listeners relative to test message talkers, and missed words. Finally, data should include ambient temperature, humidity, and wind-chill measurements, and any head covering or ear protective equipment used.

The MRT is scored by counting the number of words correct and determining the percentage of words correct, as follows:

$$\% \text{ correct} = (R - (W/5) \times 2).$$

(Where R = number right and W = number wrong.)

Calculations should be performed to provide the median, mean, and standard deviation of the speech intelligibility scores. To calculate speech interference as a function of sound levels and distances between talker and listener, plot the noise levels according to Fig. 11.5. Levels exceeding 85 dB(A) require hearing protection.

VIBRATION AND SHOCK MEASUREMENT

Mechanical vibrations and shock are dynamic phenomena that can affect the human system in many ways. The extremely complex nature of the human body makes it susceptible to a wide range of vibration and shock frequencies. When measuring and evaluating the effects of vibration and shock on humans, both psychological and mechanical effects are considered.

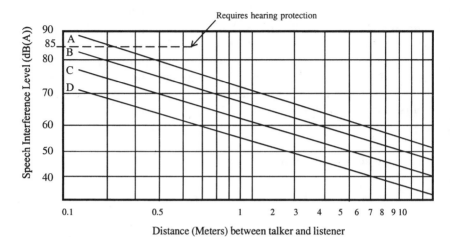

A   Shouting
B   Very Loud Voice
C   Raised Voice
D   Normal Voice

FIG. 11.5.   Speech interference levels for reliable conversation.

The most important of these effects range from kinetosis (motion sickness) in the fractional Hertz range, through whole body vibration in the range 1–80 Hz, and segmental vibration (hand-arm vibration) in the range 8–1,000 Hz (Broch, 1980). Systems exhibiting a high vibration may adversely affect performance when it causes dials, indicators, and other visual apparatus to jump and jitter, forcing us to strain our eyes and causes controls, tools, and other handheld objects to be difficult to handle, causing muscular fatigue and presenting safety hazards. There may also be other effects, such as irritability and nervousness, which can cause errors and poor judgment. Higher levels of vibration and shock can cause organ damage (Meister, 1986).

From a vibration and shock perspective, the low-frequency range (less than or equal to 80 Hz) may be considered most important (Broch, 1980). Beyond 80 Hz, the sensations and effects are dependent on clothing and other damping factors at the point of contact with the skin. Because of this, it is generally not possible to identify valid criteria outside the 1–80 Hz frequency range. A distinct resonance effect occurs in the 3–6 Hz range for the area containing the thorax and abdomen. Consequently, it is very difficult to efficiently isolate and measure vibration in a sitting or standing person. In the area containing the head, neck, and shoulders, a further resonance effect is found in the 20–30 Hz region. Disturbances are felt in the 60–90 Hz region causing eyeball resonances, and a resonance effect in the lower jaw and skull area has been found between 100–200 Hz. Also, Grandjean (1988) found disturbances in the cervical (neck) vertebrae area (3–4 Hz), lumbar area (4 Hz), and the shoulders (5 Hz). The vibration and shock data currently available is the result of studies of mostly ground- and air-vehicle personnel whose ability to perform complex tasks under adverse environmental conditions, including vibration, is particularly critical.

Whole Body Vibration

Facilities and equipment should be designed to control the transmission of whole body vibration to levels that will permit safe operation and maintenance. ISO 2631 (Internaional Standardization Organization, 1974) presents vibration and shock data as vibration criteria curves for vertical and lateral vibration over the frequency range 1 to 80 Hz. Figures 11.6 and 11.7 show the respective vertical vibration exposure criteria curves and lateral vibration exposure criteria curves as root mean square (RMS) acceleration levels, which produce equal fatigue-decreased proficiency. Exceeding the exposure specified by the curves will often cause noticeable fatigue and decreased job proficiency, mostly in complex tasks. Extra-vibration variables (i.e., personal stamina) may affect the degree of task interference, but generally, the range for onset of such interference and the time dependency observed hold constant.

Referring to Figs. 11.6 and 11.7, an upper exposure level considered acceptable with respect to both health and task performance is taken to be twice as high (6 dB above) as the fatigue-decreased proficiency boundary, whereas the reduced-comfort boundary is accepted to be about one third of (10 dB below) the stated levels. These criteria are recommended guidelines, or trend curves, rather than firm boundaries.

Segmental Vibration

Next to whole body vibration, segmental or hand-arm vibration, represents the greatest problem potential for human operators. For years, researchers have known that continued exposure to high-vibration hand tools caused pain, fatigue, and even nerve and muscle damage. Vibration levels encountered in many commonly used power tools are sufficiently high to cause pain and damage when operated for even short durations. Examples are chipping hammers, power grinders, sanders, and chain saws. Vibrations can be transmitted into the body from a handheld tool by use of one or both arms. At lower frequency levels, this may result in discomfort and reduced working efficiency. At higher levels and longer exposure periods, the blood vessels, joints, and circulation may become adversely affected. Severe, prolonged exposure may lead to progressive

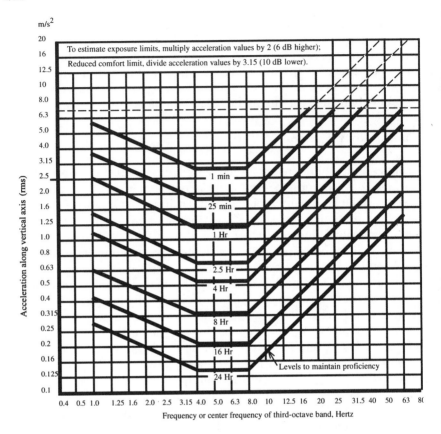

FIG. 11.6. Vertical (longitudinal) vibration exposure criteria curves. *Note:* Exposure limits are presented as a function of frequency (pure sinusoidal vibration or RMS vibration values in one-third octave bands) and exposure time.

circulation disorders in the part of the body suffering the highest level of vibration, usually the fingers and the hands (e.g., hand-arm vibration syndrome (HAVS), vibration-induced white finger (VWF) syndrome.

Vibration and Shock Test Procedures

The following instrumentation should be used for measuring shock and vibration: vibration meter, seat pad accelerometer, magnetic tape recorder, and spectrum analyzer.

Vibration Meter. To measure all but seated whole body vibration, a battery-operated vibration meter is recommended. Vibration instruments should be used in conjunction with the sound level meter and should have

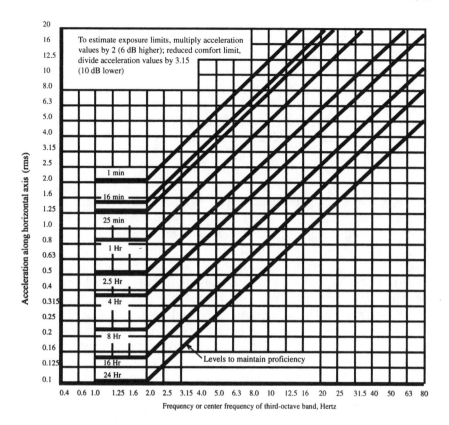

FIG. 11.7. Lateral (transverse) vibration exposure criteria curves. *Note:* Exposure limits are presented as a function of frequency (pure sinusoidal vibration or RMS vibration values in one-third octave bands) and exposure time.

the following features: a single axial accelerometer, a triaxial accelerometer, an integrator, and a one-third octave band analyzer.

Seat Pad Accelerometer. To measure whole body vibration for a seated operator, a seat pad accelerometer should be used. The accelerometer specifications should conform to those of *Standard Recommended Practice Information Report J-1013* (SAE, 1980).

Magnetic Tape Recorder. The test specialist may wish to record the signals generated by the accelerometers for later analysis by using a magnetic tape recorder, digital, computer-based recorder, or similar media.

Spectrum Analyzer.   Although not altogether necessary, a spectrum analyzer may be used to analyze the signals generated by the accelerometers.

Vibration measurements should be made at locations that are judged to
have a differential effect on vibration frequencies, amplitudes, or both.
Test conditions should include those for noise measurements. A minimum
of 5 measurements, but preferably 10, should be made under all conditions of operation. Measurements should be made as close as possible to
the source of vibration, or that point at which vibration is transmitted to
the subject's body.

For seated operators, the seat pad accelerometer is placed under the buttocks at the seat reference point. If the surface is contoured or soft, a semirigid accelerometer should be used. The pad should be carefully positioned
so that the transducer is located midway between the two bony protuber-
ances in the buttocks area and aligned parallel to the measurement axes.

Vibration measurements should also be made at the seat mounting
base. The accelerometer should be firmly attached to the seat mounting
base so that it is located at the vertical projection of the seat cushion and not
more than 100 mm (4 in.) from the vertical, longitudinal plane through the
center of the seat. The accelerometer should be aligned parallel to the
measurement axes. For standing surfaces, vibration measurements should
be made in a grid pattern along the surface. The accelerometer should be
firmly attached to the surface with adhesive or with a metal screw.

To measure hand-arm vibration, mount the accelerometer firmly parallel to the X, Y, and Z axes at each control console, panel, or surface where
the operator's hand or body makes sustained contact. To measure impact
pulses, place the accelerometer as close as possible to the point through
which vibration is transmitted to the body.

Data Analysis and Reduction

The vibration meter should yield data for vibration amplitude, in meters
per second squared (m/s^2) and converted to RMS, and one-third octaves
from 1 through 80 Hz. For nonautomated vibration analyses, draw a profile
of frequency and amplitude. Compare this profile with the appropriate fre-
quency-amplitude graphs. Problem areas will show up as areas that occur
beyond acceptable limits. Vibration data should be correlated with operator
responses to questionnaires and interviews relating to comfort and fatigue.

TEMPERATURE, HUMIDITY, AND VENTILATION

The purpose of measuring temperature, humidity, and ventilation is to
evaluate the effect of these environmental factors on occupants of enclosed areas. Test procedures apply to enclosed areas that have controls

for these factors and to enclosures that meet these provisions, with the exception of Wet Bulb Globe Temperature (WBGT), which applies to the outdoor environment and enclosed areas without any means to control these factors.

Test Issues

Heat-Stress Indices. Heat stress degrades human performance in a number of areas: physical work, tracking tasks, vigilance tasks, mental activities, and performance on industrial jobs (McCormick, 1976). Macpherson (1973) identified a great many indexes of heat stress that have been developed over the years and concluded, quite correctly, that so many indexes only meant that no one index was completely satisfactory with respect to the effects of environmental factors on human performance. It was not until the 1980s, when human factors researchers for the U.S. military settled for a composite of measurements of dry-bulb temperature, humidity, and ventilation, that the true effects of these combined environmental factors could be reasonably evaluated against human comfort and performance criteria.

Cold Stress. The effects of cold and extreme cold on human performance are well documented (McCormick, 1976; Morgan, Cook, Chapanis, & Lund, 1963). Mostly, cold stress is measured for the benefit of the personnel conducting the test outdoors so that preventive measures can be taken to avoid cold damage, particularly frostbite.

The Clo Unit. The basic heat exchange process is influenced by the insulating effects of clothing, varying from cotton underwear to a heavy parka worn in polar extremes. Such insulation is measured by the clo unit. The clo unit is a measure of the thermal insulation necessary to maintain in comfort a sitting, resting subject in a normally ventilated room at 70° F temperature and 50% relative humidity (Gagge, Burton, & Bazett, 1941). The clo unit is defined as follows: Ambient temperature (F)/(BTU/Square feet of body area). Because the typical individual in the nude is comfortable at about 86° F, one clo unit would produce the same sensation at about 70° F. Thus, a clo unit has approximately the amount of insulation required to compensate for a drop of about 16° F (McCormick, 1976).

The Wind Chill Index. The Wind Chill Index is a commonly used scale that combines ambient temperatures with wind speed to express the severity of cold environments. Table 11.8 shows an adaptation of Siple and Passel's (1945) original index. Wind chill is an important test consideration, both for the subjects involved in the test and for test personnel. Reduced wind chill equivalent temperatures were shown to lead to performance

TABLE 11.8
Wind Chill Index

Wind Speed (mph)	Air Temperature					
	40	20	10	0	–10	–20
	Equivalent Temperatures					
Calm	40	20	10	0	–10	–20
5	37	16	6	–5	–15	–26
10	28	4	–0	–21	–33	–46
20	18	–10	–25	–39	–53	–67
30	13	–18	–33	–48	–63	–79
40	10	–21	–37	–53	–69	–85

degradations: increased hand numbness (Mackworth, 1955) and increased reaction time (Teichner, 1958). Frostbite is a very real danger when conducting human factors tests in cold, windy environments, as experience in arctic conditions have shown.

Heat Stress and the WBGT Index. The WBGT index is the most practical index for characterizing the effect of heat stress on humans. Figure 11.8 shows the upper limits of exposure for unimpaired mental performance. When planning a test where human performance in temperatures

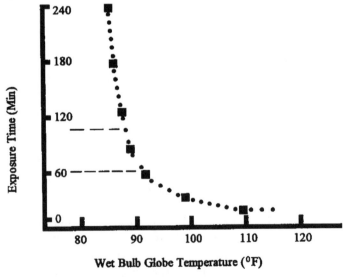

FIG. 11.8.   Upper limits of exposure for unimpaired mental performance.

above a WBGT of 77° F is involved, cautions should be observed in accordance with the information presented in Table 11.8.

A number of design criteria for temperature, humidity, and ventilation exist. The following represent the most frequently required criteria.

Heating. For mobile enclosures used for detail work or occupied for extended periods, the interior dry bulb temperature should be maintained above 50° F (100° C). A minimum of 68° F (20° C) should be maintained within occupied buildings. Check to make certain the system is designed so that hot-air discharge is not directed on personnel.

Ventilation. A minimum of 30 ft³/minute (fpm) per person should be provided into enclosures occupied by people; about two thirds should be outside air. Air flowing past any individual occupant should be at a minimum rate of 100 fpm (30 mpm). Air intakes for ventilation systems shall be located so as to minimize the introduction of contaminated air from such sources as exhaust pipes.

Cooling. For personnel enclosures where the effective temperatures exceed 85° F (29° C), air conditioning shall be provided. Air conditioning systems should be designed such that cold-air discharge is not directed on the people occupying the enclosure.

Humidity. Systems should be designed so that humidity values approximate 45% relative humidity at 70° F (21° C). As temperatures rise, this value should decrease but should remain above 15% to prevent irritation and drying of body tissues.

Test Procedures

When testing in cold and extreme-cold ambient temperatures, HFTE professionals should consider several factors that contribute to cold stress. Working without insulated gloves for prolonged periods in temperatures below 55° F (13° C) often results in stiffening of fingers and thus can degrade performance in tasks requiring manual dexterity. The test specialist should make note of whether or not surfaces that personnel touch (i.e., gearshift levers, steering wheels, seats, and other such surfaces) have a low conductivity. When testing vehicles, especially armored vehicles or vehicles with a substantial amount of metal, observations should include whether local space heating is provided and whether sufficient interior space has been provided for personnel wearing extra protective clothing. Heating should be furnished to personnel enclosures used for work or occupied for prolonged periods to maintain an interior dry bulb temperature of 68° F (20° C). A system's design should include provisions for regulating the amount of heat the heater delivers, with devices such as shutters, louvers, and fan-speed controls. Instruments for measuring temperature, humidity, and air-flow rate include dry bulb, wet bulb, and a globe, or WBGT thermometer, humidity sensors, and air-flow rate sensors.

The tester should remember that in measuring temperature, ventilation, or humidity, these factors, in combination, produce effects that are different from those produced by each factor individually. Thus, 6-ft tall and easily moveable test stands should be used, which can hold thermometers, humidity sensors, and air-flow rate sensors. These instruments should be mounted on the stand as follows: thermometers at floor level and at each 2-ft increment in the vertical direction up to and including 6 ft from the floor; humidity sensor 4 ft above the floor; air-flow rate sensors at 2 and 4 ft.

Test stands should be located and measurements taken at each operator position for vehicles and at personnel work areas for rooms and other enclosures. Measurements should be taken with the enclosure empty, minimally manned, and maximally manned. Measurements should be made under varying conditions of external temperature, humidity, and wind velocity, including snow and ice at operator stations and other personnel-occupied locations. Air-flow measurements should be made at the duct. Divide the duct opening into small grids and measure the air velocity at each grid section with a hot wire anemometer. The measurement should be averaged, $Q = VA$, where Q equals quantity of air in cubic feet per minute, A equals cross-sectional area of the duct, and V equals the average linear velocity in feet per minute. For hot surfaces where contact may be made by personnel, surface temperatures should be measured by means of a surface temperature probe. Temperatures should be averaged from among at least five points that are equidistant on the surface. Hot spots should be noted on the test report.

Data Analysis and Reduction

Compute the mean values for temperature, humidity, and air-flow rate for each of the conditions under which the test was performed. The mean values should be compared with the criteria listed earlier. Data should be compiled as degrees, percentages, and rates. These values should be compared with established criteria. Those measures that fall outside of prescribed limits should be tabulated separately under a discrepancy list. For those measures that have no standardized criteria, an evaluation by the test specialist should be made with respect to the impact on human safety and performance. The discrepancy list should be reviewed and recommendations made for each problem encountered. For convenience, all data should be presented in tabular form.

OTHER ENVIRONMENTAL TEST CONSIDERATIONS

Atmospheric Composition

Gas effluents, such as carbon monoxide, nitrous and sulfur oxides, and other toxic or noxious gases, result from the combustion process of operating internal combustion and jet engines, fuel-fired space heaters, and other fuel-burning items. Other noxious gases are known to be generated by the pyrolysis or explosive oxidation of various fuels and miscellaneous engine-driven equipment. Over 26 compounds have been identified (U.S. Army Test & Evaluation Command, 1983). Most weapons produce carbon monoxide (CO), nitrogen dioxide (NO_2), and ammonia (NH_3), and when black powder is used, sulfur dioxide (SO_2). A by-product of solid rocket engines is hydrogen chloride (HCL).

CO is considered the most deadly, if not the most pervasive, of toxic effluents. It is tasteless and odorless, thus it can concentrate quickly in confined spaces. CO leakage may be found in vehicle passenger compartments, in confined living spaces, and, in short, most any place associated with fuel-fired machines. Its principal toxic effect on the human body is the production of carboxyhemoglobin (COHb). Consequently, one of the purposes of atmospheric composition testing is to measure COHb levels, where CO is suspect, and evaluate CO concentrations. Another purpose is to measure all other gases that have a noxious effect: NO_2 is a pulmonary irritant, NH_3 is a pulmonary irritant and asphyxiant, and SO_2 is a mucous membrane irritant affecting the eyes, nose, and throat. For testing atmospheric composition the reader is referred to TECOM's *TOP 1-2-610* (1983).

Visibility

Visibility measurements should be made for vehicles and for other items where the user must view through windows, port holes, periscopes, and so forth; visually acquire and recognize components of the item; read displays and labels mounted in or on the item; and use optical devices associated with the item. Some of the factors that influence visibility include field of view (unobstructed), distance to and orientation of components requiring visibility, and viewing media (e.g., fog, rain, clear air, glass). Visibility test procedures are found in TECOM's *TOP 1-2-610* (1983b).

MEASURING THE ENVIRONMENT
USING SUBJECTIVE METHODS

As we have mentioned in our examples of lighting and noise, because of economic or other constraints, it is not always possible to measure physical phenomena. But even when you can collect objective data, these environmental phenomena, measured by instrumentation, may not be sufficient (Meister, 1986). This is because the operator's subjective impressions of the environment may not correlate precisely with objective evaluations. Thus, the test specialist has the option of relying on other, subjective means of gathering environmental information.

An earlier chapter explains how to develop and employ questionnaires. This subjective technique can help the HFTE specialist to identify and rate the severity of environmental problems, if such exist. If the person responds positively, the severity of the problem can be rated by asking how the problem affects his or her performance, according to a rating scale in which the response range is from *has no effect whatsoever on my work* to *completely disrupts my work performance*. Here, the test specialist must rely on good judgment as to whether to consider responses near the upper end of the scale (i.e., *completely disrupts my work performance*) serious enough to warrant further, instrumented measurements. Remember that subjective methods should never take the place of instrumented measurement of the environment, particularly where the environmental dimension being considered is serious enough to approach levels that may be physiologically damaging or lethal. If ratings indicate a significant degree of impairment, the specialist should follow up with instrumented measurements.

SUMMARY

This chapter described some of the more common test procedures for environmental testing in the workplace. Each procedure describes the purpose of the test, test criteria, data collection, data reduction and analysis methods, and other information. Although many of these test procedures were initially developed from military and other governmental agencies, the test methods have been refined and validated over the years and are directly applicable to commercial work environments and test programs.

REFERENCES

Beranek, L. L. (1957). Revised criteria for noise in buildings. *Noise Control, 3*(1), 19.

Broch, J. T. (1980). *Mechanical vibration and shock measurements.* Copenhagen, Denmark: K. Larson & Sons.

Burrows, A. A. (1960). Acoustic noise, an informational definition. *Human Factors, 2*(3), 163–168.

Department of Transportation, Federal Aviation Administration/CT-96/1. (1996). *Human factors design guide for acquisition of commercial-off-the-shelf subsystems, non-developmental items, and developmental systems.* FAA Technical Center, NJ: Authors.

Gagge, A. P., Burton, A., & Bazett, H. (1941). A practical system of units for the description of the heat exchange of man with his environment. *Science, 94,* 428–430.

Grandjean, E. (1988). *Fitting the task to the man* (4th ed.). London: Taylor & Francis.

International Organization for Standardization (ISO) 2631. (1974). *Guide to the evaluation of human exposure to whole body vibration.* Geneva, Switzerland: Author.

Mackworth, N. H. (1955). Cold acclimatization and finger numbness. *Proceedings of the Royal Society, B. 143,* 392–407.

Macpherson, R. K. (1973). Thermal stress and thermal comfort. *Ergonomics, 16*(5), 611–623.

McCormick, E. J. (1976). *Human factors in engineering and design.* New York: McGraw-Hill.

Meister, D. (1986). *Human factors testing and evaluation.* Amsterdam: Elsevier.

Military Standard, MIL-STD-1472. (1999). *Human engineering design criteria for military systems, equipment and facilities.* Washington, DC: Author.

Military Standard, MIL-STD-1474. (1984). *Noise limits for Army materiel.* Washington, DC: Author.

Morgan, C. T., Cook, J., Chapanis, A., & Lund, M. (1963). *Human engineering guide to equipment design.* New York: McGraw-Hill.

Siple, P. A., & Passel, C. (1945). Measurement of dry atmospheric cooling in subfreezing temperatures. *Proceedings of the American Philosophical Society, 89,* 177–199.

Teichner, W. H. (1958). Reaction time in the cold. *Journal of Applied Psychology, 42,* 54–59.

U.S. Army Test and Evaluation Command. (1983). *Test operations procedure 1-2-610.* Aberdeen Proving Ground, MD: Author.

VanCott, H. P., & Kinkade, R. (1972). *Human engineering guide to equipment design.* Washington, DC: Author.

Part II

APPLICATIONS

INTRODUCTION TO THE APPLICATIONS CHAPTERS

As we mentioned in the preface, we think the applications chapters are a key component of this book. They are included to provide examples of how the various tools and techniques are placed into practice across a wide range of applications areas. As most HFTE practitioners know, theory and practice are sometimes quite different. We hope that these examples of practitioners making it work will provide some further insights into how HFTE can be accomplished.

Further, because human factors and ergonomics professionals are generally of an inquisitive nature, these examples will appeal to that nature by showing some of the incredible range of applications areas, undoubtedly outside any one person's experience. These chapters also give HFTE practitioners a voice and an audience outside of the technical reports and briefings that are typical venue for HFTE results. We hope that providing this medium for these many fine examples of HFTE will foster greater communication about what our colleagues are doing and what our profession is capable of.

In this collection you will see some absolutely captivating examples of HFTE. We begin with a traditional HFTE area, aircraft flight deck design, but as the authors show us, there are tremendous challenges still in this application area. Our colleagues from Boeing show us not only some fine examples of the testing art but also the importance of the supporting processes and organizational foundations that must be in place. In test activities ranging from reach envelopes to displays and display symbology for a windshear training aid, this applications chapter illustrates how test and evaluation fits into, and is constrained by, the realities of a major engineering program.

Moving from the skies to the sea, we present a maritime testing example, which is an updated version of a much-appreciated chapter from our first edition. The testing of designs for shipboard effectiveness, safety, and livability, as well as the impact of automation on shipboard tasks and training, provides a revealing glimpse of the issues confronted by HFTE professionals at the U.S. Coast Guard. Their work illustrates a full range of testing methods, including structured surveys, self-administered logs, direct observation, checklists, rating scales, mock-ups, HFE standards, task analysis, cognitive analysis, skills and knowledge assessment, error analysis, subject-matter experts, and focus groups. The employment and adaptation of these techniques to the maritime environment provides yet another picture of the challenges faced by HFTE professionals in transferring techniques from the textbook to the reality of a test.

Our third application chapter brings the reader to the forest plantations of New Zealand. The forestry environment contains the full range of issues, everything from basic manual materials handling to the high technology of mechanized harvesting machines. In their chapter, the testers from the Centre for Human Factors and Ergonomics demonstrate their mastery of an extensive range of human factors and ergonomics techniques. Not content with that range, these HFTE professionals look outside the square, and indeed outside our century, to find solutions to ergonomics design problems. The development of the logger's Sabaton™ shows that many answers to our HFTE questions can be found outside the well-traveled paths of human factors and ergonomics theory.

The fourth chapter is also from New Zealand, although the issues addressed are applicable anyplace where there are roads. The chapter shows how laboratory-based technologies can sometimes be used to test road safety features that would otherwise be too costly or too dangerous to accomplish. Testing the design of speed control devices, truck driver fatigue, and the effects of the road environment on driver attention and performance illustrate applied problems tackled through the use of laboratory techniques supplemented with field observation and data collection.

The next two chapters discuss the human factors issues associated with nuclear power plant design and consumer product development; they are also updated versions from the first edition of the handbook. In the first of these chapters, the authors describe the lessons learned and revisions to the top-down HF program review model (PRM) evaluation methodology.

The evaluation of alarms, displays, controls, and decision support systems design are all based on a context of the operator tasks, system resources, and mission goals. The chapter on consumer product design shows how testing can accompany design and how the new technique of generative search can involve users in every stage of that process. Both of the approaches illustrated in these chapters are nontraditional in the sense that they move away from the evaluation associated with laboratory or field tests and move testing into the heart of the design process.

In the chapter on usability testing, we will see how the application of task analyses, user-computer interface conventions, and iterative assessments contribute to highly effective software-driver user interface designs. Here, the authors provide examples of the full range of inspection, observation, and true experimental usability testing paradigms. The reader is left with a greater appreciation of the benefits of HFTE as an essential ingredient to good usability engineering.

The chapter on command, control, communications, and information (C4I) systems provides a revealing look at the development and testing of these highly automated, highly structured, highly complex, and frequently chaotic programs. The issue of how to test the degree of human control, and even awareness, over automatic processes is a fascinating topic, of interest to nearly every HFTE professional. The balancing act between programmatic demands and human factors testing requirements described by the author underscores not only the potential contribution that testing can make but also how the needs of a system will dictate the issues to be tested.

The use of anthropometric measurement is covered in the next chapter. Here the reader is given an overview of both the theory and practice of anthropometric testing, with particular emphasis on workspace design and physical ergonomics. The chapter ends with a few brief examples of the digital ergonomics method as applied to land and sea transportation systems.

Although nearly every system test can be justifiably called unique, training systems are exceptional creatures even by that standard. The chapter describing the test and evaluation of training systems introduces the test and evaluation techniques used for each of the components of a training system: the hardware, software, courseware, procedures, and personnel. The evaluation procedures incorporate aspects of education, learning and memory, personnel selection, software interfaces, and process analysis. This chapter will introduce the reader to each of those methods, while providing a guiding hand through the potential pitfalls inherent in training systems evaluation.

Chapter 12

Human Factors Testing and Evaluation in Commercial Airplane Flight Deck Development

Alan R. Jacobsen
Brian D. Kelly
Darcy L. Hilby
Randall J. Mumaw
The Boeing Company

The goal of this chapter is to provide the reader with an understanding of effective human factors practices as applied to design, test, and evaluation of commercial air transport flight decks. As evident from earlier chapters, the application of human factors is both an art and a science because the actual application of good scientific and engineering principles often entails the art of compromise in everyday practice. The focus of this chapter is putting the right set of tools into practice and using them in the context of a real-world engineering development program; the development, design, and testing of a new airplane or a derivative of a previous airplane design.

Understanding the characteristics of the user and how that user should interact with complex systems is key to designing not only a safe system but also a cost-effective one. A well-engineered flight deck is one that has flexibility for downstream emerging functionality, is intuitive and simple, meets its intended function, and is easy to learn and operate. How one tests and evaluates the human interface for these features on modern commercial flight decks is a key part of human factors in aviation.

DESIGN ISSUES AND CONSTRAINTS

The challenge of flight deck design, as in many complex systems, is to balance between numerous dissimilar requirements, issues, constraints, and principles, many of which are difficult to quantify. For example, the pilots'

283

external vision can be affected by considerations such as crashworthy glareshield design, space for instruments on the front panel, aerodynamic shape of the nose, and structural considerations of the windows and fuselage. The instrument displays themselves, limited by panel space, must present information to the appropriate depth of detail. Novel designs, even if they simplify training, must be balanced against the benefits of common operations and existing training. Work load can be reduced by automation but possibly at the expense of situation awareness. The modern-day pilot must aviate, navigate, and communicate as well as manage complex airplane systems. Pilots must make appropriate choices when these tasks compete as well as divide tasks between themselves and the airplane systems. None of the attributes above have common metrics or criteria, but they must be appropriately balanced anyway. Figure 12.1 depicts examples of the various concepts and issues that must all be considered within the flight deck design process.

One of the less-documented design constraints for new flight decks and their systems or components is that the end product is usually constrained

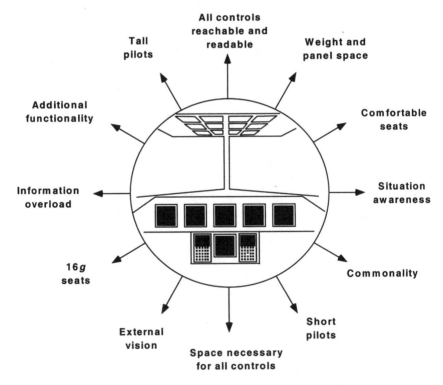

FIG. 12.1. The challenge of flight deck design is to balance between numerous dissimilar requirements, issues, constraints, and principles.

to be evolutionary, rather than revolutionary, for reasons of both safety and economy. Revolutionary changes often bring with them unintended or unknown side effects that can negatively impact safety. In addition, training costs are a significant driver in commercial aviation (Boyd & Harper, 1997), and these types of changes often require a significant increase in training time. Gradual changes tend to support more cost-effective training programs. Revolutionary changes are typically introduced only if the new functionality is worth the economic cost to the airline, in terms of hardware, software, and training.

The degree of system change also has an impact on the type of testing required of the new design. Existing in-service field experience is often used as a baseline or criteria by which to evaluate new designs. With smaller incremental changes, this testing concept is more easily accommodated. For example, when evaluating the flight crew work load of a new design, it is perfectly appropriate, as well as efficient, to judge the new design against current designs, especially if there are no radical changes. The risk of revolutionary change is that this type of evaluation would miss some unanticipated and unidentified effects. Consequently, the larger the changes, the more intensive and robust the testing program should be.

Regulations

At a very practical level, commercial air transport design is constrained by a wealth of regulatory rules and industry standards and guidelines that help define the boundaries of the human interface. The seminal documents for flight deck design in the United States are written by the Federal Aviation Administration (FAA) and are known as the Federal Aviation Regulations (FAR). These regulations are a mixture of general applicability requirements, such as intended function, and specific prescriptive implementations. Of additional significance is that the human factors requirements do not appear in any one place or section but are scattered throughout the document, indicative of the fact that human factors touch many different aspects of the flight deck. Examples of FARs pertaining to the human interface include the following:

> Each pilot compartment and its equipment must allow the minimum flight crew . . . to perform their duties without unreasonable concentration or fatigue.

> Each cockpit control must be located to provide convenient operation and to prevent confusion and inadvertent operation.

> Each instrument for use by any pilot must be plainly visible to him from his station with the minimum practicable deviation from his normal position.

All commercial transport flight deck designs must show compliance to these types of regulations; hence, they represent a significant human interface constraint. These regulations also very obviously drive the type of testing and evaluation required. Many regulatory requirements carry very subjective criteria (e.g., "convenient operation" and "easily readable"). To help designers, various means of compliance or interpretations of the FARs for specific systems or components are often defined in FAA Advisory Circulars (AC) or industry guidelines. For electronic displays, some examples of AC recommendations address the presence and perceptibility of flicker; jitter, jerkiness, or ratcheting effects; and consistent use of color coding.

It is always incumbent upon the manufacturer to demonstrate how the design meets the regulatory rules. This is usually accomplished via analysis, test, or evaluation by the certifying agency pilots. These evaluations sometimes contain measurable parameters to decide if the design meets the subjective criteria stated in the FARs. In practice, existing in-service experience is often considered during the evaluation.

GETTING FROM ANALYSIS TO REALITY

As discussed elsewhere in this book, human factors expertise is required during all phases of a development program (see Fig. 12.2). Tasks include generation of certification plans, review of acceptability of human-

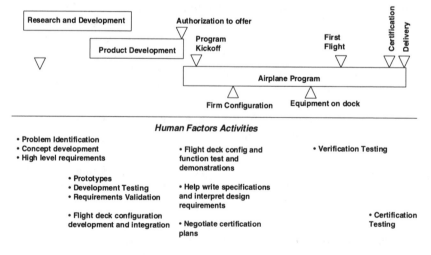

FIG. 12.2. Human factors flight deck design activities in the context of an engineering program.

machine interface during design reviews, and participation during testing stages to ensure that sufficient and necessary data are taken and analyzed appropriately. Although airplane development programs follow the typical phases of development, such as research, concept development, requirement validation, and verification, an additional phase called certification testing is always required. Significant flight testing and evaluation always occurs during this certification phase, in which the airplane, including the flight deck, is subjected to many hours of testing by regulatory agency certification pilots. This provides a comprehensive examination of appropriate human factors properties by both regulatory agency pilots and manufacturers' test pilots. It is essential at this point in the program that certification pilots fully understand the rationale, development history, flight deck design, automation, and training philosophies to appropriately judge and certify the human interface. Moreover, it is in the interest of the manufacturer that these authorities are not surprised with or asked to evaluate a system with which they are not familiar. In the press of time and resources on all parties at this point, misunderstandings can result in costly delays and design changes.

Within the context of an airplane engineering program, there is a broad range of test and evaluation tools and methods that can be employed. However, it is not always possible to employ all of these techniques all of the time. An evaluation strategy helps establish a testing plan that is consistent with the goals of a program, the degree to which it advances the state of the art, and the inevitable limitations on time and resources. In the remaining portion of this chapter we will focus on more of the details of the test and evaluation of a crew-centered flight deck design.

One of the most significant but often overlooked process tools for human factors design is the establishment of clear and concise design philosophies. In general, the design philosophies help dictate how the limitations and characteristics of the pilot are to be taken into account during the entire design and evaluation process. These philosophies also suggest *what* in the design needs evaluation and *how* potentially to test the design. Following are examples of three design philosophies of a crew-centered flight deck and how these impact testing.

First, fundamental human performance limitations must be accounted for in the design. This requires evaluating the full range of conditions that test human performance limits. For example, display format prototyping is usually carried out in rather benign office lighting conditions. Knowing that the visual system reacts differently under the lighting extremes of the flight deck environment requires that format effectiveness or usability be tested under both bright and dark ambient lighting conditions.

A second design philosophy requires that flight deck designs accommodate individual differences in pilot performance capabilities. This means that the design should be tested for usability by a population with a wide variety of skills and backgrounds. One significant issue here is cultural differences among the pilot population. These differences drive one to evaluate usability from the standpoint of pilots from a wide variety of cultures and who may or may not have the same basic precepts as the designer.

Another crew-centered design philosophy is that human-error tolerance and avoidance techniques should be applied. A wide range of methods can be used to address this most difficult issue, in accordance with the degree of novelty in the crew interface and associated risk of unexpected errors. Consideration of past practices and design, existing and future training methods, and expectations of the user population are all important. Design reviews and simulator evaluations by diverse groups are important to identify potential errors from a variety of perspectives.

These three design philosophies are top-level, and more detailed guidance is typically required to help the practitioner decide what to test and when. The following is a list of key principles in an overall flight deck design philosophy used by The Boeing Company (Bresley, 1995; Kelly, Graeber, & Fadden, 1992):

- The pilot is the final authority for the operation of the airplane.
- Both crew members are ultimately responsible for the safe conduct of the flight.
- Flight crew tasks, in order of priority, are safety, passenger comfort, and efficiency.
- Systems are designed for crew operations based on pilot's past training and operational experience.
- Systems are designed to be error tolerant.
- The hierarchy of design alternatives is simplicity, redundancy, and automation.
- Designers should apply automation as a tool to aid, not replace, the pilot.
- Fundamental human strengths, limitations, and individual differences, for both normal and nonnormal operations are addressed.
- New technologies and functional capabilities are used only when they result in clear and distinct operational or efficiency advantages and when there is no adverse effect to the crew-centered interface.

These flight deck design philosophies serve not only to develop a design consistent with good human factors principles but also to point to areas where the design needs to be tested and evaluated. These principles also serve as high-level criteria for this testing.

Tools for Flight Deck Test and Evaluation

Many of the tools used in human factors test and evaluation have been covered in previous chapters. What follows are examples of how these tools have been brought to bear on airplane flight deck design.

Computer-aided design (CAD), when used in conjunction with digital electronic anthropometric data (e.g., an electronic human model), is particularly useful when analyzing flight crew reach, body clearances, visual angles, visual occlusion, line of sight angles to displays, and external vision. An example of the effective use of CAD data and an electronic human model is in determining the lower limits of the front instrument panel. The front panel must be designed to allow shin clearance while the pilot's feet are using the rudder pedals for rudder control, braking, or both. However, panel space needs to be maximized to allow effective instrument arrangement. During the layout of the 777 flight deck, the use of a human model combined with a CAD system model of the rudder pedal system allowed the analyst to establish the pilots' shin location for a range of pilot statures. The upper edge of the shin, with some clearance room added for pilot comfort, was then used to establish the lower limit of the front panel. The use of CAD and human modeling systems allows designers to evaluate the physical integration of a flight deck before actual hardware is built and without the use of mock-ups. Some of the more sophisticated CAD modeling packages can also evaluate lighting, shadows, and, to some extent, view-ability of instruments.

Prototyping tools for display formatting, as discussed in earlier chapters, are available on the market. All of these give the human factors engineer the ability to generate new formatting concepts quickly and then either evaluate them in a stand-alone fashion or, more ideally, in a more realistic flight deck simulator environment. However, depending on a manufacturer's test schedule, nonintegrated avionics systems are often available for use in human factors evaluations in a laboratory or test bench setting. Test benches, which are used to test a system prior to installation in an airplane, do not provide a fully integrated environment but can be used to check display features. An advantage of test benches is that they are often available for use prior to any integrated testing, such as an engineering simulator or an airplane.

Flight deck mock-ups are typically used for ergonomic reach, internal vision, body clearance analyses, and, possibly, lighting and readability evaluations. The range of testing that can be performed in a mock-up is limited by the fidelity, geometry accuracy, and, in some cases, the materials used in its construction. Mock-ups can vary from the simplistic (i.e., consisting of one or more pieces of equipment constructed out of a material, such as foam-core, for example) to the comprehensive (full-size mock-up of an entire flight deck). They can include lighting and displays, but they generally do not have many moving parts (i.e., controls and switches) or flight deck effect logic. One of the advantages of using a mock-up is that they can provide a geometrically accurate test bed early in the life of a program. Decisions can be made as to the angle of certain instrument panels and preliminary reach information gathered. In summary, it is through the use of prototypes and mock-ups that many of the perceptual issues can be evaluated more thoroughly in flight deck design. Concepts that have been successfully evaluated by prototyping equipment include situation awareness, human-machine functional allocation, and decision-making strategies.

Flight deck simulators provide the most integrated and accurate representation of flight deck geometry and system performance of all the evaluation tools (except for using the flight deck of an actual airplane). Subjective evaluations in flight simulators can be performed as soon as simulators are available for the engineering program. Because the chance to perform integrated human factors evaluations in the actual flight deck of an airplane typically doesn't occur until the last stages of a design program, a simulator provides the first chance to evaluate the flight deck systems in an integrated manner. Depending on the chosen fidelity, the simulator may be used to evaluate external vision, reach, body clearance, glare, reflection, subjective response to displays, pilot comfort, and even to verify flight crew procedures. The real value of these full-mission simulators is identification of design errors earlier in the process, something that cannot be done when using only the flight test airplane. Consequently, bringing simulators online as soon as possible is an important goal for all flight deck design programs.

Lastly, the flight deck simulator is often a better device than the real airplane for testing more complex, in-situation tasks because the environment is more controllable than on the airplane. Many different types of failure modes and nonnormal situations can be run in the simulator to evaluate work load and manageability of the flight deck interfaces under both normal and nonnormal conditions.

The airplane itself, of course, represents the most realistic setting for human factors and ergonomic evaluations of the flight deck. As mentioned earlier, many evaluations can be performed using a combination of hardware or software prototypes. However, final acceptability as well as certification by the regulatory agency of the flight crew interface is most often performed on flight test airplanes. Many, if not all, of the interfaces are actually evaluated as an integral part of other flight testing because the flight crew interface is put to use on every flight of a flight test program. In addition, there are many dedicated flight crew interface tests that are required to be performed as part of the flight test program. These include basic instrument readability tests during both day and night lighting conditions.

Whatever tools are utilized, one key to ensuring that the evaluation covers the various aspects of effective performance is to develop scenarios that broadly sample the types of skills that pilots need. Evaluation scenarios for both ergonomic and cognitive evaluations need to consider the following:

- a representative set of normal tasks and procedures;
- a representative set of emergency (nonnormal) procedures;
- existing functional hazard analysis, failure-mode effects analysis, and probabilistic risk assessment scenarios that identify critical safety situations;
- minimum equipment list (MEL) conditions;
- continued safe flight and landing;
- situations in which airplane control is highly constrained;
- complex situations where task management becomes difficult; and
- other contributing factors, such as weather and ATC demands.

Testing Considerations for Evaluating Information Transfer

On the flight deck of modern commercial transport aircraft, information is presented to the flight crew through visual, tactile, and auditory modes, with the visual modality being primary. Almost all of the information that the pilot requires to aviate, navigate, and communicate is either viewed through the windshield or is presented on a host of visual displays. This requires that the flight deck-human interface designer be cognizant of the characteristics of the human visual system, including such concepts as visual acuity, color perception, light and dark adaptation, motion perception, and depth perception. Because all displays must be viewable under all known operating conditions, these displays must be visible in an illumination environment that ranges from direct sunlight to dark nighttime conditions. Many volumes have been written about human perception and sensation (Boff, Kaufman, & Thomas, 1986; Boff & Lincoln, 1988; Schiffman, 2000; Silverstein & Merrifield, 1985; Woodson, 1981) and human factors entails the practical application of this information to the modern flight deck.

One of the main design constraints affecting visual perception parameters on the flight deck is the placement of the eye reference point (ERP) for the pilots. The ERP is a point in space where the pilot is afforded, as stated in the FARs (paragraph 25.773.d), "an optimum combination of outside visibility and instrument scan." All pertinent information must be visible from this point, so the ERP is used as the pilot body location starting point for a variety of analyses. The location of the ERP impacts the geometric layout of the flight deck, including the size and placement of the windows, glare shield, main instrument panel, controls, and displays. The ERP is used when determining pilot location for reach, clearance, and strength evaluations as well as evaluations of internal and external vision. If the chosen ERP is relatively high, the down vision out the front window may be optimal; however, the effective viewable area of the front panel will be reduced because of the increased occlusion of the glare shield. If the ERP is low, the internal vision to the front panel displays may be good; however, the reach to the aft portion of the overhead panel may be impacted as well as the clear view area of the windows.

The tactile modality is important both for sending and receiving information from the pilot. Tactile feedback allows the pilot to determine switch activation without having to look at the switch. The shape of control knobs is also used to convey functionality to the pilot both visually and tactually. In fact, some control knob shapes are specified in the FARs. Sizes of switches, knobs, and other control devices is important because a large number of switches are required to be activated by the pilot in a rather confined physical geometry. These switches and knobs must be useable by a wide variety of pilots whose range of finger sizes and strength must be accommodated. Pilot strength is another critical tactile modality constraint that must be accounted for in the design of flight deck controls. Both hard stops and soft stops are used on the flight deck for rotary controls. Hard stops are fairly straightforward, and one only has to know the amount of pressure or force that can be exerted by the strongest pilot to design an adequate stop. Appropriate soft stop forces (e.g., for dual rate controls) are much more difficult to determine in that the force has to be high enough to allow the strongest pilots to be aware of its presence but not so high that it prevents the weakest pilots from moving through the stop without undue effort. These more subjective criteria are usually required to be evaluated via part-task simulations with prototype hardware devices. In performing these evaluations, it is important to include a representative sample of the user population in the test.

Another tactile constraint of flight deck design is comfort. Pilots, as well as passengers, often have to sit in their seats for extended periods of time. A real challenge is to design a seat that is comfortable for all pilots and that allows all pilots to position themselves at the ERP and yet meet 16 g crash certification criteria. Much like the automotive crash test dummies, the 16 g test involves instrumenting a crash test dummy to ensure certain head injury criteria are met. Footrests, armrests, and headrests are other examples of items that are designed to provide a measure of comfort but which must meet competing operational and regulatory requirements. Comfort testing requires that evaluators actually sit in the seat for long periods of time and identify comfort problems in a way that can direct design changes. Short stays in the seat result in misleading conclusions.

Physical (visual and tactile) ergonomics also play an important role is in the design of the flight deck layout. All pertinent displays and controls must be visible and useable by a range of pilots under all foreseeable conditions and the flight deck design must accommodate a population that ranges from 5 ft 2 in. to 6 ft 3 in., per the FARs. This represents a serious challenge to the human interface designer because the existence of stature range physically limits the location of controls and instruments. Once again, we can see that the requirements dictate that tests and evaluations of the design be accomplished with a wide range of users. By testing at just the extremes and potentially one or two other points between the limits, the design can show compliance with the requirements. This actually simplifies the test and evaluation procedures.

One of the often-overlooked information modalities on the flight deck is the auditory channel. Several classes of information are communicated aurally. Probably the most important of these are crew alerts that are time critical to the pilot. Other auditory information is that which is communicated between the airplane and the ground, between the pilots, or between the pilots and the flight attendants and passengers. Again, the flight deck represents a challenging environment for an auditory interface since the flight deck typically has a variable background noise environment, and a wide variety of native languages are found in the pilot population. The successful transmission of auditory information is therefore dependent on an adequate understanding of both the environment and the characteristics of the user population.

Testing for Situation Awareness and Workload

Some of the most significant and complex design issues on the flight deck involve cognitive processing on the part of the pilot. Safe and efficient operations rely on the flight crew's ability to maintain situation awareness (a broad concept in this setting), share control of the airplane with automated systems, make good decisions in sometimes uncertain conditions, and execute procedures appropriately for the conditions. The flight deck interface needs to support these aspects of performance, as well as keep cognitive work load and errors manageable. A design that leads to excessively high cognitive work load can prevent a crew from being successful in critical decision and management tasks.

Assessing the interface in terms of how well it supports cognition is quite different from perceptual and physical assessments in that it is more difficult to establish thresholds of performance. With cognitive performance, measures are more closely tied to the contents of a display, which requires more qualitative measures (correct behavior at the correct time) as well as an increased dependence on observations from pilots. Thus, these assessments lead back to changes in content or the organization of content within the interface.

Situation awareness has already been covered elsewhere in this book, but some specific issues regarding flight deck design may be useful at this stage. A specific issue within modern flight decks is the flight crew's awareness of the state of the automation that is used to control various aspects of aviation and navigation. In particular, there are concerns that pilots may not be monitoring the appropriate indications to maintain an awareness of changes to the state of the automation. In recent studies, eye-fixation data have been used to assess how line pilots are monitoring the indications as they fly (Endsley, 1995). These measures are linked to scenario performance measures and assessments of each pilot's understanding of how the automation state changes to determine how indications might be improved.

Another significant concept in aviation is workload. The criticality of workload has been significantly reduced with today's third-generation jet transports, especially when the principle of evolutionary design change is adhered to. However, all designs must ensure that at any given phase of flight, the pilot is neither overworked, which can cause significant errors to be made, nor underworked, which can result in the pilot becoming complacent and therefore cause him or her to miss key cues or alerts.

Assessing cognitive workload requires measuring the degree to which the pilot's mental resources are being used. Assessment of crew workload can be made for a single airplane model or can be made using a comparison of one airplane model to a previous one for which the workload is known to be acceptable. The former method requires specific tests in a controlled environment (e.g., simulator or airplane), whereas the latter method is adaptable to being performed throughout a flight test program, even without specific test conditions, if the evaluating pilots are familiar with the work load characteristics of the comparison airplane. Analytical efforts to model workload to date have been extremely labor intensive. They are usually driven by physical rather than cognitive occurrences, and the results are dependent on many choices made by the analyst when setting up the model.

REAL WORLD TESTING EXAMPLES

Following are examples of testing that have been carried out during the development and certification of major commercial airplane programs at The Boeing Company. These examples are meant to help the reader see how test and evaluation fits into and is constrained by the realities of a major engineering program.

Determination of Reach

As mentioned earlier in this chapter, flight decks are required by the FAA to accommodate pilots whose statures range from 5 ft 2 in. to 6 ft 3 in. Body clearance and reach are the two main concerns when evaluating a flight deck layout for physical fit. Although the 777 flight deck design was based on the layout of the 767 flight deck, there were enough changes (e.g., front instrument panel angle, overhead panel angles, and rudder pedal angles) to warrant evaluations of reach and body clearances both of individual panel concepts and of the integrated flight deck.

In addition to CAD-based human model evaluations (some of which were briefly mentioned earlier), a foam core mock-up was used early in the program to evaluate the reach potential of 5 ft 2 in. subjects to the new overhead panel. Reach, body clearance, and force production testing of individual design concepts (e.g., for the new rudder pedal system design and the new rudder trim control knob) were also performed early in available engineering simulators. When the 777 engineering simulators came online, evaluation of fit for both 5 ft 2 in. and 6 ft 3 in. subjects was made possible. Once the simulator was determined to be geometrically accurate, final acceptability and certification of the 777 flight deck to the applicable FARs was determined through front panel, overhead panel, and aisle stand reach engineering simulator tests using 5 ft 2 in. subjects, and body clearance tests using 6 ft 3 in. subjects.

Development of the 777 Flight Deck Window Crank

Early in the development cycle of the 777 flight deck, it was recognized that the window crank, used to open and close the flight deck window, needed to be moved to a new location. The initial location, based on the 767 flight deck layout, would have caused an interference with the planned location of new side displays. Because of structural constraints, flight deck design engineers proposed a new location slightly aft of the pilot's position when seated at the ERP. The proposed location was modeled in a CAD system by the design engineers, and ergonomics practitioners used a CAD-based human modeling system to generate figures representing the required pilot stature range (developed from certification authority and manufacturer requirements). Thus, it was possible to conduct a preliminary evaluation of the location for reach acceptability. It was also recognized that the shape of the window crank knob needed to be revised—the existing spherical shape would not lend itself to an efficient grip in the new location. CAD models, in combination with anthropometric hand and finger data, were used to design a new, flatter knob.

Although preliminary analyses were performed using the CAD and human modeling systems, it could not be determined by CAD-based analysis alone if the new location would meet the certification authority requirement that the window, if used as an emergency exit, be opened in 10 seconds or less. The pilot's potential force development, based on body position with a hand on the crank, could not be accurately determined by using only the existing electronic analysis tools. Because a geometrically accurate simulator was not online at the time, evaluations of opening time and maximum allowable rotation force were performed in a mock-up of the flight deck structure. This evaluation setting allowed the subjects to assume body positions that approximated those that would be seen in the actual flight deck. Subjects were asked to rotate the mock-up window crank a specified number of rotations as fast as possible. Cranking force was varied, and the resultant times recorded. Participants were chosen to represent not only the range of pilot statures but also the potential ranges of arm strength. It was determined that the proposed window crank location and predicted maximum cranking force were sufficiently acceptable to proceed with the design.

The completion of the engineering flight simulator and, later, the actual airplane made subsequent evaluations of the actual window crank design possible. Because the simulator was constructed using parts that were as close as possible to the current design, engineers were able to discern a knuckle clearance problem that was due to an anomaly in the CAD model. Detection of the problem happened early enough that a fix was designed, built, and put in place before any airplanes were delivered to customers. Evaluations in the airplane revealed that although the window crank location was acceptable, the actual system forces required to open and close the window throughout its full range of motion were too high. Of concern were the pilot's comfort when opening and closing the window and the pilot's ability to close the window in flight. The ability to open the window in flight is used as a means of compliance for a certification authority requirement that an alternate means of forward vision be provided, should the front window become unusable. An evaluation was performed in the engineering flight simulator to determine an acceptable force profile that accounted for the window breakout and closing forces as well as the dynamic forces throughout the window range of travel. The results were then used to modify the design of the window mechanism. The modified design was tested in the airplane and found to be acceptable and certifiable under actual flight conditions.

Attitude Display Symbols and Guidance for Windshear Escape Maneuver

In the mid-1980s an industry task force applied considerable resources toward the problem of low-altitude windshear accidents. The effort was funded in part by the FAA for development of a new windshear training aid. The training aid contained new facts and data that established an industry standard flight crew procedure for escape from an inadvertent encounter with severe windshear on takeoff and landing. Key elements of the new procedure were rapid yet precise increases in pitch attitude combined with simultaneous awareness of vertical speed performance and the margin to stall warning.

Stall warning is predominantly a function of angle of attack, which was not displayed on commercial flight decks in any way at that time. Vertical speed was displayed on a dedicated instrument. Debates ensued as to the efficacy of new, dedicated angle of attack indications in the flight deck. However, a concern was the potential difficulty for a pilot in scanning pitch attitude, vertical speed, and a new angle of attack indicator and carrying out the emergency maneuver. To address this, several competing concepts were evaluated. A pitch limit indicator, driven by angle of attack, was developed to indicate margin to stall superimposed on the attitude display. The display of flight path angle on the attitude display was felt to be a valuable substitute for vertical speed, and flight director indications that could compute and provide pitch attitude commands was yet another approach.

The purpose of the testing was to provide an operational (i.e., piloted) comparative evaluation of several display/procedure concepts. The results would be used to assist in company decisions on how best to equip airplanes in the future. The specific objectives of the test were to answer the following questions:

- Which of these display methods or combinations of them best supported the crew in accomplishing the procedure?
- Which displays would be easiest to train the pilots to use?
- Which displays were easiest to use?
- Which displays yielded the best windshear escape performance in terms of terrain avoidance?
- Which displays enabled most effective use in flying near stall warning?
- Which provided best stall protection?

For the test, several windshear scenarios and representative displays were developed and implemented in a motion base flight simulator. The test design called for a number of pilots from both flight test and flight training backgrounds to evaluate each display concept and several combinations of the concepts. Flight parameter data were collected consistent with many of the test objectives, and a questionnaire as well as a rating system were also developed. Pilot briefing materials were developed, and test procedures were developed and practiced before the actual test began. For each display, a tailored crew procedure was developed to be consistent with both the display attributes and the recommendations of the windshear training aid.

The results of this test demonstrated that the engineering data were insufficient to clearly distinguish advantages of the various display concepts. The test results did, however, show the greatest consistency of response using the computed commands displayed on the flight director pitch command. All of the displays and pilots performed reasonably well in all of the windshear scenarios. Interestingly, pilot preferences for the displays and display combinations increased steadily as information was added until the sixth combination, where all of the elements were present, when the preference ratings became suddenly bipolar. Half of the pilots rated this display at the bottom of the scale, and the other half rated it at the top. This turned out to correlate with their backgrounds, with flight crew training pilots rating this most complex display low, and flight test pilots rating it high. Combining this fact with many of the comments from the posttest questionnaire and interview process led to conclusions related to the limits of what could be expected in airline flight crew training environments. This, combined with the least consistent flying among pilots, resulted in the elimination of flight path angle display as a primary means to fly the windshear escape maneuver.

Although the flight director pitch command provided the best consistency, with all combinations being adequate in terms of performance, the pilots' observations also revealed that the pitch limit indicator was of significant value in reducing the startle effect of stall warning, which would usually occur in turbulent windshear encounters. Even though, in theory, the flight director provided all the command information a pilot needed to know, simultaneous knowledge of the stall margin enabled the pilots to trust the unusual commands experienced in windshear and enabled them to avoid overreacting to stall warning with abrupt pitch down commands that would cause altitude loss. The final configuration chosen for production was an improved flight director providing commands appropriate during a windshear encounter, augmented by the display of the pitch limit indicator.

Development and Introduction of a New Display Technology

The introduction of large format, high-resolution flat panel liquid crystal displays (LCD) onto commercial transport flight decks was accomplished during the 777 program. The FARs require that all flight deck instruments meet their intended function under all foreseeable conditions with a minimum of distraction. This had been accomplished with previous technologies, such as CRTs and electromechanical devices. However, many of the perceptual parameters of the LCDs, such as pixel definition, color, contrast, and field of view characteristics, were significantly different from previous display technologies. Yet the benefits of LCDs as a flight instrument were such that it was decided early to make a heavy investment into a test and evaluation program to ensure that they could be used on the 777 flight deck.

Several display parameters were determined to be of sufficiently high risk, that modeling and analysis alone were deemed to be insufficient to evaluate the technology. For example, basic readability using a matrix-addressed surface and field of view characteristics of LCD technology were seen as high risk for LCDs used as flight deck displays. Consequently, Boeing and several of its suppliers developed various prototype systems and devices to investigate and evaluate readability parameters, such as resolution and pixel structure, for these new displays (Jacobsen, 1990, 1995, 1996; Silverstein, Monty, Gomer, & Yeh, 1989; Silverstein, Monty, Huff, & Frost, 1987). Based on these early prototype evaluations, basic requirements for LCDs used as flight instruments were developed. These data were used both to guide and evaluate interim designs of this emerging technology. The display hardware engineers also used the data to look at the trades between parameters such as resolution, antialiasing, and pixel structure configurations, speeding the development of acceptable solutions. Although the prototyping tool was valuable in early test and development, many of the final formatting issues, such as symbol attributes, could not be decided and tested until actual hardware became available (Wiedemann & Trujillo, 1993). Final testing of these attributes was carried out in the flight deck simulator as well as during flight test. Because of the extensive use of modeling and early testing on prototypes, few, if any, problems were found during these late program tests.

To investigate the boundaries of the field of view issue on these displays, the CAD system was used to define the angles at which the pilots would see the displays on the flight deck. Knowing the characteristics of the current technology at the time, as well as what improvements would likely be made, the prototyping tool was again used to evaluate the field of view characteristics of LCDs, such as color and contrast changes. This evaluation tool proved to be invaluable for generating requirements, directing the development of the displays, and evaluating interim solutions (Jacobsen & Ketterling, 1989).

In the area of display brightness, LCDs also posed an interesting testing challenge. Because the perception of brightness is governed by a variety of display parameters that drive the cost of displays, significant modeling and prototyping effort was brought to bear on this issue as well. Because of the difficulty in simulating the high brightness conditions of actual daytime flight conditions, a display readability issue was uncovered during the flight test program that had not been identified earlier with the analysis and prototyping tools. During conditions when the sun was overhead and ahead of the airplane, the displays were difficult to read. Again, utilizing the flight deck simulator and the prototyping tools, several alternative solutions were generated. These included boosting luminance as well as creating symbology with thicker lines, both of which increase perceived brightness. While initial testing of these concepts was carried out in the simulator, final validation was carried out during dedicated flight tests.

SUMMARY

Our examples have shown that human factors testing and evaluation, even in airplane design, is as much an art as it is a science. Although many of the tools and methods are based on strong, sound scientific principles, others tend to be more subjective, and all are prone to errors of misinterpretation. Furthermore, even the most objective and scientifically sound tool requires some testing art when deciding when and how it should be used. In addition, the availability of adequate tools alone is insufficient to guarantee a successful test and evaluation program. Numerous processes and organizational foundations must be in place to ensure that the appropriate tools and methods are used in a timely fashion to support engineering development programs.

In the preceding sections we have illustrated that the organization must create a clear priority and goal of creating a crew-centered design. To do this means the appropriate tools and people will be put into the plans and schedules of the program at the front end. We have also pointed out the importance of clearly communicated design philosophies and documentation. Understanding the user's task, which in the case of flying means control, navigation, surveillance, and communication, is also necessary from both a design as well as a test and evaluation standpoint.

The art of deciding which tests to employ at which phase of the program is made more consistent when one makes a more concerted effort to match testing tools with test objectives. This requires an understanding of what types of questions need to be answered during the various phases of an engineering program. This will help ensure an effective and efficient test and evaluation program is put in place. Finally, there is often no one test or method that can cover all aspects of human factors testing. Therefore, the judicious use of multiple tactics and methods is paramount for carrying out a successful test and evaluation program.

REFERENCES

Boff, K. R., Kaufman, L., & Thomas, J. P. (1986). *Handbook of perception and human performance*. New York: Wiley.

Boff, K. R., & Lincoln, J. E. (1988). *Engineering data compendium: Human perception and performance*. Dayton, OH: Armstrong Aerospace Medical Research Laboratory, Wright-Patterson Air Force Base.

Boyd, S. P., & Harper, P. M. (1997). Human factors implications for mixed fleet operations. In *Ninth International Symposium on Aviation Psychology* (pp. 831–836). Columbus, OH: Ohio State University.

Bresley, W. (1995, April–June). 777 Flight Deck Design. *Airliner* (pp. 1–9). Seattle, WA: The Boeing Company.

Code of Federal Regulations, Title 14. *Aeronautics and Space*, Chapter 1 Federal Aviation Administration, Department of Transportation, *Part 25 Airworthiness standards: Transport category airplanes*. (2000), Government Printing Office, Washington, DC.

Endsley, M. R. (1995). Measurement of situation awareness in dynamic systems. *Human Factors*, *37*(1), 65–84.

Jacobsen, A. (1990). Determination of the optimum gray-scale luminance ramp function for anti-aliasing. *Proceedings of the SPIE International Society for Optical Engineering Human Vision and Electronics Imaging, Models, Methods, and Applications*, 202–213.

Jacobsen, A. (1995, July–September). Flat panel displays. *Airliner* (pp. 15–21). Seattle, WA: The Boeing Company.

Jacobsen, A. (1996, March–April). Flat screens and the active matrix. *World Engineering* (pp. 38–42). Hounslow, Middlesex, UK: British Airways.

Jacobsen, A. R., & Ketterling, J. A. (1989). Determining acceptable backgrounds for color LCDs. *Society for Information Display Digest of Technical Papers*, *20*, 288–291.

Kelly, B. D., Graeber, R. C., & Fadden, D. M. (1992). Applying crew-centered concepts to flight deck technology: The Boeing 777. *Flight Safety Foundation 45th IASS and IFA 22nd International Conference*, 384–395.

Schiffman, H. R. (2000). *Sensation and perception*. New York: John Wiley.

Silverstein L. D., & Merrifield, R. M. (1985). *The development and evaluation of color display systems for airborne applications* (Technical Report: DOT/FAA/PM-85-19). Washington, DC: Department of Transportation.

Silverstein, L. D., Monty, R. W., Gomer, F. E., & Yeh, Y. Y. (1989). A psychophysical evaluation of pixel mosaics and gray-scale requirements for color matrix displays. *Society for Information Display Digest of Technical Papers*, *20*, 128–131.

Silverstein, L. D., Monty, R. W., Huff, J. W., & Frost, K. L. (1987). Image quality and visual simulation of color matrix displays. *Proceedings of the Society of Automotive Engineers Aerospace Technology Conference* (Paper No. 871789).

Wiedemann, J., & Trujillo, E. J. (1993). Primary flight displays conversion: 747-400 CRTs to 777 LCDs. *Society for Information Display Digest of Technical Papers*, *24*, 514–517.

Woodson, W. (1981). *Human factors design handbook*. New York: McGraw-Hill.

Chapter 13

Maritime Applications of Human Factors Test and Evaluation

Anita M. Rothblum
Antonio B. Carvalhais
U.S. Coast Guard

One of the missions of the United States Coast Guard (USCG) is to maintain the safety and security of our nation's ports and waterways. This mission involves many different operations, including such things as licensing merchant mariners to ensure that they are qualified to operate merchant ships, inspecting commercial vessels to ensure that they meet minimum safety standards, regulating the maximum number of daily work hours for crew members to prevent excessive fatigue, operating Vessel Traffic Services in major ports to facilitate navigation in congested waterways, and investigating maritime accidents to determine the causes and to institute preventive measures. All of these operations are united by a common goal: to decrease accidents on the water. Because 75% to 96% of marine casualties (accidents) have human-related causes (U.S. Coast Guard, 1995), human factors research is essential to achieve this goal.

The USCG has an active human factors research program at its Research and Development Center in Groton, Connecticut. Human factors methods and knowledge are being applied to solve problems and increase safety in both the commercial shipping industry and in USCG shipboard and shoreside operations. Because of the diversity of USCG operations, a variety of human factors methods are routinely employed in our research projects. These methods are used to identify human factors problems, to target design and training recommendations, and to perform operational test and evaluation of new equipment, such as computer systems, small boats, and cutters.

We often need to modify traditional HFTE methods to make them compatible with our test environment. For example, some of our marine safety projects evaluate aspects of the commercial maritime system. The commercial maritime system is very complex: It is composed of commercial shipping companies (which supply the ships, cargoes, and management policies), labor unions (which typically supply the crews), U.S. and international laws and regulations, Coast Guard activities (e.g., inspection, investigation, Vessel Traffic Service), port and waterway design, and environmental variables (e.g., winds, currents, rough seas, extreme temperatures, fog). The maritime system is fraught with human factors problems (Perrow, 1984; Sanquist, Lee, Mandler, & Rothblum, 1993). Only a portion of this system is under the direct control of the USCG, yet through USCG-sponsored research it is possible to establish ways that government and industry can improve the safety and efficiency of the system. Often this type of test and evaluation cannot be carried out in a controlled laboratory or simulator environment, or even in a scripted operational test scenario: It is necessary to use real operational environments, such as commercial vessels at sea, as our laboratories. The participants in these studies are primarily concerned with performing their duties (e.g., navigating the ship, operating and maintaining the ship's engines, loading cargo, performing a law enforcement or search and rescue mission); the fact that they are participating in a research project is secondary at best. Therefore, we are challenged to develop test and evaluation methods that are compatible with the real-world operational environment and limitations.

This chapter focuses on four projects that illustrate some of the methods we use for human factors test and evaluation. The first two studies cover the OT&E of prototypes of two Coast Guard boats. OT&E is an important step in the design and development process: Testing the prototype allows us to identify and correct problems before the vessel goes into production. In the NORCREW project, the HFTE emphasis is evaluation of novel living conditions aboard the boat, whereas in the 47-ft motor lifeboat (MLB) project, the emphasis is on vessel design related to safety and effectiveness.

The final two studies illustrate methods that can be used to make automated systems more human centered. Identifying the impact of automation on shipboard tasks and training is the focus of the third study, which uses task and cognitive analyses, skill assessment, and error analysis. The final project is an evaluation of a casualty (accident) database and incorporates structured surveys, observation of work activities, and a formal human-computer interface evaluation.

APPLICATION OF HFTE TECHNIQUES
TO MARITIME SYSTEMS

Because marine systems are complex, often operate in isolated and hostile environments, and must be self-sufficient, it is critical to be able to assess the potential impact of these systems on operator performance and safety prior to deployment. Once the initial ship design and development have been completed, and a working prototype becomes available, OT&E methods are employed to focus attention on the performance of real operators using real equipment. The following section demonstrates the use of various OT&E methods to address human-related needs in marine system design. The intent of this section is not a detailed description of the methods—some of the methods are already described in detail in other chapters of this book—but to demonstrate the use of these methods in an operational setting.

The NORCREW Live-Aboard Concept

The USCG currently operates and maintains approximately 186 small-boat stations throughout the United States. The personnel who operate out of these stations provide rapid response to such mission needs as search and rescue, pollution response, aids to navigation, enforcement of laws and treaties, boating safety, and port safety and security. These facilities are costly to staff and maintain. The USCG is evaluating whether to re-furbish and continue to maintain all existing small-boat stations or investigate alternatives to some stations. One alternative that has been suggested is a live-aboard boat concept modeled after the Norwegian Society for Sea Rescue. Under the live-aboard concept (NORCREW), shoreside facilities are eliminated and replaced with a live-aboard boat. These boats are equipped with both mission-related equipment and facilities (e.g., berthing, shower and toilet, food storage and preparation equipment, recreational amenities) to accommodate crew members for extended periods of time.

An OT&E was conducted to assess the feasibility of implementing the NORCREW concept in certain small-boat station environments (Tepas & Carvalhais, 1994). The objective of the evaluation was to assess whether crews could adapt and cope with extended continuous operations from the confined NORCREW working and living environment. The primary concern was that the restricted environment would create excessive stress and compromise performance and well-being.

Methods. A number of T&E methods are available to capture crew re-actions to work and living environments. These methods include ques-tionnaires, daily logs, interviews, focus groups, and checklists, just to name a few. The questionnaire and daily logs were selected because they can collect crew reactions on an extended and prolonged basis with minimal on-site support and can exert minimal interference with operational requirements. The questionnaire is ideal for collecting information on demographics, work characteristics, personal behaviors and preferences, sleep-wake cycles, and lifestyle issues. In addition, questionnaires can be used to capture global reactions toward work and well-being and establish a profile of the respondent. Although the questionnaire provides a cost-effective means of collecting general or global information, and can assess more chronic effects, this method is too coarse to assess acute effects.

The daily log, which was administered 3 times per day, both on- and off-duty, was used to assess acute effects related to NORCREW. The self-administered daily log was selected because it offers a cost-effective means of maintaining continuous feedback on subject and measurement tool performance. By having respondents mail logs on a daily basis, one can detect whether crews are experiencing difficulties that may require intervention; ensure that the logs are being completed correctly and on a daily basis; reduce the possibility of crew members referring back to previous days' ratings; and reduce the data entry and analysis burden at the completion of data collection.

One of the major reasons for using daily logs was to assess time-of-day effects. To accomplish this, respondents were requested to complete logs at specific times of the day that corresponded to major daily events: upon waking up (before breakfast), before dinner, and at bed time. Daily events were used instead of specific times of day to lessen the burden of having to maintain a time-of-day check, reduce interference with randomly occurring operational requirements, achieve greater time-of-day consistency in the data because these events have a fairly robust schedule and the activity would provide a cue to subjects to complete their logs. Also, logs were completed before meals to minimize any food intake effects.

Prior to full implementation of the T&E methods, a pilot test was conducted to assess the acceptability, feasibility, and sensitivity of the methods and variables. The results identified some minor changes to the methods. The actual implementation occurred at two USCG small-boat stations operating in the same geographic region and exposed to similar environmental conditions, mission profiles, and mission workload levels. At one station, shore-based facilities were replaced with a live-aboard boat. A traditional small-boat station with shore-based facilities was used as the comparison station. The personnel at the comparison site matched the test group in terms of age, gender, Coast Guard experience, and rank/rate.

Summary of Results. Analysis of questionnaire and daily log data typically emphasize percentages and measures of central tendency. These indices provide the basis for comparison to other populations or to historical data. An important measure in questionnaire and log data is the response rate. This measure reveals the percentage of the population that responded to the data collection. The higher the response rate, the greater the statistical confidence that the data provide a representative sample of the population. Response rates as low as 30% are common and represent successful data collection efforts. For this effort, the questionnaire and log forms achieved a 100% and 70.25% response rate, respectively. The face-to-face delivery of the questionnaire ensured a high response rate, and the daily mailings of logs may have motivated individuals to complete the forms because they knew their actions were being monitored. Also, periodic feedback to data collection sites ensured them that their data were being processed.

Because elements of our questionnaire and log forms had been used in previous efforts (Paley & Tepas, 1994), we conducted comparisons to assess the reliability and sensitivity of our measures. For example, a typical finding is that sleepiness increases as a function of time of day, being lowest in the morning and greatest at night. This is a general finding, which can vary depending on individual characteristics, such as sleep-wake cycles and circadian rhythmicity. However, time-of-day effects are robust and can be used to assess the sensitivity of the measurement tools used in this study by comparing our data with that of previous research. These comparisons revealed similar time-of-day effects, suggesting that our data collection methodology was practical and yielded meaningful data.

Comparisons between the NORCREW concept and the traditional station did not indicate any of the differences that might be expected if the NORCREW concept was exerting any adverse effects on crew members. This evaluation did not reveal any human factors problems that would prohibit the continuation of the NORCREW concept at the present test site. However, it is important to realize that questionnaire and log form data are qualitative impressions and may not reflect the reactions of crews in other Coast Guard small-boat environments, especially if there are significant differences in geography, operations, and workload compared with the test site.

The 47-Ft Motor Lifeboat (MLB)

United States Coast Guard MLBs operate in some of the most challenging rough weather environments in the world. To meet these challenges, the Coast Guard has been evaluating a new 47-ft MLB to replace the aging fleet of 44-ft MLBs, which have been the service's primary heavy-weather rescue craft for the past 29 years. The new 47-ft MLB has been designed to operate in 20-ft breaking seas, to have self-righting capability (if accidentally capsized), to be operated for over 9 hours at a sustained top speed in excess of 25 kt, to have an operating range of 200 NM, and to have a crew complement of four people. The 47-ft MLB also has many enhanced features not included in the 44-ft MLB: increased speed, different motion characteristics, and an enclosed bridge to protect crew members from harsh environments.

In Coast Guard small-boat acquisitions, the manufacturer typically delivers a prototype vessel for initial acceptance testing. Design problems that are identified result in changes to subsequent preproduction models. With the 47-ft MLB, two significant problems became immediately evident during the prototype acceptance trials. First, certain steering maneuvers produced unrecoverable boat deviations that could result in capsizing. Second, crew performance was hampered by suboptimal instrument and control layouts, poor work space configuration, many design problems, and other human factors engineering considerations.

An OT&E was requested to conduct human factors engineering assessments of ergonomic and safety deficiencies and to assess the suitability and effectiveness of the vessel to perform Coast Guard duties. The objective of this evaluation was to ensure that the new boat met or exceeded Coast Guard specifications and needs and was operationally suitable and effective for performing various mission-related duties. To this end, two assessments were conducted to (1) identify and correct human factors deficiencies that could jeopardize crew safety and efficiency; and (2) collect crew impressions on the suitability and effectiveness of the 47-ft MLB to perform mission duties.

Methods.   For the first assessment, an OT&E protocol developed for seagoing vehicles (Malone & Shenk, 1977; Malone, Shenk, & Moroney, 1976) was used to guide evaluations of specific human factors problem areas on the prototype vessel. In addition, the American Society for Testing Materials (ASTM) 1166, *Standard Practice for Human Engineering Design for Marine Systems, Equipment and Facilities*, was used to identify areas, and provide evaluation criteria, for our assessments. The assessments concentrated on item design features, operation, and use conditions that impact human performance and safety. Checklists were prepared to guide and structure data collection. Each checklist applied seven human factors considerations to one or a set of equipment components found on the vessel (see Table 13.1).

Reviews of design and layout plans, as well as walk-throughs, were used to evaluate how well the vessel conformed to anthropometric standards and human factors guidelines to accommodate human psychophysiological tolerance limits. Some of the areas of concern included instrument/display design, visibility, and layout; controls operability; crew communications; safety and rescue equipment; hatchway design and op-

TABLE 13.1
Human Factors Considerations and Equipment Components

Seven Human Factors Considerations	Equipment Components Checklists
(1) Location/arrangement	(1) Steps, platforms, railings
(2) Size/shape	(2) Doors, hatches, passageways
(3) Direction/force	(3) Controls
(4) Information	(4) Displays
(5) Visibility	(5) Work space
(6) Use conditions	(6) Labels, manuals, markings
(7) Safety	(7) Lines, hoses, cables
	(8) Fasteners, connectors
	(9) Accesses, covers, caps

eration; stair and handrail dimensions; deck surfaces; jutting corners and equipment; reaching and lifting extremes; and obstructed views and reach; The checklists provided a formal and standardized process for conducting the reviews. During the walk-throughs, researchers inspected the vessel and accompanied experienced crew members on actual missions, documenting operational conditions and user activities for each item under consideration. Static and dynamic tests of design features were conducted, paying particular attention to the factors in Table 13.1. Interactions with crew members were recorded as they described the tasks they performed and potential problems. Video and photos were taken for later analysis and to illustrate findings. To address specific concerns of suboptimal instrument and control layouts, full-scale mock-ups of the open and enclosed control bridges were constructed and used for examining alternative arrangements of equipment. Priority schemes, based on the critical nature of task, level of use, function, and so on, were developed to generate control and instrument configuration strategies. Several alternative console layouts were mocked up and tested. This assessment was conducted on the prototype, and design changes were submitted for incorporation into the preproduction vessels (Holcombe & Webb, 1991). This assessment was then replicated, with the exception of the mock-ups, on one of the preproduction vessels to ensure changes were incorporated and benefits were realized (Holcombe, Street, Biggerstaff, & Carvalhais, 1994).

The second assessment, to collect crew impressions of the suitability and effectiveness, was conducted on all five preproduction vessels. The boats were stationed at different Coast Guard facilities to assess their performance under a variety of environmental and operational conditions. Surveys and small group discussions were used to collect crew impressions to specific operational functions that were critical to mission performance. To generate the list of critical operational functions, small group discussions were held with subject-matter experts, producing 23 operational focal points (OFP) that were critical for effective mission performance. Examples include surf operations, towing, personnel recovery, and alongside operations. A survey was developed to compare crew ratings of the 47-ft MLB to the 44-ft MLB on the OFPs. However, such a comparison only provides a relative measure of the effectiveness of the new resource. To assess how close the new MLB was to being optimal, an absolute measure was obtained by comparing the 47-ft MLB to an ideal MLB vessel. The description of the ideal MLB was individually specified. Crew members were instructed to think of the ideal MLB as the vessel with the optimal design, layout, equipment, and so forth to perform their duties.

In the absolute comparison, crew members were asked to evaluate, using a 100-point scale, how well the 47-ft MLB does compared to the ideal MLB on each of the OFPs. In the relative comparison, crew members were asked to directly estimate the relative effectiveness and suitability of the 44-ft and 47-ft MLBs with regard to each of the OFPs (Bittner, Kinghorn, & Carvalhais, 1994). These types of ratings represent a direct estimation method that has elsewhere proven valuable for rapid evaluation of responses to physical and other aspects of systems (e.g., Morrissey, Bittner, & Arcangeli, 1990; Stevens, 1975). Following the survey, small group discussions (four to seven people) were held to allow crew members to ask questions and provide additional or more detailed information on survey issues, as well as on issues not addressed in the survey. Also, the discussion gave the researchers an opportunity get clarification on issues, request recommendations for deficiencies that were identified, or both. The small group discussion can be a rich source of information especially as a follow-up to structured data collection methods.

Summary of Results. The success of the walk-through methodology was facilitated by a policy established for this OT&E to encourage and motivate participating crew members to submit recommendations that would improve the safety and productivity of the 47-ft MLB. Special lines of communication were established to receive and process recommendations, periodic committee meetings were held to review and approve or reject ideas, and final decision results were returned to the original source of the idea. This process encouraged crew members to be very active in submitting ideas. The success of this policy and our efforts can be best expressed by the fact that approximately 150 engineering change proposals were submitted for consideration, and most were accepted by the review committee for inclusion in the design of the final production vessel.

The methods used in the suitability and effectiveness evaluation were valuable in providing an estimate of not only how much the utility of the new boat increased from the previous one but also how close the new design approached the optimal. The data revealed that the 47-ft MLB ratings all approached the ideal category and were consistently superior to the 44-ft MLB. However, some deficiencies were identified that required redesign to improve the effectiveness and suitability of the vessel.

APPLICATION OF HFTE TO MARITIME AUTOMATION

The last two studies are field studies of automated systems in use by commercial mariners and the Coast Guard. Besides the traditional application of human factors standards and techniques to determine design and performance problems, T&E methods were used to identify how training would need to be designed to prepare mariners for the transition from manual to automated operation. The studies demonstrate how a wide variety of T&E tools may be applied to facilitate a thorough examination of human factors problems.

The Impact of Automation on Shipboard Tasks and Mariner Qualifications

As recently as the 1960s, merchant vessel tasks were performed manually in much the same way as they had been for hundreds of years. But in the late 1960s, automated systems began to appear aboard ships, first in the engine room with the advent of periodically unattended diesel engines and later on the deck with constant-tension winches and radar (National Research Council, 1990). Today's ships boast completely unattended engine rooms connected by computerized alarm systems to the bridge; microcomputers for accounting, general record keeping, and e-mail to land-based operations; automated satellite positioning systems (e.g., the global positioning system, or GPS); and navigation and collision-avoidance systems, such as electronic charts (ECDIS) and automated radar plotting aids (ARPA). With this boom in technology comes the concern that not all mariners understand how to use the automation effectively and safely. Indeed, there have been several automation-assisted accidents in recent years in which otherwise seasoned mariners either didn't know how to use the automated system or had trouble using it because of poor system design. To help the USCG's training and licensing requirements keep pace with automation, we began to develop methods that would elucidate the ways in which a given automated system affected the performance of shipboard tasks and the knowledge and skills needed by the crew to use the system.

Methods. Four methodologies were developed or adapted for maritime use. These are the operator function model (OFM) type of task analysis, cognitive task analysis, skill assessment, and error analysis (Sanquist,

Lee, McCallum, & Rothblum, 1996; Sanquist, Lee, & Rothblum, 1994). OFM task analysis, developed by Mitchell and Miller (1986), provides a breakdown of a function (such as avoiding collisions with neighboring vessels) into the tasks that must be performed, the information needed to perform each task, and the decisions that direct the sequence of tasks. This type of task description is independent of the automation; that is, the same tasks, information, and decisions are required, regardless of whether they are performed by a human or by a machine. For example, in collision avoidance, other vessels must be detected, their relative motions analyzed to determine whether there is a threat of collision, and a decision made regarding how a mariner should change own ship's (his or her vessel's) course or speed to avoid a potential collision. These tasks must be performed, regardless of who (human) or what (machine) does them.

The cognitive task analysis method extends the OFM by considering the mental demands that would be placed on a human operator while performing tasks. For example, for a human to detect a new ship as soon as it appears, vigilance (i.e., sustained attention) and discrimination (i.e., the ability to spot a target against the background) are required. The mental demands of analyzing the relative motion of the target vessel include plotting a series of target ranges (i.e., distance) and bearings (i.e., its angular position relative to own ship) and evaluating the ratio of change over time. Miller's (1971) task transaction vocabulary was used to categorize mental demands. Cognitive tasks are encoded using the 25 terms in Miller's vocabulary, such as *search, detect, code, interpret,* and *decide/select.* Different levels of automation can be represented by assigning the appropriate OFM tasks to humans or machines. Then the cognitive impact of automation can be identified by comparing the number and types of cognitive demands placed on the human operator under the different levels of automation. For example, when collision avoidance by manual methods was compared to the use of ARPA, it was found that virtually all of the computational demands of the manual method had been eliminated through automation.

To evaluate the impact of automation on training requirements, a skill assessment technique was developed by marrying the OFM and cognitive task analyses with the knowledge, skills, and abilities (KSA) analysis typically used in the development of training courses. This hybrid analysis focuses the knowledge and skill assessment at the task level, which allows us to distinguish changes in skill level as a result of automation. The skill assessment was performed by taking each cognitive task (from the OFM-cognitive task analysis) and determining what types of knowledge and skills were required for the proper performance of the task. For example, in collision avoidance the manual task of plotting target range and bearing was replaced in the automated scenario by knowledge of how to display target information on the ARPA. Although the basic knowledge requirements for collision avoidance do not change with automation, the procedural requirements change radically. That is, the mariner has to understand the theory behind collision avoidance, regardless of the level of automation, but the specific set of procedural knowledge and skills the mariner needs is dependent on the level and type of automation.

The way an automated system is designed can also affect the mariner's performance. Some automation hides information from the mariner, presenting only what the designer thought was needed. Unfortunately, many system designers do not fully understand the user's task, and consequently we end up with less-than-perfect, error-inducing designs. By studying the types of errors commonly made by operators, and by understanding the ramifications of these errors (i.e., are they just nuisance errors or can they cause an accident?), we gain important information that can be used in training and system redesign. Our error analysis method consisted of interviewing mariners and instructors and observing the use of automation during routine shipboard operations.

Summary of Results. The methods discussed above were applied to study the effects of automation (ARPA and ECDIS) on three navigation functions: voyage planning, collision avoidance, and track keeping. The OFM-cognitive task analysis identified how automation changed the mental demands of these functions, which, in turn, suggested a shift in emphasis (less on computation and more on interpretation) was needed on the USCG's exam for the radar observer certificate, given the wide usage of ARPA. The skills assessment and error analysis techniques identified several important types of skill and knowledge that were not fully covered in current internationally recommended training course objectives for ARPA. These same techniques also allowed the development of training course objectives for ECDIS, a relatively new piece of equipment for which no formal training courses exist. The cognitive task analysis and error analysis also proved valuable in identifying aspects of the user interface and equipment functionality that were inconsistent with the needs of the crew in the performance of the automated tasks. Taken together, these tools provide a powerful and comprehensive method of identifying the impact of automation on task and training requirements.

A Human Factors Analysis of a Casualty Investigation Database

"It was a dark and stormy night . . ." This is the popular image of conditions under which maritime casualties (accidents) take place. In reality, a large number of casualties occur under picture-perfect conditions, in calm seas with good visibility. Seventy-five to 96% of marine casualties can be traced to human error (U.S. Coast Guard, 1995), and usually to multiple errors by multiple people (Perrow, 1984; Wagenaar & Groeneweg, 1987). Many people typically think of human error as operator error or cockpit error, in which the operator makes a slip or mistake due to misperceptions, faulty reasoning, inattention, or debilitating attributes, such as sickness, drugs, or fatigue. However, there are many other important sources of human error, including factors such as management policies that pressure ship captains to stay on schedule at all costs, poor equipment design that impedes the operator's ability to perform a task, improper or lack of maintenance, improper or lack of training, and inadequate number of crew to perform a task.

As a part of its marine safety mission, the USCG routinely investigates approximately 4,000 marine casualties a year (not counting pollution cases), including events such as collisions, groundings, sinkings, fires and explosions, hull damage, and personnel injuries. A computer database was set up in 1981 for reporting the causes of marine casualties, although at that time little attention had been paid to human error causes. In 1992 the USCG completed the development of a new database system, the Marine Investigations Module (MINMOD), which includes a taxonomy for reporting human errors. The MINMOD includes several other improvements and also makes the investigating officer (IO) responsible for data entry into the computer system, rather than having clerks enter the information, as was done previously. Due to all the procedural and content changes brought about by the new MINMOD system, the Research and Development Center evaluated the data entry process and the validity and reliability of the human error data entered into the MINMOD (unfortunately, funding constraints had prohibited human factors participation in the design and development of the MINMOD). The details of this study may be found in Byers, Hill, and Rothblum (1994) and Hill, Byers, Rothblum, and Booth (1994).

Methods.   This study had two very different objectives. The first objective was to evaluate the MINMOD data entry process. To this end, structured interviews were used to question IOs about their use of the MINMOD, including strengths and weaknesses of the data entry process. IOs were encouraged to demonstrate problems as they discussed them. The interviews were supplemented by observations of the IOs as they were using the MINMOD to enter case data. This resulted in the demonstration of additional problems and the opportunity to record the frequency of errors, usage of documentation and help screens, and examples of system response time. Finally, we performed a formal evaluation of the MINMOD human-computer interface. The user interface was evaluated with respect to the screen design, user interaction, and screen navigation principles set out in design guidelines, such as in Smith and Mosier (1986).

The second objective of this study was to evaluate the validity and reliability of the human error causal data in the MINMOD. Two prior informal analyses of the MINMOD human factors data had shown that very few of the casualty reports identified any human factors causes, and those that did appeared to apply the taxonomy in an inconsistent manner. The most direct way to have measured the IOs' ability to identify and record human factors components in a casualty would have been to fabricate accident narratives and interviews with "witnesses" and to have had a large sample of IOs work through the data and record their findings in the MINMOD. Unfortunately, this was deemed too disruptive and time-consuming. Instead, we used structured interviews to determine how well the IOs understood the general concept of human error, how they tailored their investigations when a human error cause was suspected, and whether they understood the terminology in the MINMOD human factors taxonomy. We had also intended to accompany IOs on real investigations so that we could see their techniques firsthand, but there were no marine casualties during the 17 days we were interviewing IOs.

Summary of Results. The study found that the MINMOD had many serious user interface deficiencies. This wasn't surprising because the MINMOD was written to be compatible with a 10-year-old information system and thus could not benefit from the numerous interface improvements of the last decade. The combination of user interviews, observation, and the formal user-interface evaluation provided detailed and complementary feedback about features of the MINMOD and the human factors taxonomy that needed improvement. But perhaps more important, the interview data provided feedback that was not expected by management— namely, that the computer system was only part of the problem. The data provided irrefutable evidence of serious deficiencies in the promulgation of policy about how marine casualty investigations should be performed and in the training of IOs on the identification and investigation of human error causes of casualties. The richness of information that was provided by the structured interviews allowed the researchers to develop a model of the various factors that contribute to the validity and reliability of a casualty database. This model is currently being used to develop alternative methods of investigation of human error causes of casualties.

SUMMARY

This chapter discussed some of the human factors test and evaluation methods used in maritime research and development. A variety of methods were illustrated, including structured surveys, self-administered logs, direct observation, checklists, rating scales, mock-ups, HFE standards, task analysis, cognitive analysis, skills and knowledge assessment, error analysis, subject-matter experts, and focus groups. The maritime testing environment often differs substantially from the textbook case, requiring that traditional methods be adapted to accommodate the unique features of the maritime system. By continuing to refine existing methods and exploring the applicability of new ones, we can increase our knowledge of the variables that affect mariner performance and implement changes that will lead to a safer and more effective maritime system.

ACKNOWLEDGMENTS

The work discussed in this chapter was supported by the U.S. Coast Guard Research and Development Center under Contract DTCG39-94-D-E00777. The views expressed herein are those of the authors and are not to be construed as official or reflecting the views of the U.S. Coast Guard.

REFERENCES

American Society for Testing Materials (1994). *Standard practice for human engineering design for marine systems, equipment and facilities*, F-1166-94. Philadelphia: ASTM.

Bittner, A. C., Kinghorn, R. A., & Carvalhais, A. B. (1994). *Assessment of the impressions of station personnel on the effectiveness and suitability of the 47-foot motor life boat (MLB)* (AD-A285806). Washington, DC: U.S. Department of Transportation, U.S. Coast Guard.

Byers, J. C., Hill, S. G., & Rothblum, A. M. (1994). *U.S. Coast Guard marine casualty investigation and reporting: Analysis and recommendations for improvement* (AD-A298380). Washington, DC: U.S. Dept. of Transportation, U.S. Coast Guard.

Hill, S. G., Byers, J. C., Rothblum, A. M., & Booth, R. L. (1994). Gathering and recording human-related causal data in marine and other accident investigations. *Proceedings of the Human Factors and Ergonomics Society 38th Annual Meeting, 2*, 863–867.

Holcombe, F. D., Street, D., Biggerstaff, S., & Carvalhais, A. B. (1994). *Human factors assessment of the USCG 47-foot motor life boat pre-production model* (Internal Report). Groton, CT: U.S. Coast Guard Research and Development Center.

Holcombe, F. D., & Webb, S. (1991). *Human factors assessment of the USCG 47-foot motor life boat (MLB)* (NBDL-91R003). New Orleans, LA: Naval Biodynamics Laboratory.

Malone, T. B., & Shenk, S.W. (1977). *Human factors test & evaluation manual. Vol. III: Methods and procedures* (Internal Report). Falls Church, VA: Carlow International Inc.

Malone, T. B., Shenk, S. W., & Moroney, W. F. (1976). *Human factors test & evaluation manual. Vol. I: Date guide* (Internal Report). Falls Church, VA: Carlow International Inc.

Miller, R. B. (1971). *Development of a taxonomy of human performance: Design of a systems task vocabulary* (Tech. Rep. AIR-72/6/2035-3/71-TRII). Silver Spring, MD: American Institutes for Research.

Mitchell, C. M., & Miller, R. A. (1986). A discrete control model of operator function: A methodology for information display design. *IEEE Transactions on Systems, Man, and Cybernetics, 17*, 573–593.

Morrissey, S. J., Bittner, A. C., Jr., & Arcangeli, K. K. (1990). Accuracy of a ratio-estimation method to set maximum acceptance weights in complex lifting tasks. *International Journal of Industrial Ergonomics, 5*, 169–174.

National Research Council. (1990). *Crew size and maritime safety*. Washington, DC: National Academy Press.

Paley, M. J., & Tepas, D. I. (1994). Fatigue and the shiftworker: Firefighters working on a rotating shift schedule. *Human Factors, 36*(2), 269–284.

Perrow, C. (1984). *Normal accidents: Living with high-risk technologies*. New York: Basic Books.

Sanquist, T. F., Lee, J. D., Mandler, M. B., & Rothblum, A. M. (1993). *Human factors plan for maritime safety* (AD-A268267). Washington, DC: U.S. Department of Transportation, U.S. Coast Guard.

Sanquist, T. F., Lee, J. D., McCallum, M. C., & Rothblum, A. M. (1996). Evaluating shipboard automation: Application to mariner training, certification, and equipment design. In National Transportation Safety Board (Ed.), *Proceedings of the Public Forum on Integrated Bridge Systems* (Tab 5), Washington, DC: National Transportation Safety Board.

Sanquist, T. F., Lee, J. D., & Rothblum, A. M. (1994). *Cognitive analysis of navigation tasks: A tool for training assessment and equipment design* (AD-A284392). Washington, DC: U.S. Department of Transportation, U.S. Coast Guard.

Smith, S. L., & Mosier, J. N. (1986). *Guidelines for designing user interface software* (ESD-TR-86-278). Hanscom Air Force Base, MA: Electronic Systems Division, AFSC, United Sates Air Force.

Stevens, S. S. (1975). *Psychophysics*. New York: Wiley.

Tepas, D. I., & Carvalhais, A. B. (1994). *Crew adaptation evaluation of the Norwegian Crew Concept (NORCREW;* AD-A288478). Washington, DC: U.S. Department of Transportation, U.S. Coast Guard.

U.S. Coast Guard. (1995). *Prevention through people* (Quality Action Team Report, July 15, 1995). Washington, DC: U.S. Department of Transportation, U.S. Coast Guard.

Wagenaar, W. A., & Groeneweg, J. (1987). Accidents at sea: Multiple causes and impossible consequences. *International Journal of Man-Machine Studies, 27*, 587–598.

Chapter 14

Human Factors Testing in the Forest Industry

Richard J. Parker
Tim A. Bentley
Liz J. Ashby
Centre for Human Factors and Ergonomics

Forestry is one of New Zealand's largest industries. New Zealand's exotic plantation resource stands at around 1.7 million ha, representing 6% of New Zealand's total land mass (New Zealand Forest Owners Association, 1999). The forestry estate is distributed throughout New Zealand's North and South Islands, although the major concentration of plantation forest is in the Central North Island (approximately 560,000 ha). The industry is expanding rapidly, with over 60,000 ha of new planting each year. This rate of growth has major human resource implications for the industry: It is anticipated that the New Zealand forest industry will require an almost twofold increase in its skilled workforce over the next decade, as the potential harvest volume grows annually (Palmer & McMahan, 2000). Indeed, forestry operations remain labor intensive, despite the increasing mechanization of the industry. Approximately 11,000 persons were employed in the New Zealand forestry and logging industry in 1998, 30% in forestry and 40% in logging (New Zealand Forest Owners Association, 1999).

There are a number of major stages in New Zealand forestry operations, each of which involve people and human factors and ergonomics testing issues that have a significant potential to effect the ultimate value of the seedling. Silviculture involves preparation of the planting site, the planting of seedling trees, and pruning and thinning trees at various stages of growth. Harvesting, usually referred to as logging, involves the clearfelling of trees at around 28 years. Once felled, logs must be transported by heavy-goods vehicles from the forest to a variety of destinations, including sawmills and ports. The goal of this chapter is to illustrate the wide range of human factors test issues and methodologies employed in forestry operations.

FORESTRY OPERATIONS AND HUMAN FACTORS ISSUES

Silviculture operations, such as planting, pruning and thinning, are mainly manual processes. Planting work involves carrying heavy boxes of seedlings and is often undertaken on steep terrain and in extreme temperatures. Pruning involves the removal of branches from a section of the tree to encourage the growth of knot-free clearwood around the tree's core. Radiata pine (which comprises over 90% of New Zealand's plantation estate) is usually pruned in three lifts (i.e., heights): 2.2 m, 4.5 m, and 6.5 m at different stages of tree growth. Pruners must select only those trees that are defect free for pruning, as these will produce the best timber at harvest. Poor judgement at this stage can cost a forest company millions of dollars in lost timber value.

Thinning involves removal of defective trees (i.e., unpruned) and reduces competition for light, water, and nutrients. At approximately 5 years Radiata pine is thinned to waste as timber recovery is not economical. At around 14 years, trees may be production thinned. Mechanized harvesters capable of precise felling are used for production thinning to minimize damage to surrounding trees.

Most silviculture tasks are undertaken under the pressure of piece-rate conditions, and work can be tiring and monotonous. Significant hazards include the risk of musculoskeletal problems and ladder falls while pruning the higher lifts of a tree. Continuity of work in this sector can be poor, and many workers move frequently between contractors as work availability and pay rates fluctuate. Indeed, relatively high turnover in both silviculture and logging sectors has the potential to significantly impact safety and productivity, as contractors can be wary of expending resources on training for workers who may take these skills elsewhere. Moreover, frequent changes in crew membership is problematic as task safety and efficiency is in large part dependent on factors such as mutual understanding of work practices and good communication between individual crew members.

There has been an increasing mechanization of the logging operation over recent years, with many of the larger logging contractors using mechanized harvesters (see Fig. 14.1) to fell and delimb trees. Although mecha-

FIG. 14.1. Mechanized forest harvesting machine.

nization of the logging industry is increasing, the majority of trees are still felled by motor-manual methods. Many contractors cannot afford the expensive machines used by the larger logging companies, and even where felling machines are available, they generally cannot be used to fell on steep terrain, where much of New Zealand's plantation resource is situated. Falling (i.e., felling of trees using chainsaws) is a highly skilled and dangerous job. Certainly, there is little room for error when felling trees of up to 40 m in height and weighing up to 2.5 t. The faller must ensure the tree falls in the required direction without damaging surrounding trees. High winds and sloping ground make this task more difficult. Having felled the tree the faller must then trim the branches off the tree. A considerable degree of skill is needed to operate powerful chainsaws, and operators are faced with many decisions that must be made quickly and accurately.

Once a tree has been felled and trimmed, it is hauled to a clearing known as a skid site. Here log-makers grade and cut logs to lengths specified by the customer. When grading, log-makers take into account diameter and straightness of the log and knot size and number. This task requires fine judgement by the log-maker, whose grading decisions can significantly affect the value of the wood harvested by a crew.

The technological advances associated with increased mechanization have had significant consequences for the logging workforce, with large numbers of employees moving from the physically demanding job of falling into the machine cab. This change, although offering protection from the elements and dangers associated with motor-manual harvesting, has lengthened the work day for many loggers and has removed the physical/outdoor aspect of the job that was, for many, the primary attraction of working in the forest. Moreover, the sanitized atmosphere of harvesting-machine cabs has introduced new human factors problems for the industry, including long hours of mostly sedentary work increasing the risk of boredom, job dissatisfaction, occupational stress, and musculoskeletal problems among the workforce using mechanization. Other tasks involved in the logging process include hauling, in which felled trees are dragged to a processing area, skid work, in which trees are cut into shorter logs, and the loading of logs onto trucks. Much of this work takes place in relatively confined areas, with heavy plant and with people working in close proximity. The potential for accidents and injuries in areas such as skid sites (Fig. 14.2) is a major problem for the logging industry (Parker, 1998).

Heavy-goods vehicles are used to transport logs from the forest to processing centers, ports, and other destinations. The high accident incidence for log trucks involved in this process, particularly rollovers, is a major problem for the New Zealand forest industry. Indeed, accidents involving log trucks on public roads are the most visible aspect of safety failures in

FIG. 14.2.   Loggers working on skid site in close proximity to heavy machinery.

the industry. The Land Transport Safety Authority (LTSA, 1999) of New Zealand estimates from police reports that two log trucks crash every week on public roads in New Zealand, and log truck crash rates are 3 to 4 times greater than the average for all heavy combinations (Metcalf & Evanson, 2000). The most common primary causes of log truck accidents, as listed in police accident reports (LTSA, 1999), were speeding and vehicle defect.

This brief outline of the forestry process highlights the importance of people in all stages of the forestry operation and the considerable potential for people to impact on the forestry value chain. The importance of the human factor in forestry has been recognized for some time in New Zealand. Since 1974, the Logging Industry Research Association (LIRA) has been conducting applied research for the forestry industry, which has sought to optimize the forestry worker-task-environment fit in an effort to improve injury and production performance. LIRA (subsequently known as Liro Forestry Solutions, a division of the New Zealand Forest Research Institute) was established as a joint industry and government initiative to provide applied harvesting research to the New Zealand logging industry. The Centre for Human Factors and Ergonomics (COHFE) arose from the demise of Liro and is now largely funded by public money. COHFE undertakes applied research in forestry and other industries, often in collaboration with New Zealand universities and leading international experts. Figure 14.3 provides an overview of the wide range of human factors issues that have been the subject of applied research projects undertaken by the COHFE.

The high-risk nature of logging operations is reflected in much of the work undertaken by the COHFE. This work has necessarily focused on microsystem factors, with attention on the individual in accident analysis and prevention (see Fig. 14.3). Changes in theoretical perspectives on safety management (e.g., Hale & Hovden, 1998; Reason, 1990) and an increasing move to mechanization with its associated changes in working conditions within the forest industry have led to a recent refocusing of human factors research attention to macrosystem considerations, notably the organization of work, management practices, and leadership styles.

THE ACCIDENT REPORTING SCHEME (ARS)

Following the realization in the early 1980s of the need for a national forestry injury database, LIRA set about developing an accident reporting system for the industry. After a successful 15-month regional pilot trial, the scheme was extended to cover the whole New Zealand forest workforce (Prebble, 1984; Slappendel, 1995). The scheme, which still operates to-

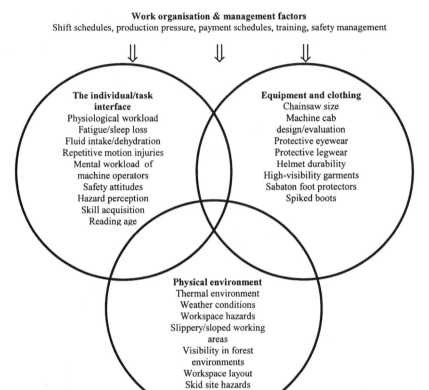

FIG. 14.3.   Overview of forestry human factors issues undertaken by the COHFE.

day, relies on logging companies to provide monthly summaries of injuries and near-miss events to COHFE. Contributing companies also supply exposure information (i.e., total man-hours worked) from which injury rates can be determined. The data collected enables the analysis of accidents in terms of a range of factors, including lost time, temporal information, activity at time of accident, and body part injured. This information is analyzed quarterly and fed back to the industry via newsletters and other COHFE publications in the form of summary statistics of trends and patterns in injury data. Secondary, but related, uses of these data are to identify key areas of concern regarding forestry safety and to provide a baseline against which interventions can be evaluated. To illustrate the issues and methods involved in forestry human factors, six case examples will be described in the sections that follow.

Felling and Delimbing Hazard Identification

Felling and delimbing operations are the most hazardous phases of the logging operation, accounting for 27% and 28%, respectively, of reported lost-time injuries between 1985 and 1991 (Parker & Kirk, 1993). This investigation recorded the frequency of potentially hazardous felling situations encountered by workers during normal working operations, comparing this data with actual injury rates as reported to the ARS. The study was part of a wider project to reduce injuries in felling and delimbing operations, which included measuring numbers and types of hazards encountered by these workers and evaluating loggers' recognition of hazardous felling situations.

A total of 14 loggers were observed throughout normal working operations, 9 for 2 days and 5 for 1 day. Job experience of the loggers varied, with nine loggers having over 5 years' felling experience, and five with less than 1-year felling. A field computer was used to record activities, and predefined hazardous felling and delimbing situations were recorded. The researchers, experienced in the tasks being observed, assessed whether a situation was hazardous (i.e., if an injury resulted or could have resulted in slightly different circumstances) using descriptions derived from Ostberg (1980). The relative frequency of observed hazards (as a proportion of all hazards recorded), which can be considered as the potential injury rate, was compared to the actual injury rate as derived from the ARS.

The inexperienced loggers were exposed to more hazards than the more experienced workers, both during felling and delimbing, as shown in Fig. 14.4. Average overall rate of exposure to hazards was 31.1 (± 3.9) hazards per 100 trees. The type and frequency of hazards varied accord-

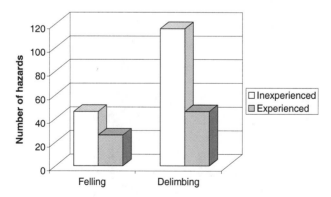

FIG. 14.4. Comparison of total number of hazards per 100 trees, as encountered by loggers.

ing to the level of logger experience. Inexperienced loggers tended to be exposed to hazards associated with procedures and techniques, such as drop starting the chainsaw (i.e., an illegal shortcut method of starting the saw) and incorrectly cutting the tree so it did not fall in the right direction. Inexperienced workers were more frequently exposed to the hazard associated with sailers (i.e., trees, debris, or branches suspended above the logger that could fall and injure or kill). Both inexperienced and experienced loggers were found to be exposed to some hazards, particularly trips, slips, falls, above-the-head use of the saw, and lost balance during delimbing.

Figure 14.5 and Figure 14.6 compare observed hazards (potential injuries) with ARS data for felling and delimbing tasks. Large differences were found for the majority of hazards. This implies that although hazard identification and recognition by workers is essential, other means of addressing safety need to be employed. For example, the potential for injury from a sailer during felling operations was much greater than the observed hazards' frequency, as shown by a higher than expected injury rate (see Fig. 14.5). Sailers are more difficult to identify during felling, because of the tree canopy, as opposed to during delimbing, where the canopy is opening and sailers can be seen (see Fig. 14.6).

Some of the hazards that workers were exposed to, identified by researchers and reinforced by ARS data, could be avoided through changes in work practice. For example, chainsaw injuries would be reduced by keeping chainsaw use below shoulder level. This emphasizes the need for effective training in appropriate techniques that are geared around actual risk reduction. Exposure to other hazards can be reduced through the use of personal protective equipment (PPE), such as eye protection; spiked boots, which reduce slips, trips, and falls; and cut resistant leg wear and foot wear.

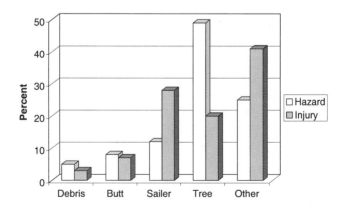

FIG. 14.5. Comparison of observed hazard frequency and actual injury rate for felling tasks.

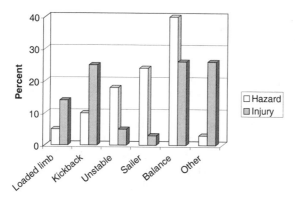

FIG. 14.6. Comparison of observed hazard frequency and actual injury rate for delimbing tasks.

Human Factors Affecting Log-making Ability

This study sought to identify the effects of perceived boredom and physiological work load on log-grading decisions made during the harvesting of trees. The study employed a combination of physiological and rating scale methodologies in a demanding field test environment.

Once felled, tree stems are graded, cut, and sorted into approximately four to eight shorter logs suitable for transport to mills. The task of grading and cutting tree stems is undertaken by a log-maker. Log making must often be carried out under conditions of high noise, extreme temperatures, and in close proximity to large mobile machines. There is evidence these variables can have a detrimental effect on mental performance in inspection tasks (McCormick, 1970).

Because the physical properties of wood within a tree vary considerably, parts of the tree need to be graded for their end use suitability. A single forestry company can produce up to 40 grades of logs with many features, including length, roundness, bend, branch diameter, damage, and decay. Each log grade has a different sale price. The classification of a stem into grade logs that give the highest value is, therefore, a complex inspection and decision task, and there is considerable variation in individual log-making performance (Cossens & Murphy, 1988). A mathematical computer model (AVIS) that determines the combination of logs that will maximize revenue for each tree has been developed (Geerts & Twaddle, 1984). The value of log-grading decisions made by the log-maker can therefore be compared with decisions made by the computer optimization model. This study investigated log-making ability, in terms of value recovery, as measured against this model.

Six logging crews considered representative of North Island logging operations were studied for 2 days each. Log-making ability was assessed at 3 times during each day (start, middle, and end). Trees to be assessed were measured for stem shape and quality characteristics required for input to the stem optimization software. The value of log-makers' log grade selections were compared with optimal log grade selections calculated by AVIS and expressed in terms of percentage of optimal value.

Log-makers' self assessment of boredom was estimated from questionnaire items that asked log-makers to rate their level of boredom on a 7-point scale 3 times during the day. Physiological workload of log making under normal conditions was estimated by recording logmakers' heart rates at 15-s intervals throughout the day using a Polar Electro PE3000 heart rate monitor. Shaded dry bulb air temperature on the landing was recorded at 15-minute intervals. Work activities of the log-makers were recorded at 15-s intervals. Data for these variables were entered into a logistic regression model to determine the relationship between each factor and log making value recovery.

Average value recovery for all log-makers was 93.4 ± 0.05% (221 stems), with a large random variation between individual log-makers, day of study, and time of day. No consistent relationship was found between value recovery and work rate (stems/hour), air temperature, heart rate, or time of day. Value recovery did show a highly significant decrease of 6% for a 2-unit increase in boredom from 2 to 4 ($p < .005$). Most of this effect occurred between ratings of 2 and 3 (on a 7-point scale: 1 = *interested*, 7 = *bored*). This suggested that even small reductions in log-maker boredom could result in a significant increase in value recovery.

A number of recommendations were made on the basis of this study. Training and selection of staff may be the most effective method to improve the log-making function. Indeed, Cossens and Murphy (1988) showed formally trained workers achieved better log-making value recovery than those with more work experience. Reduction of noise through the use of active noise reduction earmuffs and reduction in task complexity, through reducing the number of log grades and providing clearer log classification rules, should provide further benefits. Boredom might be addressed through appropriate changes to the organization of log-making work, notably, rest-break scheduling and job rotation, or enlargement. Moreover, these measures have been shown to be effective as a means to maintain worker vigilance (Fox, 1975; Krueger, 1991).

Determining Fluid Intake Requirements for Loggers

New Zealand loggers work throughout the year under physically demanding conditions and are not sheltered from the prevailing weather conditions. Felling trees with a chainsaw requires constant vigilance and concentration to avoid injury (Slappendel, Laird, Kawachi, Marshall, & Cryer, 1993). Dehydration has been linked to a decrease in cognitive performance (Adolph, 1947), potentially increasing the opportunity for injury.

A study by Paterson (1997) investigated the fluid balance, under two fluid regimes, of eight loggers who were chainsaw felling and delimbing trees. Condition A was normal fluid intake and Condition B was an additional 200 ml of sports drink (Gatorade) every 15 minutes. Heart rate was recorded at 1-minute intervals using a Polar Electro PE3000 heart rate monitor. The monitor was wired with shielded coaxial cable between the sensor unit and the receiver unit to prevent electrical interference from the chainsaw (Parker, 1992). Sweat rate was determined by weighing the subject (near nude) to the nearest 50 g before work and at the end of work each day. Weight increase due to food and fluid intake and weight loss due to feces and urine output were recorded each day to correct final body weight. A Husky Hunter field computer running a continuous time study program (Siwork3) was used to collect time and motion data for each subject (Rolev, 1990).

Other measures recorded were (1) skin temperature via thermistors attached to the skin at four sites (upper chest, bicep, thigh, and calf), (2) self-assessed rate of perceived exertion (RPE; Borg, 1982), (3) thermal comfort (Cotter, 1991), and (4) body part discomfort (Corlett & Bishop, 1976). These latter three measures were collected using a questionnaire administered at hourly intervals throughout the day.

Under the normal fluid condition subjects consumed an average of 2.7, ± 1.2, L of fluid per day, compared with 5.5, ± 1.3, L of fluid per day under the additional fluid condition. The normal fluid condition did not completely replace fluid lost, and subjects were losing in excess of 1.2% ± 1.3%, body weight per day. With additional fluid (800 ml/hour) subjects were losing 0.7% ± 1.0%, body weight per day. The rate of work (number of trees felled and delimbed) between the normal fluid intake condition and the additional fluid intake condition was unchanged. However, there was a significantly lower mean heart rate for subjects under the additional fluid condition, and this difference increased in the afternoon. There were no differences in body core temperature or skin temperature between the two conditions.

This field test demonstrated the beneficial effects of drinking sufficient fluid during the day. This study set the stage for further work on fluid intake and ultimately resulted in a fluid and nutrition education package for loggers.

High Visibility Clothing for Loggers' Safety

The relative conspicuity of loggers' clothing was tested using a laboratory-based eye-tracking system to establish which colors are most visible in pine forests. The test was designed to reproduce the dual visual requirements of logging machine operators: high-precision visual demands in operating the machine and peripheral vision demands necessary for avoiding other workers in the vicinity. Logging operations involve use of heavy machinery, which can be in close proximity to workers who are on foot, for example, during the processing of logs on the skid site. From 1991 to 1993, the ARS indicated four fatalities as a result of collisions between machines and workers who were not seen (see Fig. 14.7); a total of 27 lost-time injuries resulted in an average of 34 days off work (Isler, Kirk, Bradford, & Parker, 1997). Optimizing the visibility of workers by determining which garments are most easily seen should help reduce these incidents.

The laboratory-based experiment required the participants to perform two tasks. The central tracking task, simulating machine operation, involved tracking a moving dot on a computer screen by moving a joystick. The computer screen was embedded in a slide screen, on which slides of either a pine forest scene or camouflage patterns were projected. A peripheral task involved locating images of six test shirts, which were set into the projected picture. Test shirt designs are shown in Fig. 14.8.

During the experiment, the luminance of the screen was gradually increased, and decreased, during the task to simulate variable lighting conditions. Additionally, the test shirt images were moved through 10 positions, so that by the end of the experiment each design had been observed in each position. Monitoring of the subjects' eyes (using an eye-tracking system) allowed ranking of tests showing which shirts were detected and in which order.

Fluorescent yellow was the most conspicuous garment against the pine forest background, being detected first in 43% of the trials. Black was only detected first in 1% of the trials and was detected last, or not at all, in 35 out of the 99 tests. The fluorescent chevron on the black garment did not perform much better than the plain black (detected first in 5% of tests). Fluorescent orange, which is widely used in other industries, was detected

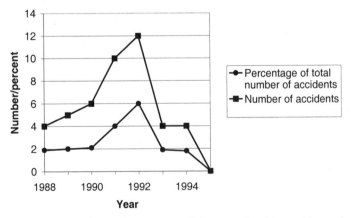

FIG. 14.7.   Number and percentage of all reported accidents with people that were not seen.

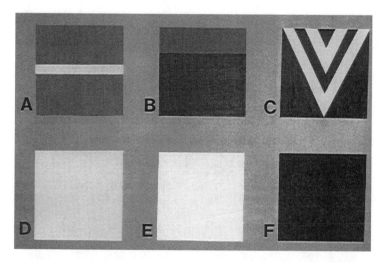

FIG. 14.8. High-visibility shirt designs. Top row: red with yellow stripe, red and green, black with yellow stripes. Bottom row: yellow, white, black.

first significantly less than the yellow (14% of tests) and was undetected or last in 11% of the tests.

In conjunction with the laboratory trials, prototype garments were also evaluated in the field, for comfort, practicality, durability, and cost. This involved providing loggers with test garments of various designs and obtaining feedback after a period of use (Fig. 14.9). Loggers currently use the resultant high-visibility clothing throughout the industry, garments that have been adapted to incorporate their company name or logo. The subsequent designs are a compromise of the blocks of florescent yellow, allowing material to breathe through the use of chevrons and logos. These also serve to encourage a bond between team members, and in some cases the contractors award high-visibility clothing to workers who are performing well. By 1993, following the study, most forest companies required the use of high-visibility clothing. The proportion of not-seen accidents significantly dropped (following an initial increase probably through more accurate reporting and increased awareness) with none reported by 1995 (Sullman, Kirk, & Parker, 1996).

The Use of Spiked Boots During Felling and Delimbing Activities

Harvesting and silvicultural work in New Zealand forests increasingly involves working on steep terrain, with various soil types and weather conditions exacerbating the risk of slips, trips, and falls. Analysis of 6 years of data from the Forest Industry ARS showed 17.5% of the lost-time injuries

FIG. 14.9.   Loggers wearing high-visibility clothing.

during felling and delimbing activities were a result of slips, trips, and falls. These injuries resulted in the loss of 2,870 days over the 6-year period. This study field tested the impact of wearing spiked boots on physiological workload, feller productivity, and safety (Kirk & Parker, 1994). Unlike a standard rubber sole, the sole of a spiked boot has approximately 20 screw-in spikes, each 1-cm long, inserted into it.

Four experienced workers were studied during normal work activities. The study areas consisted of moderate to steep terrain in transition pine with mean butt diameters of 48 cm (*SD* = 8.2). A weather station measured air temperatures (ambient, wet and dry bulb and black globe) at 15-minute intervals during the study days, from a position as close as reasonably practicable to the work site. The activities of the workers were standardized and recorded using a continuous time study method. A Husky Hunter field computer recorded 24 separate elements of the work cycle. Examples of activities included walk, backcut (i.e., the final saw cut when felling), and delimbing, with the total cycle time being the time taken to fell and delimb the tree. The workers were studied over 3 days, wearing the conventional boots. Each worker then wore a pair of spiked boots for a 4-week period, during which time normal work activities were continued. Following the 4-week period, the continuous time study method was repeated in the same stand and with similar terrain and conditions. In both conditions, the number of times the workers slipped was recorded into the field computer by the observer.

The observer also recorded additional information on tree characteristics, ground slope (at the site of felling), access, and delimbing slope, for each cycle time. The occurrence of a slip was categorized into three groups according to the location of the worker at that time: working from bare ground, working from standing on the stem, and working from standing on slash (branches, bark, stumps, and foliage on the ground). This way each slip could be related to the ground conditions as well as the specific work activity. A slip was recorded if workers lost their footing, resulting in loss of balance or sliding of the feet.

Physiological workload was measured using a heart-rate data-collection technique (Astrand & Rodahl, 1977) during the wearing of conventional boots and spiked boots to estimate fallers' work loads. The heart rate of each faller was recorded at 15-s intervals using a Polar Electro Sport Tester PE3000 heart rate monitor. Additionally, task activities were recorded at the same 15-s interval to relate heart rate recordings to the specific activities being undertaken.

Productivity was not reduced during the use of spiked boots, and one faller showed a significant decrease in the cycle time when wearing the spiked boots. Analysis of recorded slips showed a significant ($p < 0.01$) decrease in slips on dirt, slash, or the stem when spiked boots were worn (see Table 14.1). There was a particularly large reduction in slipping when working on slash.

Average working heart rates were calculated, excluding rest periods or other delays in activities. The recording of climatic conditions allowed comparisons of work loads between the two methods, with no significant differences between wet-bulb globe temperatures. Three out of the four workers showed no significant differences in average heart rates between the two boot types. One worker had a higher heart rate when wearing the spiked boots, which may be due to steeper ground slope during that phase of the test.

The use of the spiked boots indicated a reduction in the rate of slipping, which in turn has the potential of significantly reducing lost-time injuries associated with slips, trips, and falls. Workers provided subjective opinions supporting the findings, reporting increased confidence when using the spiked boots and claiming that they would continue to wear the boots. Spiked boots are now commonly used in the industry, and ARS data indicates a significant reduction of slipping injuries among loggers and silvicultural workers following their introduction.

TABLE 14.1
Location and Frequency of Slips per 100 Stems Delimbed

Location of Slip	Bare Ground	Stem	Slash
Rubber soled boot	43	56	76
Spiked boot	19	13	15

Chainsaw Size for Delimbing

The purpose of this study was to compare the cardiovascular work load imposed by delimbing with small, medium, and large capacity chainsaws under controlled conditions (Parker, Sullman, Kirk, & Ford, 1999). In logging operations, after trees are felled, the branches must be removed (delimbed) before further processing can take place. New Zealand loggers commonly use large chainsaws, weighing up to 10 kg (e.g., the Stihl 066 and the Husqvarna 394). From an ergonomic perspective, a heavy chainsaw is a poor choice because it will impose a greater physiological work load on the operator and place greater loadings on the lower spine than a lighter chainsaw (Tatsukawa, 1994).

Delimbing by chainsaw (i.e., motor-manual delimbing) is both physically arduous and potentially dangerous (Parker & Kirk, 1993). The ARS noted that chainsaw felling and delimbing accounted for 46% (i.e., 93 people) of all New Zealand lost-time logging injuries in 1995 (Parker, 1996). Gaskin, Smith, and Wilson (1989) reported 46% of the logging workforce suffered from back problems. A contributing factor to this high injury rate could be the high physiological and biomechanical load of motor-manual felling and delimbing (Gaskin, 1990; Kukkonen-Harjula, 1984; Parker & Kirk, 1994).

Test participants included 11 males and 1 female from the logging industry. All were experienced chainsaw operators who had worked under normal logging conditions in New Zealand forests. Chainsaws of 60-, 70-, and 90-cc displacement were used to delimb eight test stems 6 m long and 0.2 to 0.3 m in diameter, mounted with their long axis 0.7 m above the ground (see Fig. 14.10). Branches were simulated by using dowel branches 18 mm in diameter, in whorls of four branches, 0.7 m apart—an estimate of whorl separation in *Pinus radiata* trees.

A timing device, mounted on the chainsaw operator's belt, was linked to an earpiece inside the protective earmuffs, which beeped at 5-s intervals. The chainsaw operator delimbed branches in a methodical way at a rate of one whorl (four branches) every 5 s. At the completion of delimbing, (approximately 5.5 minutes) the chainsaw operator rested for 10 minutes. The procedure was repeated twice with chainsaws of different sizes. The order of chainsaw use was determined by a Latin square design for the three chainsaw sizes and 12 (subsequently 11) chainsaw operators.

All heart rate data were expressed as a proportion of heart rate range (% Δ HRratio; Morton, Fitz-Clark, & Banister, 1990; Trites, Robinson, & Banister, 1993), calculated as follows:

$$(HR_{work} - HR_{prework}) / (HR_{maximum} - HR_{prework}) \times 100$$

Where:

HR_{work} = Heart rate while working, last 3 minutes

$HR_{prework}$ = Heart rate before working, while resting

$HR_{maximum}$ = 220 − age (years), maximum heart rate adjusted for age

Cardiovascular strain (log % Δ HRratio) increased with greater chain-saw mass ($p < 0.001$) indicating the greater effort required to delimb with a larger chainsaw. Delimbing with a chainsaw is a physically demanding task, and the results indicate that cardiovascular strain increased as a function of increased chainsaw weight.

FIG. 14.10.   Chainsaw operator delimbing test stem.

The delimbing stem was set 70 cm from the ground, which disadvantaged the shorter chainsaw operators, whose elbows were closer to the ground. The proportion (log % Δ HRratio) utilized decreased as elbow height increased. Elbow height was inversely related to cardiovascular strain, indicating that the delimbing task was more physically demanding for chainsaw operators with elbows closer to the ground ($p < 0.05$). However, the proportion increased with increased arm length ($p < 0.01$). Arm length was positively related to cardiovascular strain, indicating that the delimbing task (for all saws) was more difficult with longer arms. Perhaps shorter arms offer a greater mechanical advantage to the operator.

Under simulated and controlled conditions the results of this study indicate that chainsaw weight has a significant effect on cardiovascular strain for chainsaw operators delimbing simulated tree stems. The experimental design allowed the identification of order of chainsaw use, chainsaw operator elbow height, and operator arm length as factors contributing to cardiovascular strain. Further work is required to determine the influence of chainsaw size and power under the normal operational conditions existing in New Zealand plantation forest harvesting operations.

Development of the Sabaton Foot Protector

In this final study, we show an example of human factors testers looking outside their application area (and, in this case, their century) to find a novel solution to a very common problem: chainsaw laceration of loggers' feet. Data from the ARS indicated there were nine injuries in 1998, resulting in an average of 15 days off work for each injury. Most (95%) chainsaw lacerations to feet were to the left foot, and a survey of 60 injured loggers identified the area behind the big toe as the most at risk.

No fully practical cut-resistant boot currently exists for loggers. Current chainsaw cut-resistant footwear consists of either rubber boots or fabric kevlar foot covers. Rubber gumboots are hot, do not allow perspiration to escape, and offer little ankle support. Kevlar foot covers are a single-use solution; they must be replaced once cut. In an attempt to develop rugged and practical chainsaw cut-resistant foot protection for loggers, COHFE researchers investigated protective footwear from the past. Foot armor worn by medieval knights was developed to be flexible, to allow combat on foot, and to deflect blows from weapons (see Fig. 14.11 and 14.12). These same qualities are required by loggers.

An armorer skilled in the medieval style was engaged to construct sabatons from modern materials. On the advice of metallurgists, stainless steel, aircraft aluminum, and titanium were tested for chainsaw cut resistance. Titanium proved the most cut resistant, but it proved too expensive for commercial manufacture. Stainless steel, although heavier, proved to

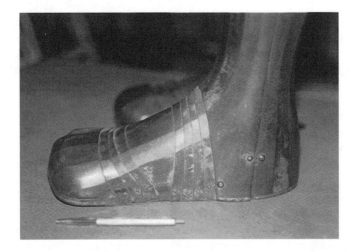

FIG. 14.11. English sabaton 1520—Tower of London, England.

FIG. 14.12. Logger's sabaton 1999—Kaingaroa Forest, New Zealand.

be sufficiently chainsaw cut resistant. Prototype sabatons were tested in field trials with loggers, and design modifications were made until a practical foot protection device was developed.

CONCLUSION

In common with other workers who undertake human factors test and evaluation, COHFE has to work within the methodological constraints associated with field research. Notable among these are the absence of statistical controls in some experimental work, inability to control all possible interfering (nonexperimental) variables, and problems associated with evaluation of interventions over time. Despite these limitations, COHFE has sought to provide relevant and timely research solutions to the New Zealand forest industry. Close links with the forest industry and use of the ARS to identify test and evaluation priorities have proven invaluable in achieving these aims. We have attempted to illustrate the wide range of test methodologies required and some of the forestry issues investigated to date. Recent significant industry changes, including increasing mechanization, longer worker hours, and concerns over the industry's ability to attract and retain sufficient numbers of skilled workers, have presented new issues for COHFE to test. Two major projects are currently considering organizational-motivational issues important to the problem of attracting and retaining skilled workers to the industry. Sources of stress and job satisfaction in the forest workforce are being investigated, with a range of qualitative survey techniques being used to support construction of questionnaire items to measure worker stress and satisfaction. It is intended that this study will be able to determine the association between key factors and external variables, such as turnover, absenteeism, and injury involvement. A second study is considering the influence of contractor-supervisor leadership style on a range of factors. Current projects to address problems related to increased mechanization and working hours include the use of electromyography biofeedback as a training tool to reduce musculoskeletal problems among mechanized operators, methods for improving vision in logging vehicles, and shift scheduling and fluid and nutritional requirements for logging and silvicultural workers. In the future, huge logging machines, which are capable of moving over steep terrain, extracting tree stems, and transporting them to a clearing, and the increased use of information technology to support human decision making in various forest operations suggest but two new human factors challenges.

REFERENCES

Adolph, E. F. (1947). *Physiology of man in the desert*. New York: Interscience.

Astrand, P. O., & Rodahl, K. (1977). *Textbook of work physiology*. New York: McGraw Hill.

Borg, G. A. V. (1982). Psychophysical bases of perceived exertion. *Medicine and Science in Sports and Exercise, 14*(5), 377–381.

Corlett, E. N., & Bishop, R. P. (1976). A technique for assessing postural discomfort. *Ergonomics, 19*(2), 175–182.

Cossens, P., & Murphy, G. (1988). Human variation in optimal log-making: A pilot study. *Proceedings of the International Mountain Logging and Pacific North West Skyline Symposium*, 76–81.

Cotter, J. D. (1991). *Hypothermia: An epidemiology and an assessment of a garment for preventing immersion hypothermia.* Unpublished masters thesis, University of Otago, Dunedin.

Fox, J. G. (1975). Vigilance and arousal: A key to maintaining inspector performance. In C. G. Drury & J. G. Fox (Eds.), *Human reliability and quality control* (pp. 89–96). London: Taylor & Francis Ltd.

Gaskin, J. E. (1990). An ergonomic evaluation of two motor-manual delimbing techniques. *International Journal of Industrial Ergonomics*, *5*, 211–218.

Gaskin, J. E., Smith, B. W. P., & Wilson, P. A. (1989). *The New Zealand logging worker—A profile* (Project report No. 44). Rotorua, NZ: Logging Industry Research Association.

Geerts, J. M., & Twaddle, A. A. (1984). A method to assess log value loss caused by crosscutting practice on the skidsite. *New Zealand Journal of Forestry*, *29*, 173–184.

Hale, A. R., & Hovden, J. (1998). Management and culture: The third age of safety. A review of approaches to organisational aspects of safety, health and environment. In A-M. Feyer & A. Williamson (Eds.), *Occupational injury: Risk, prevention and intervention* (pp. 129–165). London: Taylor & Francis.

Isler, R., Kirk, P., Bradford, S. J., & Parker, R. (1997). Testing the relative conspicuity of safety garments for New Zealand forestry workers. *Applied Ergonomics*, *28*(5/6), 323–329.

Kirk, P., & Parker, R. (1994). The effect of spiked boots on logger safety, productivity and workload. *Applied Ergonomics*, *25*(2), 106–110.

Krueger, G. P. (1991). Sustained military performance in continuous operations: Combat fatigue, rest and sleep needs. In R. Gal & A. D. Mangelsdorff (Eds.), *Handbook of military psychology* (pp. 244–277). Chichester, UK: Wiley.

Kukkonen-Harjula, K. (1984). Oxygen consumption of lumberjacks in logging with a powersaw. *Ergonomics*, *27*, 59–65.

LTSA (1999). New study links crash rates to truck stability. Available: *http://www.ltsa.govt.nz/news/news/990826 01.html*

McCormick, E. J. (1970). *Human factors engineering.* New York: McGraw-Hill.

Metcalfe, H., & Evanson, T. (2000). Causal factors in heavy vehicle accidents. In *Liro Project Report No. 89.* Rotorua, NZ: Logging Industry Research Organisation.

Morton, R. H., Fitz-Clark, J. R., & Banister, E. W. (1990). Modeling human performance in running. *Journal of Applied Physiology*, *69*, 1171–1177.

New Zealand Forest Owners Association. (1999). *New Zealand forest industry, Facts and figures 99.* Wellington, NZ: Author.

Ostberg, O. (1980). Risk perception and work behaviour in forestry: Implications for accident prevention policy. *Accident Analysis and Prevention*, *12*, 189–200.

Palmer, W., & McMahon, S. (2000). Attracting and retaining a skilled work force. *Liro Report*, *25*(3), ISSN 1174-1234.

Parker R. J. (1992). Ergonomics research in the New Zealand logging industry. *New Zealand Ergonomics*, *7*(3).

Parker, R. J. (1996). *Analysis of lost time accidents—1995, logging* (Project report Vol. 21, No. 21). Rotorua, NZ: Logging Industry Research Association.

Parker, R. J. (1998). *Analysis of lost time injuries—1997. Logging ARS* (Project report Vol. 23, No. 15). Rotorua, NZ: Logging Industry Research Association.

Parker, R. J., & Kirk, P. M. (1993). *Felling and delimbing hazards* (Project report Vol. 18, No. 22). Rotorua, NZ: Logging Industry Research Association.

Parker, R. J., & Kirk, P. M. (1994). *Physiological workload of forest work* (Project report Vol. 19, No. 4). Rotorua, NZ: Logging Industry Research Association.

Parker, R., Sullman, M., Kirk, P., & Ford, D. (1999). Chainsaw size for delimbing. *Ergonomics*, *42*(7), 897–903.

Paterson, T. (1997). Effect of fluid intake on the physical and mental performance of forest workers. *Liro Project Report*, *66.* Rotorua, NZ: Logging Industry Research Organization.

Prebble, R. L. (1984). A review of logging accidents in New Zealand. *Proc. LIRA Seminar Human Resources in Logging* (pp. 65–71). Rotorua, New Zealand: LIRA.

Reason, J. (1990). *Human error*. New York: Cambridge University Press.

Rodahl, K. (1989). *The physiology of work*. London: Taylor & Francis.

Rolev, A. M. (1990). *Siwork 3 version 1.2: Work study and field data collection system*. Denmark: Danish Institute of Forest technology, Amalievej 20, DK-Frederiksberg C.

Slappendel, C. (1995). *Health and safety in New Zealand workplaces*. Palmerston North, NZ: Dunmore Press.

Slappendel, C., Laird, I., Kawachi, I., Marshall, S., & Cryer, C. (1993). Factors affecting work related injury among forestry workers: A review of the literature. *Journal of Safety Research*, *24*(3), 19–32.

Sullman, M., Kirk, P., & Parker, R. (1996). The accident reporting scheme: What happens to the data and why. *Liro Report, 21*(32). Rotorua, NZ: Logging Industry Research Organisation.

Tatsukawa, S. (1994). An ergonomic evaluation of the cutting work with a light-weight chain saw on steep terrain. In J. Sessions (Ed.), *Proceedings of the IUFRO/NEFU/FAO Seminar on Forest Operations Under Mountainous Conditions* (pp. 166–172). Harbin, China: Northeast Forestry University.

Trites, D. G., Robinson, D. G., & Banister, E. W. (1993). Cardiovascular and muscular strain during a tree planting season among British Colombia silviculture workers. *Ergonomics*, *36*, 935–949.

Chapter 15

Human Factors Testing Issues in Road Safety

Samuel G. Charlton
Brett D. Alley
Peter H. Baas
Jean E. Newman
Transport Engineering Research New Zealand Ltd.

Traditional approaches to improving the safety of our road transportation system have typically focused on each of the components of the system in isolation: building more crashworthy vehicles, engineering roads to improve traffic control devices and road geometry, and public education programs designed to ensure drivers have appropriate attitudes towards speed and drink driving. Although this approach has enjoyed a modicum of success in increasing some aspects of transportation safety, the continuing toll of accidents and injuries (approximately 650,000 worldwide deaths annually) suggests that a more integrated approach is needed.

One such approach has been taken by the people of Sweden. In 1997 the Swedish parliament enacted a revolutionary piece of legislation known as Vision Zero. Simply stated, Vision Zero holds that it is ethically unacceptable for anyone to be killed or severely injured by the road transport system. Further, it places the responsibility for the level of safety in the system on the designers of the system. To achieve the required reductions in road fatalities and injuries (50% reductions by 2007) human factors research is being used in the integration of road engineering, vehicle design, and educational initiatives (Tingvall, 1998).

Human factors engineers working in the area of road safety have argued that analysis of the transport system must include the entire system: vehicle, road, and driver factors (Charlton & Baas, 1998; Persaud, Bahar, Smiley, Hauer, & Proietti, 1999). Although approximately 90% of road crashes are attributed to driver error (Treat et al., 1977), that figure must be taken to include the effects of poorly designed roads and vehicles. In other words, driver error does not occur in a vacuum. There are many contributing factors to any road crash, and simply citing human error as the probable cause has the effect of hiding the design challenge inherent in the circumstances of the crash and thus occasioning a similar crash in the future.

We feel strongly that an integrated approach to road safety is required to achieve greater levels of road safety. Engineering a safer transportation system requires testing the components of the system in concert, examining a full complement of road, vehicle, and driver factors that occasion driver errors, crashes, and injuries (see Fig. 15.1). This chapter will describe some of the human factors testing methodologies being applied to the road transport system in New Zealand with that aim. We describe these methods and the results they have produced to date in three interrelated areas: the design of rural-urban thresholds, driver fatigue testing, and modeling driver-vehicle-road interactions.

TESTING RURAL-URBAN THRESHOLDS

The rural-urban threshold has consistently been identified as a speeding "black spot." Many motorists appear to find it difficult to slow down from highway or open road speeds to a slower speed when entering an urban or

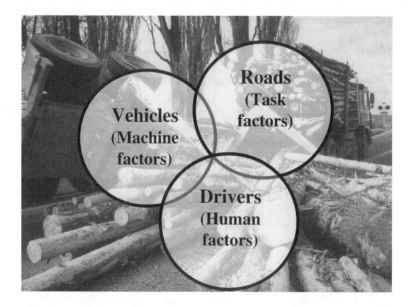

FIG. 15.1.   Crashes are a function of vehicle, road, and driver interactions.

semiurban area. Commonly, the motorist is faced with a situation in which they are required to decelerate to 70 kph or 50 kph after having traveled at 100 kph for an extended amount of time. In this situation, the motorist tends to underestimate the speed at which they are traveling and, as a result, they drive too fast (Denton, 1976). Essentially the same situation presents itself as drivers approach roundabouts, construction zones, bridges, motorway off-ramps, and corners (Denton, 1966; Griffin & Reinhardt, 1996). The driving maneuver common to all of these situations is deceleration, and it becomes a problem when visual cues induce drivers to underestimate their speed and thus fail to decelerate to an appropriate speed.

To combat this problem, rural-urban threshold treatments are being introduced around New Zealand in increasing numbers, but their effectiveness seems to be variable. Some sites prove to be very effective, whereas some seem to have the opposite effect from that intended. Threshold treatments range from simple 70-kph roundels to complex treatments involving a combination of road narrowing, hatching, traffic islands, and oversized signs (see Fig. 15.2). The treatments share the common goal of slowing drivers, but the interpretation of the best method to achieve the goal have ranged from enhancing the attention-capturing capability of the speed signage, combating the effects of visual adaptation, and simple driver intimidation. Responding to concerns voiced by the Land Transport Safety Authority of New Zealand (LTSA), and with the cooperation of Transit New Zealand and a local roading authority, we undertook a test program designed to identify the threshold features that are

FIG. 15.2.   A typical rural-urban threshold treatment in simulation.

effective in slowing drivers to urban and suburban speeds. The remainder of this section describes that test program.

It has been estimated that 90% of the information needed for driving is visual (Bartmann, Spijkers, & Hess, 1991; Kline, Ghali, Kline, & Brown, 1990; Sivak, 1996). However, drivers concentrate their eye movements on only a limited portion of the visual field, which becomes narrower and deeper with increased vehicle speeds (Bartmann et al., 1991). On rural roads the percentage of eye fixations on driving-irrelevant objects (e.g., buildings, advertising signs, shrubs) is substantially higher than for urban freeways and roads and is characterized by longer duration fixations or gazes (Bartmann et al., 1991). Research has shown that traffic signs and other traffic control devices typically consume a total of 15% to 20% of a driver's attentional capacity (Hughes & Cole, 1986). The amount of attention given to traffic signs, however, is not enough to ensure that any more than 1 in 10 of the signs is noticed by drivers (Hughes & Cole, 1984; Shinar & Drory, 1983). A range of other studies have also noted generally low levels of attention and recall for warning signs and traffic control devices and have questioned the effectiveness of the current system of traffic and warning signs (Fischer, 1992; Johansson & Backlund, 1970; Macdonald & Hoffmann, 1991; Summala & Hietamaki, 1984). Thus, common sense would suggest that introducing larger speed signs, to increase the likelihood of capturing driver attention, might be effective in getting drivers to slow down.

Another approach to understanding drivers' failure to decelerate to appropriate speeds can be based on the effects of visual adaptation. A perceptual system that is presented with a constant stimulus for an extended amount of time will suffer from the effects of adaptation, where the sensation invoked reduces in magnitude as a function of time (Gibson, 1937). There are abundant data that show that after an extended exposure to a certain speed, drivers will underestimate the speeds at which they are traveling (Denton, 1966, 1973, 1976, 1977, 1980; Evans, 1970; Recarte & Nunes, 1996; Snider, 1967). Further, as drivers decelerate, the effects of sensory adaptation are compounded by a visual motion after-effect (VMAE), a sensation in the opposite direction to that invoked by the stimulus, once again causing the motorist to underestimate his or her speed. These perceptual phenomena can be counteracted through increases in the structure and complexity of the visual environment. It has been suggested that that local edge rate (i.e., the speed at which specific portions of the environment flow past the moving observer) provides the most salient egospeed information for drivers (Owen, Wolpert, Hettinger, & Warren, 1984; Smith, 1994; Warren, 1982). It follows that an increase in edge rate leads to the perception of increased speed. Regarding the perceptual issues of adaptation and VMAE, the most effective rural threshold design would be that which increases the structure and complexity of the visual environment, thus increasing the edge rate presented to the motorist. In practical terms, this can be interpreted to mean that for a rural-urban threshold to be effective the design has to produce an increase in the edge rate in the visual environment.

Alternatively, Fildes and Jarvis (1994) cite research that indicates that increases in the structure of the environment (increasing the edge rate) tend to be most effective in slowing motorists down in situations where the driver feels intimidated. When an individual feels threatened in a driving situation, he or she will rely more on sensory input (as opposed to cognitively mediated rules) to perceive and maintain the speed at which he or she is traveling. So, in the case of rural-urban thresholds, oversized signs can act in two ways to slow motorists down. First, they provide a small but critical amount of edge rate information to peripheral vision. Second, the oversized signs intimidate the motorist, forcing him or her to rely on sensory input alone to judge speed, creating a perceptual situation in which the countermeasure is most effective.

A series of laboratory tests and field measurements were undertaken to evaluate 11 different rural threshold designs in terms of their effectiveness in slowing drivers to the prescribed speed. In the laboratory, simulated roads were presented to participants on a medium-fidelity driving simulator comprised of a partial automobile cab with steering wheel and foot pedal controls, a fully instrumented dashboard, and driving scenes projected full size on a wall screen in front of the simulator. In a complimentary test, a proposed rural-urban threshold was constructed and tested in the simulator from design blueprints prior to its placement on the road. Field data were collected at the site before and after its placement and compared with the effects obtained for the same threshold in the simulator. One test employed an additive factors method (where a progression in design entails adding more features to the previous design) to investigate the effectiveness of alternate typical rural threshold designs. Another test was designed to isolate the effects of attention and speed adaptation by removing the speed restriction information from the various simulated rural threshold designs so that any effects of sign design and lateral proximity of the traffic islands could be attributed to perceptual factors alone.

When the effectiveness of different rural threshold designs were tested in the simulator, it was found that as the complexity of the rural threshold designs increased, there was a concomitant increase in the effectiveness of the treatment. It was further observed that the single most effective aspect of rural threshold design was the inclusion of oversized signs. Although this observation suggested that capturing a driver's attention was the key to threshold effectiveness, when we removed the speed restriction information from the oversized signs the rural threshold designs were still effective. That is, even when there was no statutory reason for drivers to slow down at the signs (i.e., no lowering of the speed limit), they still decreased their speed, presumably due to perceptual factors associated with increased edge rate information.

One of the interesting results of the test program experiment was that, although there were significant differences in speed through the threshold, by 250 m downstream all of these significant differences had disappeared. Further, in the case of signs without speed restrictions, there was a tendency for drivers to return to a speed of a greater magnitude than that at which they were originally traveling. Ultimately, these threshold designs were making participants speed up, and this is problematic. It is notable that some on-road implementations of rural thresholds have been seen to have similar effects on motorists. This suggests that the effects of a rural-urban threshold are transitory and that to achieve full effectiveness complimentary countermeasures may need to be introduced further downstream. Another finding of note was a habituation effect such that the earlier a particular rural threshold design was presented in the test sequence, the more effective it was. With repeated exposure, the thresholds' effectiveness diminished. This trend too has been noted in on-road implementations, some thresholds producing an average speed reduction of 10.4 kph at 6 months but only 5.2 kph at 12 months after implementation. Finally, the simulator test of the proposed threshold conducted prior to its implementation on a local road predicted only slight or marginal effectiveness. Early indications from actual traffic speeds obtained after the threshold was completed suggest that the prediction was accurate.

The test program demonstrated that the introduction of oversized signs and a restricted lateral proximity were effective in slowing drivers down. The data suggest that the introduction of oversized signs worked to both intimidate the motorist and to produce a critical increase in edge rate in peripheral vision, causing the observer to perceive an increase in speed. Further, we noted the presence of habituation effects such that the effects of rural-urban thresholds did not persist downstream of the treatments and that even their immediate effects will diminish with repeated exposure. The results of this testing program are being used by the LTSA and Transit New Zealand to draft design guidelines for urban-rural thresholds. A follow-on test program to investigate complimentary countermeasures designed to maintain speed reductions downstream of the initial rural threshold treatment is also under discussion with LTSA and Transit New Zealand.

DRIVER FATIGUE TESTING

The prevalence and effects of driver fatigue in New Zealand have been the focus of another human factors test effort employing a variety of testing methodologies. Performance deficits of various sorts have been well documented in fatigued drivers: Drivers are less able to detect and report street signs as driving time increases (Naatanen & Summala, 1976). Long distance bus and truck drivers report poorer awareness of both traffic signs and other traffic when fatigued (Feyer & Williamson, 1993) and poor gear changing and slowed reaction time to traffic and other driving situations (Hartley et al., 1996). Errors of commission and omission have also been reported as driving performance decrements caused by fatigue (Brown, 1994). Errors of commission made by experienced drivers may initially be action slips (e.g., driving to a familiar but incorrect destination) resulting from an increasing reliance on well-practiced and automatic motor sequences. At more extreme levels of fatigue, errors may take the form of mistakes or poor decision making (e.g., misjudging braking distances). Increased errors and long responses in a reaction-time task were found to occur in long truck journeys (Hartley et al., 1996). Similarly, increases in risky driving maneuvers have been demonstrated to increase by up to 50% with increased time on task (Holding, 1983).

The National Transportation Safety Board (NTSB) of the United States has long regarded the role of fatigue in transportation as a significant concern for the road, air, sea, and rail industries (National Transportation Safety Board, 1995). The actual incidence of fatigue-related accidents, however, remains unknown. As fatigue is a subjective psychological experience, the momentary presence or absence of fatigue cannot be readily measured after the fact; its role in a crash can only be inferred from indirect evidence and potentially unreliable reports from individuals involved (Summala & Mikkola, 1994). Furthermore, the effects of fatigue can vary across individuals and situations, from mild performance decrements to nearly complete incapacitation (National Transportation Safety Board, 1995). Consequently, crash statistics typically include only the more easily measured or inferred definitions of accident causation and thus may underestimate the true scope of fatigue-related accidents.

In New Zealand, as with many other countries, fatigue in the transport industry is regulated through legislation and restrictions based on hours of service. Hours of service regulations (HSRs) commonly limit the maximum number of hours driven per day or per week and specify minimum rest periods; alternatively, they may regulate actual driving hours or working hours that include nondriving activities. The New Zealand Transport Act requires commercial truck drivers to maintain a logbook recording details of driving and nondriving work periods and rest periods. Drivers' log books are used to check compliance with prescribed driving hours and duty cycles. The monitoring of the driving hours of truck drivers is at present the only method of measuring (or inferring) the incidence of fatigue on New Zealand roads. Yet the fact that log books contribute little to fatigue management and are widely abused was recognized in the *Report of the Transport Committee on the Inquiry into Truck Crashes* (New Zealand

House of Representatives, 1996). In conjunction with the Road Safety Trust, we undertook the test effort described in the remainder of this section in an attempt to find how common driver fatigue is in New Zealand and the degree to which New Zealand truck drivers suffer from fatigue-related effects on their driving performance.

One of the first activities in the test program was the development of a paper-and-pencil driver survey. The goals of the driver survey were (1) to identify key demographic and work-rest patterns of truck drivers in New Zealand, (2) to collect information on drivers' attitudes about fatigue and propensity toward daytime sleepiness for comparison to other studies of driver fatigue, and (3) to obtain self-assessments on drivers' momentary levels of fatigue. To minimize the disruptive effects of the testing protocol on the drivers' schedules, it was also desirable to make the questionnaire short enough to complete in 10–15 minutes. The questions were adapted from a variety of other surveys related to fatigue, truck driving, or both, and the resulting draft questionnaire was reviewed by independent researchers in the field.

The next task involved the acquisition and configuration of a performance-based assessment instrument. The hardware and software chosen was a specially adapted version of the Truck Operator Proficiency System (TOPS), commercially available from Systems Technologies, Inc. TOPS is based on the combination of a standard driving task, a dual-axis subcritical tracking task (maintaining speed and steering in a controlled but unstable environment, a virtual roadway affected by the appearance of random wind gusts requiring steering correction), and a secondary or divided attention task requiring driver monitoring and periodic responses (symbols presented in the side mirrors to which the driver responded by indicating for a left turn or right turn, or by pressing the horn button). The TOPS driving scenarios were modified to make them relevant to New Zealand drivers (i.e., road markings, left-side driving, display of metric rather than imperial speedometer units). The resulting performance test consisted of an 8-minute testing session composed of a straight-road scene and 27 to 30 divided attention events. The test scenario was divided into four 2-minute data collection blocks for analysis purposes. Data from the first 2-minute block was excluded from the analysis. Twenty driver performance metrics (e.g., average lane deviation, average vehicle heading error, average throttle activity, average time for a divided attention response) were collected throughout the test scenario and were used to calculate a pass-fail score by the TOPS performance index algorithm.

Drivers were sampled at a variety of sites along long-haul truck routes in New Zealand, including truck stops and depots in Northland, Auckland, Bay of Plenty, Gisborne, Hawkes Bay, Taradale, Wanganui, and the Wellington ferry terminal. Data were collected from a total of 606 truck drivers. Drivers taking the test a second time and occasional drivers (e.g., farmers moving stock) were removed from the data set prior to analysis, leaving a total sample of 596 truck drivers. During a typical data collection session a trailer containing the testing equipment was parked so that it was visible and accessible to drivers as they moved between their vehicles and the dispatchers' office, break room, or dining room. Individual drivers were approached and given a brief overview of the purpose of the study and the time required to complete the test. Drivers expressing a willingness to participate were asked to sign an informed consent form that, in writing, guaranteed confidentiality of their simulator performance and responses to the survey questions. The paper-and-pencil survey was verbally administered and was followed by the tester showing the driver to the trailer for the driving simulator performance portion of the test. Following a 2-minute orientation the drivers were given a final opportunity to ask questions, and then the 8-minute performance test was conducted. At the end of the performance test each driver was informed as to whether or not they had passed the test, and, in the case of a failure, the nature of the failure (i.e., the component of the test not meeting the criterion) was explained to the driver in some detail. All drivers were then thanked for their participation, provided with a LTSA fact sheet on driver fatigue, and given a complementary chocolate bar.

Drivers were tested across the complete range of typical shifts, ranging from 0 to 19 hr of driving immediately prior to participating in the test. When asked their typical number of days worked per week, the participating drivers reported an average of 5.35 (ranging from 0.5 to 7 days). The drivers also reported an average shift length of 11.11 hr (ranging from 3 to 16 hr, $SD = 2.02$ hr). Examining the activity data from the 48 hr prior to the survey, however, shows that the number of hours spent driving in the previous 24 hr actually ranged from 0 to 23 hr (an average of 8.98 hr, $SD = 3.99$ hr). This latter statistic is shown in Fig. 15.3, and as can be seen over 30% of the drivers sampled had exceeded their 11 hr service maximum in the previous 24 hr. The total number of hours of driving in the previous 48 hr ranged from 0 to 45 (an average of 15.895 hr, $SD = 7.28$ hr). The full activity data from the drivers shown in Table 15.1 contain several other findings of interest. Although the average amounts of sleep appear fairly reasonable, there were drivers reporting as little as 3 hr of sleep in the past 48 hr. Also of note are the relatively low numbers of meals reported by the drivers (a snack counted as 0.5 meal). Finally, some drivers reported substantial hours spent in physical work or desk work, perhaps reflecting a pattern of duty rotation (10% of the sample reported more hours of physical work or desk work than driving over the previous 48 hr).

One other finding from the survey worth noting in this brief review concerned eight questions on the degree to which the drivers were likely

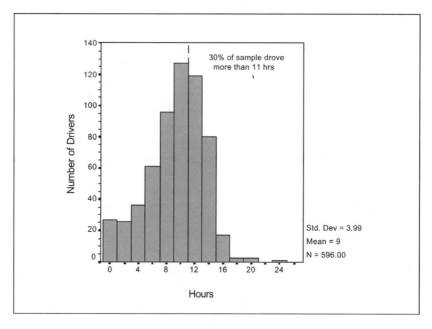

FIG. 15.3.   Hours of driving in past 24 hours.

TABLE 15.1
Driver Activity Data

	Mean	SD	Minimum	Maximum
Hours driving in past 24 hr	8.978	3.993	.00	23.00
Hours driving in past 48 hr	15.895	7.283	.00	45.00
Hours sleeping in past 24 hr	7.241	1.723	.00	16.00
Hours sleeping in past 48 hr	14.688	2.947	3.00	27.00
Meals in past 24 hr	1.901	0.711	.00	4.00
Meals in past 48 hr	3.676	1.268	.00	8.00
Physical work/exercise past 24 hr	1.242	2.300	.00	17.00
Physical work/exercise past 48 hr	2.702	4.217	.00	23.00
Desk work in past 24 hr	0.418	1.576	.00	15.00
Desk work in past 48 hr	0.810	2.759	.00	27.00
Relaxing in past 24 hr	3.940	2.925	.00	17.00
Relaxing in past 48 hr	8.746	5.760	.00	28.00
Partying in past 24 hr	0.216	0.937	.00	8.00
Partying in past 48 hr	0.555	1.844	.00	17.00

to feel sleepy in various situations (e.g., how likely are you to doze or fall asleep when watching TV, sitting quietly after lunch?). These questions, known collectively as the Epworth Sleepiness Scale (Maycock, 1995), were included to provide a point of comparison with studies linking the likelihood of daytime sleepiness with accidents by car drivers and heavy-goods vehicle drivers. The average Epworth score in our sample (6.13) was somewhat higher than that observed for heavy-goods vehicle operators in the United Kingdom (5.7).

As regards the performance test, of the 596 drivers in the sample, 450 met the performance standards associated with the TOPS test (a failure rate of 24%). Inspection of the data, however, revealed several areas where the failure rates were particularly high. As shown in Fig. 15.4, some types of freight were associated with higher than average rates of failure. The figure shows that logs, stock, and furniture freight categories are substantially worse than average (the dashed line figure shows the overall failure rate). Interestingly, these failure rates paralleled the distribution of crash rates across transport industry segments in national crash statistics. Another factor that played a significant role in the pass-fail rates associated with the performance test was the average distance driven per shift reported by the drivers. Of drivers reporting 250 km or fewer driven per shift, 27% failed the performance test. The amount of sleep reported also

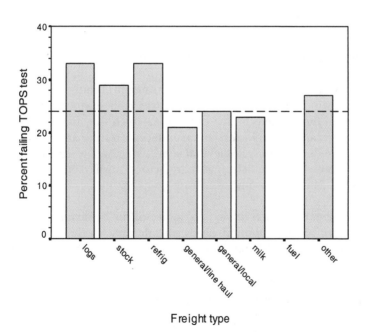

Freight type

FIG. 15.4.   Psychomotor test failure rates for each freight category.

appeared to be a factor in that drivers reporting fewer than 10 hours of sleep were more likely to fail the test (a 28% failure rate). Perhaps the greatest influence on failure rates, however, was the age of the drivers. Thirty-four percent of the drivers over the age of 37 (the mean driver age) failed the performance test, as compared with only 17% of drivers age 37 or younger. Looking at this finding from another perspective, 62% of the drivers who failed the performance test were over 37 years old.

The results of the test provide the most comprehensive look to date at driver fatigue in New Zealand truck drivers. The results of the activity survey show that a considerable number of drivers are operating in excess of the hours of service regulations and paint a rather discouraging picture of the workaday world of the New Zealand truck driver. Further, the results of the Epworth Sleepiness Scale show that New Zealand drivers have somewhat higher levels of daytime sleepiness than do heavy-goods vehicle operators in the United Kingdom. The performance testing results serve to amplify these findings with high failure rates for drivers in certain freight categories and short-haul routes. Whether the age difference associated with the performance test reflects some underlying resistance to fatigue effects or whether it is an artifact of the computer-based testing procedure cannot be ascertained at this point. The future of this work in calibrating the performance test to New Zealand road conditions and other performance standards such as those for blood alcohol levels is currently under discussion.

DRIVER-VEHICLE-ROAD INTERACTIONS

As regards the human factors of the road transportation system, a truly integrated testing approach must address drivers' behavioral and cognitive processes in the context or environment in which they function (i.e., representative roads and vehicles). In an ambitious testing and research program we are developing a model that represents the interrelationships between various driver, vehicle, and road factors. It will enable us to look at issues such as the following:

- The extent to which drivers are aware of the behavior of the trailers they are towing. Opinion varies considerably on this issue. If driver feedback is found to be a problem, it may be possible to develop appropriate warning devices for drivers or address it through weights and dimension changes.
- Whether truck speed advisory signs would be effective in reducing the number of rollovers.
- Road geometry features that cause problems for drivers.

- Improved productivity through the better matching of heavy vehicle weights and dimensions and road design.

The various components of the model will be linked to form an overall model that comprises a series of driver, vehicle, and road modules. Validation of each module is achieved incrementally through on-road measurements using an instrumented vehicle. The vehicle performance module uses complex multibody simulation software, and vehicle models typical of New Zealand vehicle configurations are being developed. These vehicles will be able to "drive" over three-dimensional road surfaces that simulate actual New Zealand roads. Measured road geometry data are already available for all of the major New Zealand highways. These data include the following at 10-m intervals: three coordinates (North, East, and elevation), three angles (heading, gradient, and camber), and horizontal and vertical curvature.

The driver module is based on driver performance profiles for various driving events and driving situations. Specific driving events such as braking, reaction to road hazards, and the maintenance and adjustment of vehicle speeds will be used to determine how drivers' response times vary as a function of the drivers' attentiveness and involvement in the driving task. Reviewing the statistics on road crashes from an information-processing perspective, a disproportionate amount of crashes appear to be the result of the perception stage of the perception-decision-action cycle governing driver behavior (Charlton, 1997). Further, crashes do indeed result from poor driving skills (action stage) and faulty risk perception (decision stage), but they are overshadowed by the number of crashes attributable to loss of attention and information processing overload (Charlton, 1997; Treat et al., 1977). The role of attentional factors (e.g., situation awareness and cognitive workload) in contributing to safe transportation systems (land based, air, and maritime) is well established and receiving increased attention by regulatory bodies, enforcement agencies, and industry groups (Baas & Charlton, 1997; Charlton & Ashton, 1997). Further, it has been hypothesized that driver attentiveness is a key variable affecting the time course of a driver's perception-decision-action cycle (White & Thakur, 1995).

We have employed a combination of observation, experimentation, and field testing in the initial development of the driver-vehicle-road interaction model. One early test manipulated the cognitive workload of drivers by introducing a secondary task into the driving situation. The measures of primary interest were the effects of varying levels of workload on driving performance (lane positioning, braking, speed maintenance, stopping times, and occurrence of collisions). The testing apparatus was a medium-fidelity driving simulator comprised of a partial automobile cab with steering wheel and foot pedal controls, a fully instrumented dashboard, and driving scenes projected full size on a wall screen in front of the simulator. A typical driving scene from the simulator is shown in Fig. 15.5. Acceleration, deceleration, and steering dynamics were based on vehicle dynamics equations derived from a vehicle performance model of a 3,000-kg truck.

Three different driving scenarios, based on actual roads and containing representative traffic levels, were developed for presentation on the driving simulator described in the previous paragraph. The driving scenarios

FIG. 15.5.   Road scenes from the driving simulator.

represented typical rural and semirural road conditions with typical road geometry, lane markings, road signs, and roadside objects, such as trees and houses. Each driving scenario also contained three braking vehicles that required the participants to stop to avoid a collision. The speed of these leading vehicles at the time of braking was 80 kph. Their braking was indicated via brake lights and began when the participants approached to a range of 50 m. The deceleration rate of both the lead vehicle and the simulator vehicle was determined by the vehicle dynamics equations. The brake perception-reaction times of the participants to the three hazards were the principal measures of driving performance. During the driving scenarios, the participant's mental workload was manipulated by the inclusion of a secondary embedded task (periodically retuning the car radio) to systematically control and monitor the levels of cognitive workload experienced while driving. The radio signal (tuned to the local university radio station) used for the embedded secondary task degraded to static on an average of every 11 s (high workload), every 35 s (moderate workload), or once every 2 minutes (low workload). Participants retuned the radio by pressing a push button on the dashboard to the left of the steering wheel (in the space normally occupied by automobile radios). One of the three levels of the radio tuning task (low, medium, or high) accompanied each driving scenario, the task level was counterbalanced across the three scenarios, and the presentation order counterbalanced across all participants.

Contrary to expectations, brake perception reaction times tended to be slowest under the medium workload condition and fastest under the low and high workload conditions (see Fig. 15.6). Brake perception-reaction times do not by themselves, however, necessarily constitute safe driving. As Fig. 15.6 also shows, the medium workload condition was associated with the fewest number of collisions with other vehicles. One possible reason for the somewhat paradoxical finding that medium workload is associated with both slower brake perception-reaction times and a lower rate of collisions is, as can be seen in the figure, the average speed the participants were driving at the time the lead car initiated braking was greater for the low and high workload conditions. This finding was unexpected and would appear to indicate that maintenance of appropriate vehicle speed requires commitment of attentional resources. When attentional resources are over committed (as in the high workload condition) or are diverted elsewhere (low workload, driving consigned to automatized processes), drivers may drive too fast.

Participants also displayed much better lane position under the medium workload condition, as evidenced by their lateral displacement scores. Driving in the high workload condition led to erratic lane positioning. In contrast, drivers in the low workload condition did not appear to

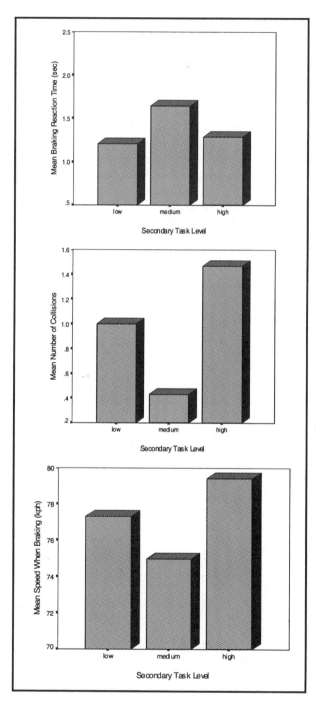

FIG. 15.6.   Brake perception-reaction times, collision rates, and speed as a function of cognitive workload.

be attending to their lane position, as they took few corrective actions to improve it. Off-accelerator reaction times appeared to be linearly related to the level of cognitive workload. Drivers in the low workload condition were first to react to a vehicle stopping ahead of them, whereas drivers in the high workload condition took the longest to react. Coupled with their brake perception-reaction time data, medium workload drivers appeared very deliberate in their reactions to traffic, yet their slower speed allowed them to avoid collisions. In contrast, drivers in the high workload condition hit the brakes very quickly (once they had detected the vehicle stopping ahead of them), but their high speed and the fact that they detected the vehicle rather later than in low and medium workload conditions resulted in a high rate of collisions.

A field test of the findings obtained in the laboratory was conducted on a 7.3 km circuit of rural and semirural public roads south of Hamilton, New Zealand. Ten volunteer drivers participated in the field trial. The test vehicle was a 1998 Mitsubishi Canter with two axles, a tare weight of 3,130 kg, and a 2.745-m wheelbase (see Fig. 15.7). It was instrumented to collect engine speed, brake pedal and accelerator activity, following distance, lane displacement, ground speed, steer angle, lateral acceleration, pitch, and roll rate. The driving scene was also recorded by a video camera mounted on the dashboard of the truck, and an event marker was used to mark the beginning of each lap of the circuit. All data were sampled at 25 Hz. An analogue filter system was used to filter all channels; the cut-off frequency was 10 Hz. The filtered signal was acquired using LabVIEW running on a 486 DX 33 desktop PC and a National Instruments ATMIO 16 A/D Board. As Fig. 15.7 shows, the data acquisition system was mounted in the cargo bay of the truck.

Before starting the trial the participants were given verbal instructions regarding the nature of the driving task, the secondary task, and the braking maneuvers required. The participants drove the on-road test circuit once as practice and were given an opportunity to ask any questions about the field trial tasks. The participants then drove the circuit three times, with a 1 to 2 minute rest break between each lap. The secondary task for the road trials consisted of responding to a prerecorded 3-s tone presented on a variable interval schedule against background music recorded from local radio stations. Participants were instructed to tap the dashboard whenever they heard the tone. One of the three levels of the secondary task (none, medium—average of one tone every 36 s—or high—average one tone every 11 s) was presented on each lap of the test circuit, with the order of presentation counterbalanced across the participants. The lead vehicle (driven by a member of the research team) braked quickly 3 or 4 times during each lap of the test circuit, requiring the participant to stop the test vehicle (usually between 70–80 kph).

FIG. 15.7.　Instrumented field test vehicle and on-board data acquisition system.

To reduce exposing other motorists to hazards associated with the field trials, the test vehicle was followed by a member of the research team in a vehicle displaying a sign advising following motorists that the group of vehicles may stop unexpectedly. The test vehicle displayed an identical sign also directed at following motorists. In addition, the research team members in the lead and tail cars were in constant communication via radio telephones to ensure that all emergency stops were conducted at a time when there were no other motorists in the vicinity and on a section of the circuit where there was plenty of space for other motorists to see and safely overtake the three field trial vehicles as necessary.

The test vehicle was instrumented to collect several measures of driving performance during each driving scenario. As with the simulator test, the primary measure of interest was brake perception-reaction time. This was measured, in milliseconds, from the moment the lead car's brake lights were illuminated, and it began decelerating to the time the experimental participant placed his or her foot on the simulator's brake pedal. In addition, the speed of the participant's vehicle, the average lateral deviation from the center of the lane, and the lateral deviation variability were recorded.

Overall, brake perception-reaction times were somewhat shorter in the field test than those in the simulator test (0.35 s shorter when averaged across all conditions), with the greatest difference occurring in the medium and high workload conditions. A different pattern was observed for the off-accelerator reaction times, with the low workload condition in the simulator being shorter than those observed in the instrumented vehicle. Under high workload conditions the results for the two tests were nearly identical in mean and variance. In the medium workload condition the results were quite similar with the reaction times for the field test being more variable and slightly slower than the simulator test. Averaging the off-accelerator reaction times across all workload conditions, the difference between the two tests was approximately 0.10 s (0.64 s for the field trial, as compared with 0.53 s for the simulator). Finally, drivers in the instrumented vehicle showed greater variation in average lateral displacement (lane deviation), as compared with the simulator test, but essentially the pattern of results was produced.

In summary, the results obtained from the two tests were very similar, and where there were differences across the two situations (brake perception-reaction times for medium and high workload conditions, off-accelerator reaction times for low workload, and vehicle speeds under high workload conditions), it was the case of drivers in the field trial behaving somewhat more conservatively. These differences are not altogether surprising when one considers some of the differences in the conditions of the two tests. Participants in the field test were driving a large unfamiliar vehicle with the foreknowledge that they would have to execute rapid braking maneuvers at unpredictable points in the drive. Additionally, the prominence of the observer/data collector in the passenger seat may have further contributed to the cautious driving behavior of the field test participants. Finally, there were also differences in vehicle feedback presented to the drivers resulting from the simplified vehicle dynamics in the simulator (which did not alter vehicle speed or engine noise based on road geometry). Given these differences in driving conditions, somewhat different driving performance might be expected. For the few performance differences observed, it remains an arguable point which of the tests is more representative of typical driving conditions. Viewed in terms of the larger goal of obtaining a variety of driver performance data across a range of drivers at different levels of driver attentiveness and quantifying elements of the perception-decision-action cycle for inclusion in the road-vehicle interaction model, the tests were highly successful. Additional simulator and field trials are being used to represent other aspects of driver behavior, such as speed maintenance, vehicle following distances, overtaking (passing), and speed selection at curves. These tests also include alternative secondary tasks such as navigation, mileage estimation, and cell phone use.

SUMMARY

The three test programs described above continue at the time of this writing. They were chosen to represent the range of human factors test methods and approaches we are using to improve the safety of our road transport system. Other projects underway, or in the early stages of development, include a test of alternative passing lane designs using the laboratory simulator and observations in the field, a large-scale survey of risk perception and patterns of road use for various road user demographic groups, the development of a testing procedure for examining the comprehensibility and attentional conspicuity of various road signs and traffic control devices, and an assessment of the environmental and community impact of projected road use trends in New Zealand. Underlying all of these test programs is our conviction that a system perspective (road, vehicle, and driver factors) must be adopted to improve the safety of our roads, and further that we, as human factors professionals, have an imperative to do our utmost to make them safe.

REFERENCES

Baas, P. H., & Charlton, S. G. (1997). Heavy vehicle safety in New Zealand. *Proceedings of the International Large Truck Safety Symposium*, 121–125.

Bartmann, A., Spijkers, W., & Hess, M. (1991). Street environment, driving speed and field of vision. In A. G. Gale, I. D. Brown, C. M. Haselgrave, & S. P. Taylor (Eds.), *Vision in Vehicles—III*. Amsterdam: Elsevier Science B.V.

Brown, I. (1994). Driver fatigue. *Human Factors, 36,* 298–314.

Charlton, S. G. (1997). *Time course of drivers' perception-decision-action cycle as a function of cognitive workload and situation awareness: Final report* (TARS Technical Report 97-111; Report contracted by Transport Engineering Research NZ Ltd.). Hamilton, NZ: Traffic and Road Safety Group, University of Waikato.

Charlton, S. G., & Ashton, M. E. (1997). *Review of fatigue management strategies in the transport industry* (TARS Technical Report 97-91; Report contracted by Land Transport Safety Authority NZ/Road Safety Trust). Hamilton, NZ: Traffic and Road Safety Group, University of Waikato.

Charlton, S. G., & Baas, P. H. (1998). The interaction of driver, vehicle, and road contributing factors in truck crashes. *Proceedings of road safety: Research, policing, education conference, 2* (pp. 209–213). Wellington, NZ: Land Transport Safety Authority/New Zealand Police.

Denton, G. G. (1966). A subjective scale of speed when driving a motor vehicle. *Ergonomics, 9,* 203–210.

Denton, G. G. (1973). *The influence of visual pattern on perceived speed at Newbridge M8, Midlothian* (TRRL Report LR). Department of the Environment. Crowthorne, Berkshire: Transport and Road Research Laboratory.

Denton, G. G. (1976). The influence of adaptation on subjective velocity for an observer in simulated rectilinear motion. *Ergonomics, 19,* 409–430.

Denton, G. G. (1977). Visual motion aftereffect induced by simulated rectilinear motion. *Perception, 6,* 711–718.

Denton, G. G. (1980) The influence of visual pattern on perceived speed. *Perception, 9,* 393–402.

Evans, L. (1970). Speed estimation from a moving automobile. *Ergonomics, 13,* 219–230.

Feyer, A-M., & Williamson, A. M. (1993). The influence of operational conditions on driver fatigue in the long distance road transport industry in Australia. *Proceedings of the Human Factors Society 37th Annual Meeting,* 590–594.

Fildes, B. N., & Jarvis, J. (1994). *Perceptual countermeasures: Literature review* (Report CR4/94). Canberra, New South Wales: Road Safety Bureau, Roads & Traffic Authority (NSW).

Fischer, J. (1992). Testing the effect of road traffic signs' informational value on driver behavior. *Human Factors, 34,* 231–237.

Gibson, J. J. (1937). Adaptation with negative after-effect. *Psychological Review, 44,* 222–244.

Griffin, L. I., III, & Reinhardt, R. N. (1996). *A review of two innovative pavement marking patterns that have been developed to reduce traffic speeds and crashes.* Available: http.phoenix. webfirst.com/aaa/Text/Research/pavement.html

Hartley, L. R., Arnold, P. K., Penna, F., Hochstadt, D., Corry, A., & Feyer, A.-M. (1996). *Fatigue in the Western Australian transport industry. Part one: The principle and comparative findings.* Perth, Western Australia: Murdoch University.

Holding, D. (1983). Fatigue. In R. Hockey (Ed.), *Stress and fatigue in human performance.* Chichester, England: Wiley.

Hughes, P. K., & Cole, B. L. (1984). Search and attention conspicuity of road traffic control devices. *Australian Road Research, 14,* 1–9.

Hughes, P. K., & Cole, B. L. (1986). What attracts attention when driving? *Ergonomics, 29,* 377–391.

Johansson, G., & Backlund, F. (1970). Drivers and road signs. *Ergonomics, 13,* 749–759.

Kline, T. J. B., Ghali, L. M., Kline, D. W., & Brown, S. (1990). Visibility distance of highway signs among young, middle-aged, and older observers: Icons are better than text. *Human Factors, 32,* 609–619.

Macdonald, W. A., & Hoffmann, E. R. (1991). Drivers' awareness of traffic sign information. *Ergonomics, 34,* 585–612.

Maycock, G. (1995). *Driver sleepiness as a factor in car and HGV accidents* (TRL Report 169). Crowthorne, Berkshire, UK: Safety and Environment Resource Centre, Transport Research Laboratory.

Naatanen, R., & Summala, H. (April, 1976). *Road user behavior and traffic accidents.* Amsterdam: North-Holland.

National Transportation Safety Board. (1995). *Factors that affect fatigue in heavy truck accidents. Volume 1: Analysis* (Safety Study NTSB/SS-95/01). Washington, DC: Author.

New Zealand House of Representatives. (August, 1996). *Report of the Transport Committee on the inquiry into truck crashes.* Wellington, NZ: Author.

Owen, D. H., Wolpert, L., Hettinger, L. J., & Warren, R. (1984). Global metrics for self-motion perception. *Proceedings of the Image Generation/Display Conference III, 3* (pp. 406–415). Phoenix, AZ: The IMAGE Society.

Persaud, B., Bahar, G., Smiley, A., Hauer, E., & Proietti, J. (1999). Applying the science of highway safety to effect highway improvements—A multi-disciplinary approach. *Proceedings of the Canadian Multidisciplinary Road Safety Conference XI.* Halifax, Nova Scotia.

Recarte, M. A., & Nunes, L. M. (1996). Perception of speed in an automobile: Estimation and perception. *Journal of Experimental Psychology: Applied, 2,* 291–304.

Shinar, D., & Drory, A. (1983). Sign registration in daytime and nighttime driving. *Human Factors, 25,* 117–122.

Sivak, M. (1996). The information that drivers use: Is it indeed 90% visual? *Perception, 25,* 1081–1089.

Smith, M. R. (1994). *Perceived aftereffects in simulated rectilinear motion as a function of edge rate and flow velocity.* Unpublished manuscript, University of Canterbury.

Snider, J. N. (1967). Capability of automobile drivers to sense vehicle velocity. *Highway Research Record, 159,* 25–35.

Summala, H., & Hietamaki, J. (1984). Driver's immediate responses to traffic signs. *Ergonomics, 27,* 205–216.

Summala, H., & Mikkola, T. (1994). Fatal accidents among car and truck drivers: Effects of fatigue, age, alcohol consumption. *Human Factors, 36*(2), 315–326.

Tingvall, C. (1998). The Swedish 'vision zero' and how parlimentary approval was obtained. *Proceedings of Road Safety: Research, Policing, Education Conference, 1* (pp. 6–8). Wellington, NZ: Land Transport Safety Authority/New Zealand Police.

Treat, J. R., Tumbas, N. S., McDonald, S. T., Shinar, D., Hume, R. D. Mayer, R. E. Stansfin, R. L., & Castellen, N. J. (1977). *Tri-level study of the cause of traffic accidents* (Tech report DOT-HS-034-3-535-77 TAC). Bloomington: Indiana University.

Warren, R. (1982, October). *Optical transformation during movement: Review of the optical concomitants of egomotion* (Tech. Rep. AFOSR-TR-82-1028). Bolling AFB, DC: Air Force Office of Scientific Research. (NTIS No. AD-A122 275)

White, D., & Thakur, K. (1995, June). *Stability in the real world—Influence of drivers and actual roads on vehicle stability performance.* Paper presented at the Fourth International Symposium on Heavy Vehicle Weights and Dimensions, Ann Arbor, MI.

Chapter 16

Human Factors Evaluation
of Hybrid Human-System Interfaces
in Nuclear Power Plants

John M. O'Hara
William F. Stubler
James C. Higgins
Brookhaven National Laboratory

Nuclear power plant personnel play a vital role in the productive, efficient, and safe generation of electric power. Operators monitor and control plant systems and components to ensure their proper operation. Test and maintenance personnel help ensure that plant equipment is functioning properly and restore components when malfunctions occur. Personnel interact with the plant through a wide variety of human-system interfaces (HSIs), including alarm systems, information systems, controls, diagnostic and decision support systems, and communications systems. The primary HSIs are located in a main control room; however, there are specific HSIs distributed throughout the plant. Personnel actions are guided by plant operating procedures and by extensive training for both normal and emergency operations using plant-specific training simulators.

In the first edition of this handbook, we described the development of our approach to HFE evaluations of advanced nuclear power plants (O'Hara, Stubler, & Higgins, 1996a). The evaluation methodology, referred to as the HFE program review model (PRM) is a top-down approach. Top-down refers to an evaluation that starts at the top with high-level plant mission goals that are broken down into the functions necessary to achieve the mission goals. Functions are allocated to human and system resources and are broken down into tasks. Personnel tasks are analyzed for the purpose of specifying the alarms, information, decision support, and controls that will be required to allow plant personnel to accomplish their functions. Tasks are arranged into meaningful jobs assigned to individuals. The detailed design of the HSI, procedures, and training to support job task performance is the bottom of the top-down process. The evaluation methodology addresses each of these aspects of design.

In this chapter, we will describe the revisions we have made to the evaluation based on two developments occurring since the original publication: lessons learned from applying the methods and extension of the scope of the methodology to plant modernization using new computer-based interfaces. First, we used the PRM to support the Nuclear Regulatory Commission's review of several advanced reactor design certifications and for Department of Energy Operational Readiness Reviews. In addition, the evaluation methods have been adopted internationally by many design organizations and regulatory authorities. Some aspects of the methods have also been used in nonnuclear industry domains and by standards organizations, such as ISO, International Electrotechnical Commission (IEC), and IEEE Standards Association (IEEE-SA). These experiences have provided many lessons learned regarding HFE design evaluation and have provided insights into the types of revisions that are needed to improve the methodology.

The second development was the need to extend the scope of the methods from new, advanced plant applications to the evaluation of plant modernization programs. We began to examine the approach taken by existing plants to modernize their HSIs by integrating new, advanced technologies with the existing conventional technologies. Also, we discovered that there were significant human performance issues that were not addressed by our methodology because they were unique to the modernization environment. In the next section, we will briefly discuss hybrid HSIs and will follow with a discussion of the revisions to the evaluation method stimulated by them and by lessons learned from applying the methodology.

CONTROL ROOM MODERNIZATION: HYBRID HUMAN-SYSTEM INTERFACES

The HSIs in many nuclear power plants are being modernized for several reasons, including (1) the need for equipment replacement because of high maintenance costs or lack of vendor support for existing equipment; (2) enhancement of HSI functionality, plant performance, and reliability; and (3) enhancement of operator performance and reliability. Another reason is that HSIs are impacted by upgrades and modifications in other plant systems as well. The instrumentation and control system, for example, is the principal link between the HSIs and plant processes, systems, and components. As instrumentation and control systems change, there is an impact on the types of HSI technologies available to plant personnel.

The modernization is being accomplished by integrating advanced HSI with the original equipment. The term *advanced HSI* refers to technologies that are largely computer based, such as software-based process controls, computer-based procedure systems, and large-screen, overview displays. As advanced HSI technologies are integrated into control rooms based on conventional technologies, hybrid HSIs are created. Hybrid HSIs are characterized by the mixture of conventional and advanced technologies.

The potential benefits of advanced technologies are compelling. New digital systems can provide personnel with information they did not have with conventional systems. Improved instrumentation and signal validation techniques can help ensure that the information is more accurate, precise, and reliable. The high stability of digital control systems can reduce the need for operators to adjust control settings to compensate for drift. In addition, data processing techniques and the flexibility of computer-based information presentation offer designers the ability to present information in ways that are much better suited to personnel tasks and information processing needs. For example, alarm processing techniques can be applied that help reduce the number of alarms during process disturbances so that operators can attend to those that are most significant. Further, the alarm information can be integrated with other plant information to make the alarm more meaningful. The HSI can also provide operators with alarm management tools so that operators can readily, as their information needs dictate, sort alarms, obtain detailed information about the technical basis of the alarms, and obtain appropriate procedures for responding to the alarms. These technologies should result in improved power plant availability and safety through the avoidance of upsets to normal operation, forced outages, and unnecessary shutdowns.

Although advanced HSIs can greatly improve operator and plant performance, it is important to recognize that, if poorly designed and implemented, there is the potential to negatively impact human performance, increase human errors, and reduce human reliability (Galletti, 1996; O'Hara, Stubler, & Higgins, 1996b). Galletti (1996), for example, reviewed five events involving digital system upgrades in U.S. nuclear plants. He concluded the following:

> The discussed events focus attention on the fact that the finest details of system implementation, such as minor modifications to the HMI and operating procedures or subtle indications of system abnormalities, are often overlooked during operator acclimation to new systems. These subtleties however, have led to plant transients and delays in the mitigation of such transients. (p. 1161)

The U.S. National Academy of Sciences (NAS) reached a similar conclusion. The NAS identified upgrades of existing nuclear power plants as a specific area of concern because the design process must integrate the upgrade system with existing plant equipment and systems (NAS, 1995). In addition, consideration must be given to ensuring that an adequate technology and training base supports the modification, the retraining of operators, and any environmental sensitivities of the upgrade equipment.

Significant interest in human performance aspects of hybrid HSIs exists in the international nuclear community. The International Atomic Energy Agency (IAEA, 1998) recognized the following:

> the approaches and concerns for the implementation of advanced technologies in new plants are very different from those who perform backfits to existing operating plants. . . . New plants allow full-scale implementation of new advanced technologies, but backfitting is constrained by existing experiences and practices, particularly in the area of human-machine interface. (p. 16)

These observations led us to conduct an investigation of the trends in plant modernization and the identification of the human performance issues associated with them. Based on an analysis of literature, interviews, site visits, and event data, we found that there is an evolution from large control rooms with spatially dedicated displays and controls to compact, workstation-like consoles with computer-based controls and displays (O'Hara, Stubler, & Higgins, 1996b). Some of the specific trends and related issues identified are discussed in the following sections.

Changes in Automation

Plant control system upgrades can provide higher levels of automation than the original design (e.g., full-range feed-water control systems). Changes in automation can have a major effect on the operator's role, defined as the integration of the responsibilities that the operator performs in fulfilling the mission of plant systems and functions. Because the implementation of automation has been mainly driven by advances in technology, changes in automation often fail to provide a coherent role for operators. In addition, concerns have been identified regarding the design of the HSIs that link personnel to automated systems. Even when a process is fully automated, personnel must still monitor its performance, judge its acceptability, and, when necessary, assume control. HSIs have been lacking in their ability to support these personnel task demands. Automated systems have generally been designed with inadequate communication facilities, which make them less observable and may impair the operator's ability to track their progress and understand their actions. In one case this problem led to operators defeating or otherwise circumventing a properly operating automated system because they believed it was malfunctioning.

The overall effect of technology-centered automation and inadequate HSI design is that human performance may be negatively affected. Automation can increase the complexity of the plant, and problems can arise because the operator lacks a good mental model of the behavior of the automated system (i.e., how and why it operates). Many of the problems associated with human interaction with automated systems have been attributed to poor situation awareness, which is difficult to maintain when the operator is largely removed from the control loop. In such conditions operator alertness may suffer. Conversely, the workload associated with transitions from monitoring automated systems to assuming manual control during a fault in an automated system can be very high. Further, there may be an erosion of the required skills for tasks that have been automated. Because nuclear power plant designs still require the operator to assume control under certain circumstances and to act as the last line of defense, the consequences of poor integration of the operator and the automated system can be significant.

User-System Interaction

User-system interaction refers to the means by which operators interact with the system. There are two kinds of tasks that need to be considered: primary and secondary. The primary tasks of operators in complex systems involve the generic cognitive activities of monitoring and detection, situation assessment, response planning, and response implementation. A computer-based HSI may contain hundreds of displays and controls that are viewed through a limited number of video display units (VDUs). To access the controls and displays necessary to perform primary tasks, operators engage in secondary tasks (e.g., navigating through displays, searching for data, manipulating windows, choosing between multiple ways of accomplishing the same task, and deciding how to configure the interface). There may be a cognitive cost for accessing information (O'Hara, Stubler, & Nasta, 1998; Wickens & Carswell, 1995; Wickens & Seidler,1997). The amount of secondary task workload can be high in designs where there is much information, the information is presented in a limited number of displays, and the HSI provides for great configuration flexibility. Thus, the potentially negative and distracting effects these tasks can have on performance needs to be carefully addressed. These secondary task effects are a common theme in many of the topics discussed in the following.

Alarm System Design and Management

The alarm system is one of the primary means by which process abnormalities come to the attention of plant personnel. Advanced, computer-based alarm systems are available as upgrades and include such characteristics as alarm reduction processing, techniques for making alarms available to operators (dynamic prioritization, suppression, and filtering), a wide variety of display devices and presentation formats, increased controls for interacting with alarm systems, and flexibility of alarm features (such as setpoints). The operational experience and research on the effects of these features on operator performance have revealed mixed results (Brown, O'Hara, & Higgins, in press; O'Hara, Brown, Higgins, & Stubler, 1994). For example, when computer displays were first used to present alarms, operators were unable to effectively use the alarm systems under periods of high alarms because of scrolling message lists and limited display area. Further, although considerable effort has been devoted to alarm reduction techniques, studies are not conclusive on their effectiveness. One problem occurs when designers set alarm reduction goals in terms of percent of alarm reduction. Although this might seem reasonable, it may not relate to the operator's use of the system and, therefore, might not noticeably improve crew performance.

Information Design and Organization

In conventional control rooms, the large, spatially fixed arrangement of indications present individual parameters as separate pieces of information. Data integration and interpretation are accomplished by the operator based on pattern recognition, training, and experience. However, computer-based display systems provide many ways for data and information to be processed and presented in an effort to make the information more immediately meaningful. The human performance considerations can be grouped into two broad categories: display formats and display management and navigation.

Display Formats. Graphic displays provide different representations of the plant's functions, processes, and systems. To make information more meaningful, efforts have been made to map displays to underlying cognitive mechanisms, such as perceptual processes and mental models. If achieved, such displays should approach properties of direct perception (i.e., immediately understood with little need for interpretation). However, this places a significant burden on the designer both to anticipate the information needs at various levels of abstraction (e.g., physical to functional) and to map them into appropriate display formats that correspond to the operator's mental models. Although research has shown that display formats can have a significant effect on performance, the identification of the appropriate format for different tasks needs to be addressed during the design process.

Graphic displays also contain many more display elements (e.g., abbreviations, labels, icons, symbols, coding, and highlighting) than do conventional displays. As displays convey more information at multiple levels of abstraction, the complexity of both the elements and the displays becomes greater. Therefore, generalization of guidelines, such as for the use of color, needs to be assessed for alphanumeric and simple displays and more complex graphic forms, such as configural and ecological displays (Bennett, Nagy, & Flach, 1997; O'Hara, Higgins, & Kramer, 1999; Vicente & Rasmussen, 1992). The relationships between graphical and coding elements may be complex and the effects of relatively new graphical techniques, such as animation, need to be addressed.

Nuclear power plant display pages typically are composed of a variety of formats. Considerations include how these formats should be combined to form display pages, how different display formats might interact within a display page, and how complex information should be divided to form individual pages.

Display Management and Navigation.   Digital instrumentation and control systems usually have the capability to present much more information than was available in analog systems. This information is typically presented on a limited number of VDUs. To address the mismatch between available information and display area, information is usually arranged into a hierarchy of display pages. As stated above, new display systems may contain thousands of display pages, creating additional workload due to secondary tasks. For example, when the information needed by operators exceeds the available display area, the operator may be required to make rapid transitions between screens, try to remember values, or write values on paper. The human performance concerns include the demands placed on the operator's memory for remembering display locations, the inability to quickly access needed information, and delays in accessing needed displays. In addition to increased workload and distraction, these problems can increase operator response time. Factors that contribute to the display navigation demands include trade-offs between density of information contained in an individual display and the number of displays in a display network, the arrangement of the displays within the network, the number and types of paths available for retrieving data, the cues for directing information retrieval, and the flexibility of software-driven interfaces, which allows information to be displayed in a variety of formats and locations.

Soft Controls

In conventional control rooms, the predominant means for providing control input are by individual, spatially dedicated control devices. They are always in the same location and always provide the same control function. Soft controls are control devices that are defined by software; thus, a control may change as a function of the display screen presented. Several concerns have been identified including ease of using soft controls as compared to hard controls, absence of tactile and aural feedback, loss of dedicated spatial location of controls, limited display surfaces on which to present soft controls, navigation to controls during configuration tasks and emergency situations, accuracy in mapping between the displays that depict plant components and displays containing the soft controls that operate them, response time and display update rate of complex systems, reliability of soft interfaces for critical tasks, and environmental factors (e.g., glare).

Soft controls can impose additional interface management demands for finding and accessing controls and providing input (Stubler, O'Hara, & Kramer, 1999). For example, the need to coordinate separate displays for selecting plant components and providing input may impose display navigation and window management burdens. "Wrong component" errors occur when operators select the wrong component from a mimic display or provide input to the wrong input display. These interface management tasks often impose sequential constraints (i.e., displays and controls must be accessed one at a time). Potential problems include slow overall operator response because of high navigation burdens, missed actions because of interruptions, and capture errors—confusion between control operations with similar steps. Soft controls are also associated with data entry errors. These include accidental actuation because of inadequate guards or inappropriate actuation logic, large-magnitude input errors because of typing errors, ineffective verification steps, and improper use of automatic blocks for inappropriate input values. The consequences of these errors can be aggravated by digital control systems that respond quickly to the incorrectly entered data.

Computer-Based Procedures

Plant procedures provide instructions to guide operators in monitoring, deciding on appropriate actions, and controlling the plant. Procedures have traditionally been presented in paper format. Computer-based procedures (CBPs) provide a range of capabilities that can support the operator's tasks and reduce demands associated with the paper-based medium. However, some human performance concerns have been identified for

CBPs. These include the loss of situation awareness in terms of losing the big picture of operations because only a portion of the procedure can be observed at one time and losing a sense of location with regard to the total set of active procedures (O'Hara, Higgins, Stubler, & Kramer, 1999). Navigation within one procedure, or between multiple procedures, and related supporting information can also be time-consuming and error prone.

The appropriate level of automation of CBP systems is another question. Although paper procedures require the operator to monitor plant indications, CBPs can contain current plant values in their displays. The operators may become overreliant on the CBP information and may not feel the need to look at other sources of information in the control room. As a result, important information presented by other displays that are not part of the CBP system may be missed.

During a total loss of the CBP system, special demands may be imposed on operators as they attempt to control the plant using backup systems. The transfer from CBP to paper procedures during such failures can be challenging, especially if their presentation formats are different, such as switching from text to flow chart formats.

Computerized Operator Support Systems

Computerized operator support systems (COSSs) assist operators with cognitive tasks, such as evaluating plant conditions, diagnosing faults, and selecting response strategies. They are often knowledge-based systems. Although most of these applications have been used offline, systems for performing diagnostics in real time are emerging. Despite the development of many COSSs, there has not been a great deal of experience with operational aspects of their use. Several experimental evaluations of the value of expert systems to reactor operators have been inconclusive. Problems have been identified related to the task relevance of the information provided, level of explanatory detail, the complexity of the COSS information processing, lack of visibility of the decision process to the operator, and lack of communication functions to permit operators to query the system or obtain level of confidence information regarding the conclusions that have been drawn. In addition, poor integration with the rest of the HSIs has limited the usefulness of COSSs.

Maintenance of Digital Systems

The maintenance of digital equipment places new demands on personnel capabilities and skills (Stubler, Higgins, & Kramer, 1999). Recent events in nuclear power plants and other complex systems indicate the importance of digital system maintenance on system performance and safety. A review of fault-tolerant, digital control systems indicated that inadvertent personnel actions, especially during maintenance, was one of the two leading causes of failure of these systems. Recent failures of digital systems in U.S. nuclear power pants illustrate how inadequate integration of software-based digital systems into operating practices and inadequate mental models of the intricacies of software-based digital systems on the part of technicians and operators caused systems to become inoperable. The events also show the susceptibility of software-based systems to failure modes different from analog systems. Errors associated with such activities as troubleshooting, programming, and loading software are significant problems. This topic may become increasingly important as the online maintenance capabilities of digital control systems are used to reduce the duration of plant outages. A greater understanding is needed of the task demands and error modes associated with maintenance on digital systems.

Configuration Control of Digital Systems

For many digital control systems, logic configuration can be performed rapidly via an engineering workstation. The use of such workstations for configuration poses questions regarding the types of safeguards that are needed to maintain the integrity of the control system. Errors due to a lack of awareness of the mode of the configuration (e.g., configure versus test) workstation are possible. Undesirable changes may be made without detection. Also, safeguards for access to the configuration workstation and related equipment need to be considered.

Staffing and Crew Coordination

There is a trend in process control industries toward reduced overall staffing levels through the implementation of multiunit control rooms, compact computer-based control consoles, and automation. However, although these technologies can reduce operator workload in some situations, they may increase workload in other situations. Therefore, a greater understanding is needed of the effects of the technologies on staffing levels of U.S. nuclear power plants. These technologies significantly change the ways in which crew members interact, communicate, and coordinate their activities (Roth & O'Hara, 1998).

Training and Acceptance

When plants are modified, personnel may have to adapt to changes in their role in the plant, knowledge and skill requirements, task characteristics, and behavior of the plant process that results from new equipment.

Two factors are important in making the transition to new technology: training and operator acceptance. In addition, during the upgrade process, interim periods may occur in which the HSI has characteristics that are more challenging to operators than either the original configuration or the end configuration. The process by which training programs incorporate these new requirements is important to maintaining safe operations. The training process may need to become more adaptable to the more frequently changing and complex computer-based technologies and the diversity of HSI technology in hybrid plants. It will also need to address a broader range of skills needed to use hybrid interfaces and the potential for negative transfer of skills between the conventional and advanced technology.

Upgrade Implementation and Transition

The method by which HSI modifications are implemented may affect human performance, especially during the transition from the old design to the new hybrid design. During the modification process, interim periods may occur in which the HSI has characteristics that are more challenging to operators than either the original configuration or the final configuration.

In summary, hybrid HSIs have a broad impact on human actions that are important to the following:

- *Personnel role.* A change in functions and responsibilities of plant personnel (e.g., caused by a change in plant automation in a control system)
- *Primary tasks.* A change in the way that personnel perform their primary tasks (i.e., process monitoring, situation assessment, response planning, and response execution and control)
- *Secondary tasks.* A change in the methods of interacting with the HSI
- *Cognitive factors.* A change in the cognitive factors supporting personnel task performance (e.g., situation awareness and workload)
- *Personnel factors.* A change in the required qualifications or training of plant personnel

HFE EVALUATION METHOD

The overall purpose of the evaluation is to ensure that HFE considerations have been integrated into the plant development and design; the HSIs, procedures, and training reflect state-of-the-art human factors principles and satisfy all other appropriate regulatory requirements; and the HSIs,

procedures, and training promote safe, efficient, and reliable performance of operation, maintenance, test, inspection, and surveillance tasks.
The evaluation process was divided into 10 elements reflecting four stages
of design: planning, analysis, detailed design, and verification and validation (see Fig. 16.1). The development of the evaluation approach is described in O'Hara et al. (1994).

When evaluating safety, evidence is collected and weighted toward or
against an acceptable finding. Different types of information are collected,
each with its strengths and weaknesses. The reviewer seeks convergent validity, that is, consistent findings across different types of information.
The types of information evaluated include HFE planning, including an
HFE design team, program plans, and procedures; design analyses and
studies, including requirements, function and task analyses, technology
assessments, and trade-off studies; design specifications and descriptions;
and verification and validation analyses (e.g., compliance with accepted
HFE guidelines and operation of the integrated system by operators under actual or simulated conditions). The general purpose and objectives of
the HFE PRM evaluation elements are summarized in the following sections. The individual evaluation criteria for each can be found in O'Hara
et al. (in press).

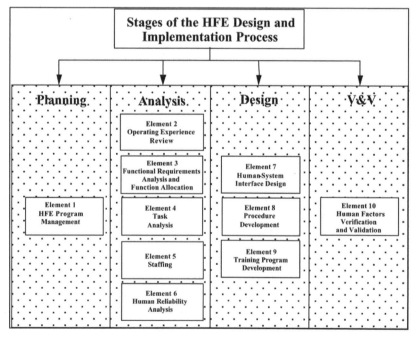

FIG. 16.1.   Human factors evaluation elements.

Element 1: HFE Program Management

The first part of the evaluation addresses the designer's program planning (i.e., the qualifications and composition of the design team, the goals and scope of the HFE activities, the design processes and procedures, and the technical program plan). This evaluation will help to ensure that the plan describes the technical program elements in sufficient detail to ensure that all aspects of HSI are developed, designed, and evaluated based on a structured top-down systems analysis using accepted HFE principles.

Element 2: Operating Experience Review

This aspect of the evaluation focuses on the designer's use of operating experience. The issues and lessons learned from previous operating experience should provide a basis for improving the plant design in a timely way (i.e., at the beginning of the design process). To make use of these insights, the designer must identify and analyze HFE-related problems and issues from similar previous designs. This allows negative features associated with earlier designs to be avoided in the current design and positive features to be retained. In addition to design similarity, the designer should become familiar with the operating experience related to new technological approaches being considered in the current design. For example, if touch-screen interfaces are planned, HFE issues associated with their use should be reviewed. A third area of valuable operating experience is the process by which previous changes were implemented. Plants are continuously changing and a great deal of experience exists regarding what did and did not work in the past.

The sources of operating experience can include general literature, event reports, and interviews with plant operations and maintenance personnel who have familiarity with the systems and technologies. Obtaining information from the user community is quite important for several reasons. First, personnel may use interfaces in ways that are different from what designers had planned. Second, with any complex system, not all human performance is documented in available sources. We have found that interviews with personnel yield very good information to supplement the other sources.

Element 3: Functional Requirements Analysis and Allocation

Modernization programs can change plant systems and HSIs in ways that alter the role of personnel. The purpose of the this aspect of the evaluation is to ensure that the applicant has sufficiently defined and analyzed the plant's functional requirements, so that the allocation of functions to human and machine resources can take advantage of human strengths and avoid human limitations.

The analysis of functions and their assignment to hardware, software, and personnel is where the human role in the system begins to be defined. The term *function* may be used in different ways by designers. It can refer to high-level plant functions, such as preventing or mitigating the consequences of postulated accidents that could cause undue risk to the health and safety of the public. It can also refer to the functioning of an individual piece of equipment, such as a valve or a display system. At the level of analysis addressed here, the former, high-level definition is intended. A function is an action that is required to achieve a desired goal. These high-level functions are achieved through the design of plant systems and through the actions of personnel. Often a combination of systems is used to achieve a higher-level function.

The functional requirements analysis is conducted by designers to determine the objectives, performance requirements, and constraints of the design, define the high-level functions that must be accomplished to meet the objectives and required performance, and define the relationships between high-level functions and the plant systems responsible for performing the function. For each function, the set of system configurations or success paths responsible for or capable of carrying out the function should be clearly defined. It may be helpful to describe functions initially in graphic form (e.g., functional flow-block diagram). Function diagramming should be done at several levels, starting at top-level functions, where a general picture of major functions is described, and continuing to lower levels, until a specific critical end-item requirement emerges (e.g., a piece of equipment, software, or personnel). The functional decomposition should link high-level functions and specific plant systems and components. The description of each high-level function includes the following:

- purpose of the high-level function;
- conditions that indicate that the high-level function is required;
- parameters that indicate that the high-level function is available;
- parameters that indicate the high-level function is operating (e.g., flow indication);
- parameters that indicate the high-level function is achieving its purpose (e.g., reactor vessel level returning to normal); and
- parameters that indicate that operation of the high-level function can or should be terminated.

Once functional requirements are defined, they can be allocated to personnel (e.g., manual control), system elements (e.g., automatic control and passive, self-controlling phenomena), and combinations of personnel and system elements (e.g., shared control and automatic systems with manual backup). Function allocation seeks to enhance overall plant safety and reliability by exploiting the strengths of personnel and system elements, including improvements that can be achieved through the assignment of control to these elements with overlapping and redundant responsibilities. Function allocation should be based on HFE principles using a structured and well-documented methodology that seeks to provide personnel with logical, coherent, and meaningful tasks.

Function allocation should be performed using a structured, documented methodology reflecting HFE principles. The technical basis for functional allocation can be any one or a combination of the evaluation factors. For example, the performance demands to successfully achieve the function, such as degree of sensitivity required, precision, time, or frequency of response, may be so stringent that it would be difficult or error prone for personnel to accomplish. This would establish a basis for automation (assuming acceptability of other factors, such as technical feasibility or cost). In some cases, the allocation of some functions may be mandated by regulatory requirements (i.e., the function must be performed by personnel or must be performed by an automatic system). An example functional allocation process and considerations is shown in Fig. 16.2.

As noted earlier, this analysis should consider not only the primary functions of personnel but also the secondary functions. These include responsibilities of personnel in supervising and ensuring the performance of automatic functions (e.g., the requirement for personnel to monitor automatic functions and to assume manual control in the event of an automatic system failure). A description of the personnel role integrated across functions and systems should be developed. It should be defined in terms of personnel responsibility and level of automation.

Element 4: Task Analysis

Task analysis is conducted to identify the task requirements for accomplishing the functions that have been allocated to personnel. It defines the HSI requirements for supporting personnel task accomplishment. Also, by exclusion, it defines what is not needed in the HSI. Task analysis should address primary tasks because plant modifications can significantly alter them. It should also analyze secondary tasks because changes in technology can alter the way primary tasks are performed and can introduce new secondary tasks.

The objective of this aspect of the evaluation is to ensure that the task analysis identifies the system and behavioral requirements of personnel tasks. The task analysis should provide one of the bases for making design

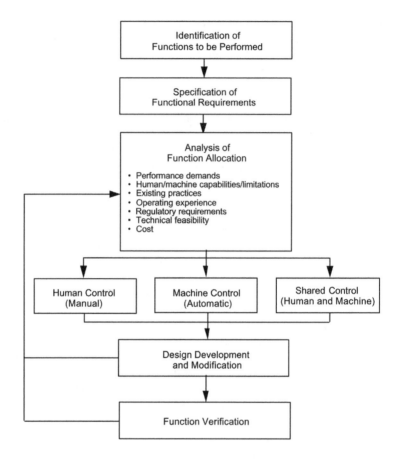

FIG. 16.2. Allocation of functions to human and machine resources.

decisions; be used as basic input for developing procedures; be used as basic information for developing staffing, training, and communication requirements of the plant; and form the basis for specifying the requirements for the displays, data processing, and controls needed to carry out tasks. The task characteristics that may be analyzed, depending on the specific tasks involved, are identified in Table 16.1.

Element 5: Staffing

Initial staffing levels may be established as design goals early in the design process, based on experience with previous plants, customer requirements, initial analyses, and government regulations. However, staffing goals and assumptions should be examined for acceptability as the design

TABLE 16.1
Task Considerations

Type of Information	Example
Information requirements	Parameters (units, precision, and accuracy)
	Feedback required to indicate adequacy of actions taken
Decision-making requirements	Decisions type (relative, absolute, probabilistic)
	Evaluations to be performed
Response requirements	Type of action to be taken
	Task frequency, tolerance and accuracy
	Time available and temporal constraints (task ordering)
	Physical position (stand, sit, squat, etc.)
	Biomechanics
	movements (lift, push, turn, pull, crank, etc.)
	forces required
Communication requirements	Personnel communication for monitoring information or control
Workload	Cognitive
	Physical
	Overlap of task requirements (serial vs. parallel task elements)
Task support requirements	Special and protective clothing
	Job aids or reference materials required
	Tools and equipment required
Workplace factors	Ingress and egress paths to the work site
	Work space envelope required by action taken
	Typical and extreme environmental conditions, such as lighting, temperature, noise
Situational and performance-shaping factors	Stress
	Reduced manning
Hazard identification	Identification of hazards involved (e.g., potential personal injury)

of the plant proceeds. Further, changes in technology can have significant effects on staffing, especially in crew interactions and qualifications.

The objective of the staffing review is to ensure that the applicant has analyzed the requirements for the number and qualifications of personnel in a systematic manner that includes a thorough understanding of task requirements and applicable regulatory requirements.

Element 6: Human Reliability Analysis

To assess the impact of system failures on plant safety, a probabilistic risk assessment (PRA) is performed. Human actions are modeled in the PRA and human reliability analysis (HRA) is used to determine the potential for and mechanisms of human error. HRA has both qualitative and quantitative aspects, both of which are useful for HFE purposes. The analyst can consider not only potential errors of omission (i.e., failure to perform required actions) but also possible errors of commission (e.g., operating a system in the wrong mode or making a wrong diagnosis because the indications of a specific disturbance are very similar to another disturbance for which the crew is well trained). Once these types of errors are identified, the analyst can examine what operators are likely to do in light of their incorrect situation assessment.

From a quantitative standpoint, the PRA can be used to identify the human actions that are critical to plant safety. That knowledge is valuable in focusing design efforts to make the system error tolerant (i.e., an effort can be made to minimize critical errors through design of the HSIs, procedures, and training or through changes in plant automation). The design can also focus on providing support for error detection and recovery in case critical errors do occur.

HRA is most useful when conducted as an integrated activity in support of both HFE and HSI design activities and risk analysis activities. In that way the risk analysis benefits from detailed information about the operational tasks and the design of the HSI, procedures, and training. The design activity also benefits from being able to focus design resources on those human actions that are most critical to plant safety.

Element 7: Human-System Interface Design

The HSI design evaluation assesses how well the function, task, and staffing requirements are translated into the detailed HSI. Whether the design employs off-the-shelf equipment or a new design, the systematic application of HFE principles and criteria should guide the design. The following aspects of HSI design are evaluated: design methodology, detailed design and integration tests and evaluation, and design documentation.

HSI Design Methodology

Requirements Development. HSI requirements define the functional characteristics that the individual HSI resources, such as alarms, displays, and controls, must reflect. The major inputs to the HSI requirements are as follows:

- the requirements of personnel functions and tasks (e.g., the specific information and controls that are needed to perform crew tasks);
- staffing and job analyses (e.g., the layout of the overall workplace and the allocation of controls, and displays, and so forth, to individual

consoles, panels, and workstations based on the roles and responsibilities of individual crew members);

- the need for a safe and comfortable working environment;
- system requirements (e.g., characteristics and constraints imposed by the instrumentation and control hardware and software);
- regulatory requirements; and
- specific customer requirements.

HSI Concept Design. A survey of HSI technologies can be conducted to support the development of alternative design concepts, provide assurance that proposed designs are technically feasible, and support the identification of human performance concerns and trade-offs associated with various HSI technologies.

The alternative design concepts are evaluated in terms of how well they meet design requirements, and an HSI concept design is selected for further development. The evaluation ensures that the concept selection process is based on a criteria such as personnel task requirements, human performance capabilities and limitations, HSI performance requirements, inspection and testing considerations, and maintenance requirements. Another consideration is the use of proven technology and the operating experience of predecessor designs. The types of evaluations that may be especially useful in concept design are trade-off analyses and performance-based tests (see later section titled "Tests and Evaluations").

HSI Detailed Design and Integration

This aspect of the evaluation addresses how the designer sets out to develop the detailed design specification from the conceptual design. Topics addressed in the evaluation include development of design-specific HSI guidelines, tailoring of the design to the user population, critical personnel actions, automatic systems, integration of operations across personnel locations, staffing and shifts, environmental conditions, and accommodation of nonoperations activities.

Development of Design-Specific HSI Guidelines. The detailed design should support personnel in their primary role of monitoring and controlling the plant while minimizing personnel demands associated with using the HSI (e.g., window manipulation, display selection, and display system navigation). HFE guidelines are available from many sources (e.g., O'Hara et al., 1996b) to support designers in accomplishing this goal. The general guidance should be tailored to the specific detailed design. Use of HFE guidelines can help ensure that the design reflects good ergonomics principles and practices and can promote consistency and standardization of the HSIs during the design process.

Guidelines are most usable when they are expressed in concrete, easily observable terms; are detailed enough to permit use by design personnel to achieve a consistent and verifiable design; are written in a style that can be readily understood by designers; and support interpretation and comprehension, which is supported by providing graphical examples, figures, and tables.

Tailoring the Design to the User Population. Two considerations that should be addressed when designing the HSI to accommodate a specific user population include user stereotypes, customs, and preferred practices and user anthropometry.

Critical Personnel Actions. As noted in the earlier HRA discussion above, critical actions should be addressed by designing the system to be as error tolerant as possible.

HSI to Automatic Systems. The design of interfaces for automatic systems should be based on the information needed by personnel for proper monitoring of the automated system, for deciding whether the automated system is properly functioning, and for assuming manual control in the event of automated system failure or malfunction.

Integration of Operations Across Personnel Locations. HSI design should consider the integration of crew activities across workplaces, such as the main control room and local HSI distributed throughout the plant. The design should support the communication and coordination of crew members.

Staffing and Shifts. The design process should take into account the use of the HSIs during minimal, nominal, and high-level staffing and the use of the HSI over the duration of a shift, where decrements in performance due to fatigue may be a concern.

Range of Environmental Conditions. HSI characteristics should support human performance under the full range of environmental conditions, such as normal and credible, extreme conditions. For the main control room, requirements should address conditions such as loss of lighting, loss of ventilation, and main control room evacuation. For the remote shutdown facility and local control stations, requirements should address constraints imposed by the ambient environment (e.g., noise, temperature, contamination) and by any protective clothing that may be necessary.

Accommodation of Non-Operations Activities. The HSI should be designed to support inspection, maintenance, test, and repair of the equipment. This is a necessary and ongoing activity in plants.

Tests and Evaluations

Testing and evaluation of HSI designs are conducted throughout the HSI development process. Evaluations are conducted to help refine design details and ensure its usability. Some general types of evaluations include trade-off evaluations, subjective evaluations, and performance-based tests.

Trade-Off Evaluations. Trade-off evaluations should be considered when selecting between design options. Relevant human performance dimensions should be carefully selected and defined so that the effects of design options on human performance can be adequately considered in the selection of design approaches.

Subjective Evaluations. Designs can be subject to evaluation by knowledgeable subject-matter experts. In the early stages of design, the HSI can be evaluated based on drawings. As the design proceeds, mock-ups, functional prototypes, and simulations can be used to provide a basis for the subject-matter experts to provide their evaluations.

Performance-Based Tests. Performance-based tests involve the collection of data on human-system performance. Such an approach provides a basis on which to evaluate design concepts and make design decisions using objective criteria. For example, when considering which of two display formats to use, the design can be compared by having a sample of operators use the two designs in simulated task activities. Performance data, as well as subjective opinions, can be collected. The data can then be analyzed and the two designs compared along relevant performance criteria.

HSI Design Documentation

Documentation of the HSI design includes the technical basis for HSI requirements; the detailed HSI description, including its form, function, and performance characteristics; and the basis for the HSI design characteristics with respect to operating experience and literature analyses, trade-off studies, engineering evaluations, and experiments.

Element 8: Procedure Development

Procedures are an essential component of plant design. Personnel use procedures to guide normal, abnormal, and emergency operations. Emergency operating procedures are especially important in that they reflect the designer's detailed analysis of design-based accidents and the various strategies personnel can use to maintain and recover functions that are critical to plant safety.

The purpose of this aspect of the evaluation is to help ensure that the procedures are technically accurate, comprehensive, explicit, and easy to use. To accomplish this, they should be developed from the same design process and analyses as the HSIs and subject to the same evaluation processes. Procedures should be modified to reflect changes in plant systems that may have resulted for modernization. In addition, the procedures should provide guidance for the interim transition periods between old and new designs during which time the plant will be operational.

Element 9: Training Program Development

Training of plant personnel is an important factor in ensuring safe and reliable operation. Training design should be based on the systematic analysis of job and task requirements. The HFE analyses provide a valuable understanding of the task requirements of operations personnel. Therefore, the training development should be coordinated with the other elements of the HFE design process.

In our examination of plant modernization programs, the failure to train personnel for the transition periods from old to new design was identified as a significant problem. Training programs often address the new design but may not adequately account for the temporary configurations crews must operate before the final design is fully implemented. The training should cover all changes that will occur in plant systems, personnel tasks and roles, HSIs, and procedures.

Element 10: Human Factors Verification and Validation

Verification and validation together seek to comprehensively determine that the design conforms to human factors design principles and that personnel can successfully perform their tasks safely and efficiently. The main verification and validation evaluations include HSI task support verification, human factors design verification, and integrated system validation.

HSI Task Support Verification.   The objective of HSI task support verification is to ensure that the design provides all information and controls necessary to support personnel tasks. This is accomplished by comparing the HSI (alarms, displays, controls, etc.) to the task requirements developed during task analysis. The analysis should seek to identify task requirements that are not available in the HSI and procedures. When discrepancies are identified, the design should be modified to include the missing requirements.

Human Factors Design Verification. Even when the HSI provides all necessary task information and controls, it is also important to verify that task requirements are represented in the HSI and the procedures in a well-designed, usable, and error-tolerant manner. This verification is accomplished by evaluating the physical and functional characteristics of the HSIs using HFE guidelines, standards, and principles. Deviations should be either corrected or justified on the basis of trade study results, literature-based evaluations, demonstrated operational experience, or tests and experiments.

Integrated System Validation. The final design should be validated to show that the integration of hardware, software, and personnel can perform effectively to achieve all operational requirements. Integrated system validation is performed using an actual system, prototype, or high-fidelity simulator. A representative sample of crews is used to perform a representative sample of normal and failure scenarios. Measures of safety functions, systems, and personnel performance are obtained. The data should be compared to performance criteria derived from operational and engineering analyses to assess overall performance.

ACKNOWLEDGMENTS

This research is being sponsored by the U.S. Nuclear Regulatory Commission (NRC). The views presented in this paper represent those of the authors alone, and not necessarily those of the NRC.

REFERENCES

Bennett, K., Nagy, A., & Flach, J. (1997). Visual displays. In G. Salvendy (Ed.), *Handbook of human factors and ergonomics* (2nd ed.). New York: Wiley.

Brown, W., O'Hara, J., & Higgins, J. (in press). *Advance alarm systems: Guidance development and technical basis* (BNL Report W6290-4-1-9/98). Upton, NY: Brookhaven National Laboratory.

Galletti, G. S. (1996). Human factors issues in digital system design and implementation. *Proceedings of the 1996 American Nuclear Society International Topical Meeting on Nuclear Plant Instrumentation, Control, and Human-Machine Interface Technologies* (pp. 1157–1161).

IAEA. (1998). *Safety issues for advanced protection, control, and human-machine interface systems in operating nuclear power plants* (Safety Report Series No. 6 - STI/PUB/1057). Vienna, Austria: International Atomic Energy Agency.

NAS. (1995). *Digital instrumentation and control systems in nuclear power plants: Safety and reliability issues.* Washington, DC: National Academy Press.

O'Hara, J., Brown, W., Higgins, J., & Stubler, W. (1994). *Human factors engineering guidelines for the review of advanced alarm systems* (NUREG/CR-6105). Washington, DC: U.S. Nuclear Regulatory Commission.

O'Hara, J., Higgins, J., & Kramer, J. (1999). *Advanced information systems: Technical basis and human factors review guidance* (NUREG/CR-6633). Washington, DC: U.S. Nuclear Regulatory Commission.

O'Hara, J., Higgins, J., Stubler, W., Goodman, C., Eckinrode, R., Bongarra, J., & Galletti, G. (1994). *Human factors engineering program review model* (NUREG-0711). Washington, DC: U.S. Nuclear Regulatory Commission.

O'Hara, J., Higgins, J., Stubler, W., Goodman, C., Eckinrode, R., Bongarra, J., & Galletti, G. (in press). *Human factors engineering program review model* (NUREG-0711). Washington, DC: U.S. Nuclear Regulatory Commission.

O'Hara, J., Higgins, J., Stubler, W., & Kramer, J. (1999). *Computer-based procedure systems: Technical basis and human factors review guidance* (NUREG/CR-6634). Washington, DC: U.S. Nuclear Regulatory Commission.

O'Hara, J., Stubler, W., & Higgins, J. (1996a). Human factors evaluation of advanced nuclear power plants. In T. O'Brien & S. Charlton (Eds.), *Handbook of human factors testing and evaluation* (pp. 275–285). Mahwah, NJ: Lawrence Erlbaum Associates.

O'Hara, J., Stubler, W., & Higgins, J. (1996b). *Hybrid human-system interfaces: Human factors considerations* (BNL Report J6012-T1-12/96). Upton, NY: Brookhaven National Laboratory.

O'Hara, J., Stubler, W., & Nasta, K. (1998). *Human-system interface management: Effects on operator performance and issue identification* (BNL Report W6546-1-1-7/97). Upton, NY: Brookhaven National Laboratory.

Roth, E., & O'Hara, J. (1998). *Integrating digital and conventional human system interface technology: Lessons learned from a control room modernization program* (BNL Report J6012-3-4-7/98). Upton, NY: Brookhaven National Laboratory.

Stubler, W., Higgins, J., & Kramer, J. (1999). *Maintenance of digital systems: Technical basis and human factors review guidance* (NUREG/CR-6636). Washington, DC: U.S. Nuclear Regulatory Commission.

Stubler, W., O'Hara, J., & Kramer, J. (1999). *Soft controls: Technical basis and human factors review guidance* (NUREG/CR-6635). Washington, DC: U.S. Nuclear Regulatory Commission.

Vicente, K., & Rasmussen, J. (1992). Ecological interface design: Theoretical foundations. *IEEE Transactions on Systems, Man, and Cybernetics, 2,* 589–606.

Wickens, C., & Carswell, C. (1995). The proximity compatibility principle: Its psychological foundation and its relevance to displays design. *Human Factors, 37*(3), 473–494.

Wickens, C., & Seidler, K. (1997). Information access in a dual-task context: Testing a model of optimal strategy selection. *Journal of Experimental Psychology: Applied, 3*(3), 196–215.

Chapter 17

Generative Search in the Product Development Process

Marty Gage
Elizabeth B.-N. Sanders
Colin T. William
SonicRim

This chapter explores the expanding role of the human factors professional in consumer product development. The chapter opens with a short discussion of changes that are taking place within the emerging field of user-centered product development. The main section then presents a framework for conducting generative search (i.e., predesign research) to support the product development process. Finally, there is a discussion on the emerging role of the human factors practitioner within the development process.

THE EMERGING CONTEXT IN CONSUMER PRODUCT DEVELOPMENT

Much of the effort of the human factors practitioner in the past has been on the evaluation of product concepts or prototypes. Evaluative research addresses the usability or the perceived desirability of concepts, prototypes, or both. Because it requires concepts or products to evaluate, it is often conducted too late in the product development process to make substantive changes based on its findings.

There has been a shift in the last few years, however, toward conducting user-centered research much earlier in the product development process. The shift has been moving away from evaluative research toward predesign research, or generative search. Generative search is research done for the purpose of generating ideas, opportunities and concepts that are relevant to the users, as well as the producers, of products. Generative search occurs very early in the design development process. Its purpose is to search for and discover one or all as-yet unknown, undefined, or unanticipated user or consumer needs.

Part of the shift from evaluative research to generative search is a shift in focus from the designing of things to a focus on the experiences people have and seek to have. Today we are beginning to hear about experience design, the aim of which is to design users' experiences of things, events, and places. In this new experience-based paradigm, the role of the human factors practitioner expands beyond usability and cognition to include emotion—how someone feels or wants to feel while engaged in the experience—which represents an opportunity to expand the domain of human factors. Usability is merely one part of the experience. Human factors must now address the experiences people have with products. The way a product makes a person feel is just as important as its usability.

It is important to remember, however, that we can never really design experience because experiencing is a constructive activity. That is, a user's experience (e.g., with communication) is constructed of two equal parts: what the communicator provides and what the receiver or user brings to the interaction. Where the two parts overlap is where the actual communication occurs. Knowing about users' experiences, then, becomes vital to the process of designing the communication. If we have access to both what is being communicated and what experiences are influencing the receipt of communication, then we can design for experiencing communication. In fact, if we can learn to access people's experiences—past, current, and potential—then we can make user experience the source of inspiration and ideation for design. And by making user experience the source of inspiration, we are better able to design for experiencing. This represents an opportunity for human factors practitioners to expand their role in the development process.

GENERATIVE SEARCH IN CONSUMER PRODUCT DEVELOPMENT

What follows is a framework to aid in conducting generative search. It is not meant to prescribe an approach but to be a starting point for developing your own approach to generative search. This framework will be compared to the typical manner of conducting scientific research.

Introduction

When doing scientific research one usually begins with a review of the literature. Similarly, when beginning to conduct generative search one usually starts with a review of the literature. In this instance, the literature be-

EMOTIONAL

What are the different types of meaning users attach to their phones (i.e., status symbol, to be connected, security, etc.)?

What inadequacies, frustration, inconveniences do people associate with their phones?

How is the phone perceived in terms of time management and productivity?

What element of fun, if any, is associated with cellular phone use?

USE

Other than the primary user who uses the phone?

What other products is the phone used in conjunction with?

What is the process each person went through to learn their phone numbers after they were purchased?

How does each person manage and store their phone numbers?

How do people take notes when using their phones?

How do people use their phone in relation to a pager and a two way pager?

Where do people typically use their phones, and in what way does that impact how the phone is used?

When people carry their phones with them, where do they carry it and where do they want to carry it?

What influences them to carry it the way they carry it?

What are the culture and gender specific uses of the phone? How do the markets differ?

What are the culture and gender specific inhibitions associated with phones?

Along with battery charging what other types of maintenance do people do, such as cleaning, etc.?

What are the different ways people use their phones (i.e., outgoing vs. incoming calls, voice mail, etc.)?

FIG. 17.1.   Research questions relevant to experiences surrounding cell phone use.

ing reviewed can include previous market research, trade or industry publications, newsstand publications, and so forth. This activity is usually referred to as secondary research. Other preliminary activities can include a review of competitive products, store audits to understand how the products are sold, as well as immersion, or experiencing the product firsthand.

In scientific studies, the introduction usually ends with a hypothesis stated in relation to the subject matter of the literature. In generative search, on the other hand, the introduction usually ends with a list of questions or issues to be addressed in the research. Take, for example, the case study of a generative search into people and their cell phones. As shown in Fig. 17.1, when one begins addressing the whole experience surrounding a product, the initial list of questions to be addressed can be broad in scope, in this case addressing aspects of emotion and use.

Participants

Of all the areas in a scientific research report, the subjects section is typically the shortest because many academic studies tend to use undergraduate psychology students. In generative search, as in most human factors research, much effort is made to ensure that the research participants are representative of the target user population. Participant recruiting is usually conducted by professional agencies that specialize in market research. These agencies have a database of consumers to call on. Participants are usually recruited over the phone using a questionnaire referred to as a screener (see Figure 17.2).

Once the design team has agreed on who would be the ideal participants in the study, the human factors practitioner must work closely with a market researcher to develop a participant screener. Participants who pass the screener and agree to participate in the research are usually paid for their participation.

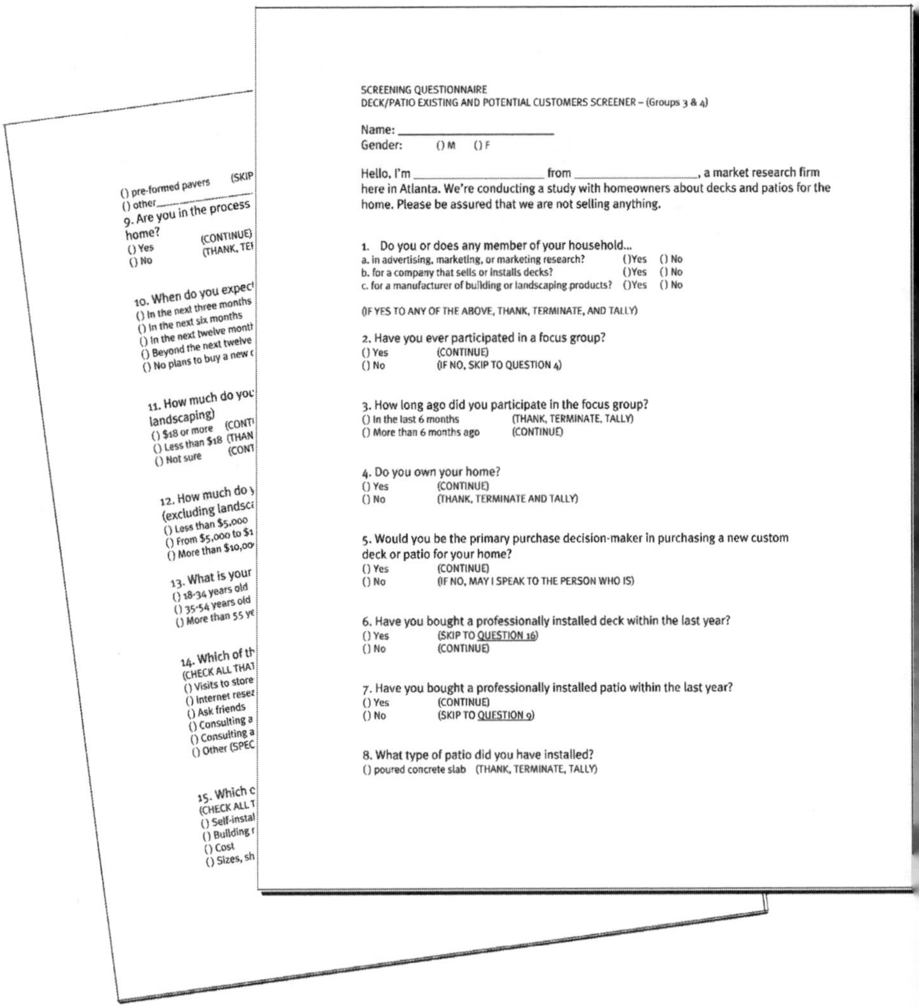

FIG. 17.2.   A sample screener for use in recruiting participants.

Apparatus

In scientific and evaluative research, the tools for data collection can range from simple to elaborate. This is also true for generative search. The overriding goal of generative search tools is to access user experience. The development of new tools for generative search represents a unique opportunity for human factors practitioners to expand their role in product development.

The different ways of accessing experience have evolved over time. Traditional design research methods have been focused primarily on observational research. Researchers watch what people do via task analysis, often now called ethnography. This might include videotaping what people do, keeping logs of mouse clicks, or having users report what they do through interviews or diaries. These methods are good for tracking people's experiences in the present. Traditional market research methods, on the other hand, have been focused more on what people say and think, as reported through focus groups (i.e., meetings with small groups of people for feedback), interviews, and questionnaires. These methods also work well for learning about people's experiences in the present, with some exploration of the past also available as well as limited opportunity for looking into the future.

The new tools are focused on what people make (i.e., what they create from the tool kits we provide for them to use in expressing their thoughts, feelings and dreams). As shown in Fig. 17.3, these domains of learning—what people say, do, and make—overlap and complement each other, and so human factors professionals should be prepared to make use of all three.

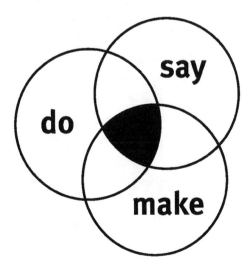

FIG. 17.3. All three domains of learning can contribute to understanding people's experiences.

The make tools alluded to previously are the most recent development in design research. Make tools are primarily visual so that they can serve as a common ground for connecting the thoughts and ideas of people from different disciplines and perspectives. Because they are projective, the make tools are particularly good in the generative phase of the design development process, for embodiment of both abstract concepts (e.g., branding ideas) and concrete ones (e.g. instructions for using a product). Two sample tools are shown in Fig. 17.4a and Fig. 17.4b.

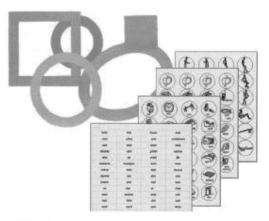

FIG. 17.4a.   A sample cognitive mapping tool kit.

FIG. 17.4b.   A sample Velcro-modeling tool kit.

When make tools are used in the generative phase of the design development process, user-generated artifacts result. There are many different types of make tool kits that facilitate the expression of a wide range of artifacts, models, or both, ranging from emotional to cognitive, looking from the past to the future, and two- or three-dimensional in structure. With emotional tool kits, people make artifacts that show or tell stories and dreams. As an example, a collage can be a two-dimensional emotional tool kit for exploring the past, present, and future (see Fig. 17.5). Tools such as this work well in accessing people's unspoken feelings and emotional states, feelings they may not be able to articulate in words.

On the other hand, cognitive tool kits can be useful for accessing cognitive processes and models. As with the emotional tool kits, cognitive tool kits can come in many forms. One example is the Velcro™ modeling tool kit, which can be well suited for accessing cognitive ideas for the future in three dimensions. With cognitive tool kits, people make artifacts such as maps, mappings, three-dimensional models of functionality, diagrams of relationships, flow charts of processes, and cognitive models. Figure 17.6a and Fig. 17.6b show completed artifacts from the tool kits shown earlier.

Every artifact tells a story, and it is important to ask the creator of the artifact to tell the story behind it. The stories associated with the artifacts from the emotional tool kits tell of feelings, dreams, fears, and aspirations. The stories associated with the artifacts from the cognitive tool kits tell us how people understand and misunderstand products, events, and places.

FIG. 17.5. A father completes a collaging exercise as his son explains what he has created.

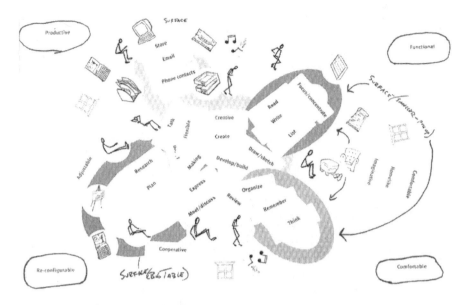

FIG. 17.6a. A completed cognitive mapping exercise using the tool kit in Fig. 17.4a.

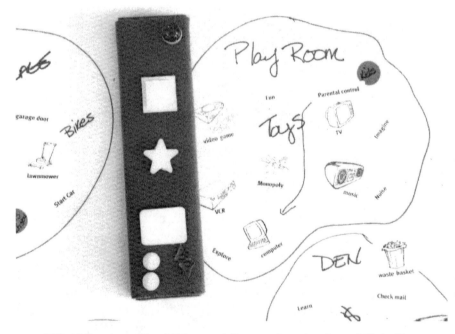

FIG. 17.6b. A completed Velcro-modeling exercise using the tool kit in Fig. 17.4b.

The cognitive tool kits also can reveal the intuitive relationships between product components. People can't always explain in words alone about their unmet needs. If they could, they would probably no longer be unmet. The new tools are an emerging visual language that people can use to express feelings and ideas that are often difficult to express in words. This new language relies on visual literacy and begins to bring it into balance with verbal literacy.

The shift toward these make tools represents a larger shift toward an emerging participatory mindset. More and more, the end user is called on to take a proactive role in the development process. Participatory activities are beginning to be used to open up the generative phase of the development process to all the stakeholders involved. By knowing how to access people's feelings and ideas and emotions and thoughts, one is able to facilitate respect between a company and its customers. Such respect can lead to the merging of the dreams and aspirations of customers with the strategic goals of a company (see Fig. 17.7). This intersection is required for companies to achieve sustainability in the rapidly changing business climate.

Procedure

Both scientific research and generative search pay close attention to how the apparatus is used or how tools are executed with participants. In gen-

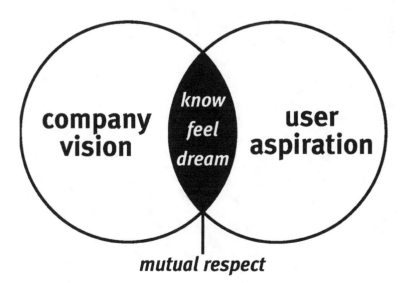

FIG. 17.7. Mutual respect emerges when companies learn what people know, feel, and dream.

erative search, data is usually collected during a face-to-face interview or group discussion.

In face-to-face interviews up to three members of the search team visit the participant within his or her home, work environment, or natural context, depending on the area of inquiry. A familiar context for the participant is ideal for this type of research.

A second method is to collect data with small groups of 6–8 participants. This is usually done within the context of a focus group facility. These facilities are available in most cities and can be found in the phone book or by searching the Web using the keywords *focus group facilities*. Focus group facilities have a number of rooms, each containing a large table where the moderator (i.e., the person who leads the discussion) and the participants sit. The room shown in Fig. 17.8 has a one-way viewing mirror where other team members can observe from the outside.

Both types of interviews are audiotaped and videotaped. A discussion guide or outline is used as a tool to help the interviewer manage the discussion. This guide outlines discussion questions, follow-ups, and probes to explore the answers to these questions, the instructions for each tool, and the specific questions to ask while a participant is explaining the tool. The process of conducting the research is based on allowing people to ex-

FIG. 17.8.   A focus group discussing a participant's collage. Most focus group facilities feature two-way mirrors so that team members and clients can watch.

FIG. 17.9. A father and son collaborating on creating an ideal product using a Velcro tool kit.

press themselves, both through the participatory tools and through discussion of their experiences (see Fig. 17.9).

Pretools can be used before the research begins to bring the subject of inquiry to the forefront of the participant's mind. After the participant has been recruited, but before the actual interview, exercises can be shipped to them to work on prior to the meeting. The most typical pretools are a workbook and disposable camera. The workbook usually contains a set of open-ended questions and scales. The disposable camera is accompanied by a set of instructions for photography. Often participants can be given instructions to mount the pictures from the disposable camera in the workbook and title them. The participants will need enough time to get their film developed prior to the discussion (although an alternative is to supply them with an instant camera instead). Use of these pretools allows people to become more attuned to areas of life that they might normally take for granted.

Using or executing the tools is one of the finer points of generative search; one should be careful not to center the meeting too exclusively on the tools, as they are components of an overall discussion or conversation centered on the person. Consideration must be given to the instructions for each tool and where that tool fits within the context of the overall discussion. A pilot test is strongly recommended because the discussion guide and tool instructions usually require refinement. The order in which the tools are used during the discussion is another key issue. A total of three to four tools can be used within a given interview or group discussion, as long as the tools themselves do not become overwhelming. Each tool must build the groundwork for the next tool and progress through the goals of the discussion. Sometimes a tool is more important for putting the participant into a state of mind for the next tool or later conversation than it is for the actual data the tool itself generates.

Analysis

Generative search usually results in massive amounts of data, including the artifacts created by participants, transcripts, videotapes, and outputs from databases and statistical analyses. The responsibility for making sense of the data belongs not only to human factors practitioners but also to the entire design team. This presents another opportunity to expand the role of the human factors practitioner to one of team integrator.

The opportunities for analyzing data from participatory tools are infinite. The overriding goal in generative search is to find patterns in the data. This includes both patterns within a specific tool or exercise and patterns that are shared across tools or exercises. Merging multivariate statistics with the tool data provides for numerous opportunities to identify patterns for new user segments through more thorough tool analysis and interrelationships between tools.

Each member of the product development team should play an active role in creating the tools, collecting the data, and analyzing the data. The larger goal in analyzing the data is to look for patterns, both within any given tool and across tools. The human factors practitioner is responsible for encouraging the participation and feedback of all team members. When generative search is executed correctly, each discipline will find meaningful information.

One foundation of analysis is to put information in a multirelational database, which makes it easier to sort data and observe patterns. Responses to open-ended questions can be entered and sorted. Titles or captions for photos can be analyzed in this manner as well. Scaled data can be pulled from this database and analyzed by descriptive and multivariate statistics. Collages are often analyzed using the frequency data of each word or image, which can be placed into a summary collage to provide a visual data array of the most frequently selected words and images.

As stated in the apparatus section, the tools alone are not adequate. It is also important to understand the descriptions that people give to the visual artifacts they create. Audiotapes should be transcribed. These transcriptions can then be attached to each participant's data to understand the meaning that the artifacts embody.

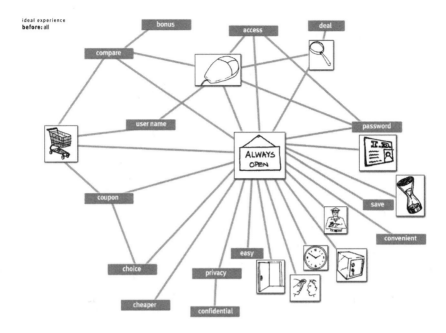

FIG. 17.10. A Pathfinder diagram is one method for visually exploring patterns in data.

As previously mentioned, the visual nature of the tools is critical. The importance of visualization also applies to presentation of the data. Reports and statistics alone are not adequate; the human factors practitioner also may have to communicate visually. Edward Tufte's books (1990, 1992, 1997) on visual representation of information are strongly recommended as resources for exploring new, more meaningful ways of presenting data. Statistical tools such as Pathfinder, a multidimensional scaling tool that presents data in a node diagram, also should be considered. Pathfinder analyses provide representations in which concepts are represented as nodes, and relations between concepts are represented as links connecting the related nodes (see Fig. 17.10; Schvaneveldt, 1990).

Discussion

In scientific research the findings from the data are related back to the hypothesis. The findings are then interpreted relative to the subject of inquiry. Observations relative to the apparatus or experimental procedure are analyzed. Finally the hypothesis is reviewed and modified. External validity is usually taken for granted, and internal validity receives scrutiny.

In generative search, the external validity receives more scrutiny than the internal validity. The goal is to translate the new understanding of user experience into design implications, business opportunities, brand attributes, and so forth. This presents another new challenge for the human factors practitioner: translating generative search findings into actionable results. Learning to do this takes time and experience. To facilitate this learning it is recommended that you make an effort to understand how each discipline works. Listen to what your team members say and watch what they do. Learn to anticipate their questions and concerns. Observe what information they find useful and how they apply it.

The output of generative search is a wide range of hypotheses. A hypothesis can be a product concept, a business strategy, or a product positioning. Ultimately, the hypothesis will be evaluated by users. Will people buy the product? Will they recommend the product to others? Will competitors copy the product? Will the product be used for a long period of time, or quickly wind up in thrift stores and garage sales?

When a generative search procedure is developed and executed correctly, it can yield findings that have implications ranging from high-level business strategy decisions to very specific design criteria. The human factors practitioner must make an effort to develop and communicate these findings. One way to do this is through an experiential model of user aspiration, which is an abstract visual representation of an ideal experience. An experiential model, such as that shown in Fig. 17.11, describes how people want to feel and the means by which they can realize their desired experience. Consider developing an experiential model that expresses the ideal experience the user is seeking.

Experiential Model

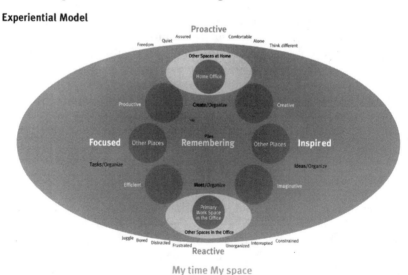

FIG. 17.11. An experiential model of people's perceptions of different workplaces.

THE ROLE OF THE HUMAN FACTORS PRACTITIONER IN CONSUMER PRODUCT DEVELOPMENT

Regardless of the industry or product category, consumer product development today is an interdisciplinary team activity. As the process emerges in many companies, the human factors practitioner if often faced with the challenge of creating a role within the product development team. To that end, the following suggestions may be useful.

The role of the human factors practitioner in new product development has been to conduct research to understand and draw conclusions about users and their perceptions of the products being developed, usually via interviews and observations. In fact, many consumer product development companies do not have a long history of using human factors professionals. They tend to have a longer history of using market researchers as the focus of user research efforts. Today the human factors practitioner must work closely with market research. Both human factors and market research practitioners are focused on understanding people; as such, human factors professionals must learn how to merge their focus on use with the market research focus on purchase behavior. Both groups can benefit if each side understands the principles that the other is applying.

Human factors professionals also must learn to work closely with industrial designers to get their ideas and findings visually embodied. Because designers are often the people responsible for integrating the thinking into a range of solutions, the human factors professional must learn how to communicate with them. You may want to draw diagrams, rough sketches, or even make a model that you can use to express your thoughts. Don't worry about how your ideas look; just be concerned with how well you are able to communicate your thinking.

Interface designers also are good partners for the human factors practitioner. Working closely with them enables integration of human factors principles into their interface designs. Again, this can be facilitated by communicating with them in their context—hand-drawn screen concepts often will work much better than written reports.

By developing a role that communicates across and respects all disciplines in the development process, you as a human factors practitioner can enable communication between these disciplines, which ultimately can benefit end users. For instance, you could find out from the electrical engineers what sensors may be on a given product. If you identify any designed for system functioning that could be used to communicate useful information to the user, you can share this information with the interface designer so as to make this information available and usable.

SUMMARY

There is both an opportunity and a challenge to expand the role of the human factors practitioner in consumer product development. Expanding from a focus on use to the whole of the experience will enhance the strategic importance of the human factors practitioner to the organization. A focus on generative search in addition to evaluative research will support the shift to experience and increase the impact the practitioner can have on new products. Moving from scientific reports to visual communication will help to enable the seamless transition of research into design.

REFERENCES

Schvaneveldt, R. W. (1990). *Pathfinder associative networks: Studies in knowledge organization.* Norwood, NJ: Ablex.

Tufte, E. R. (1990). *Envisioning information.* Cheshire, CT: Graphics Press.

Tufte, E. R. (1992). *The visual display of quantitative information.* Cheshire, CT: Graphics Press.

Tufte, E. R. (1997). *Visual explanations: Images and quantities, evidence and narrative.* Cheshire, CT: Graphics Press.

Chapter 18

Usability Testing

Stephen D. Armstrong
United Services Automobile Association

William C. Brewer
Lockheed Martin, Inc.

Richard K. Steinberg
Schafer Corporation

Creating good design is like baking from a recipe. Despite following recipe directions, perfectly and deftly applying all the best cooking techniques, a good chef will still taste the batter before putting it in the oven. Usability testing fulfills the same purpose as tasting the batter: It verifies that a system or product design meets our expectations for usefulness and satisfaction before we move into production. Like the taste test, usability testing is generally quick and cheap yet likely to reveal much about the goodness of our design. Conducting a usability evaluation is so critical to ensuring quality, no development effort should be considered complete until someone has "tasted" the final design. To assist usability evaluators, this chapter offers practical guidance on understanding the goals and strategies of usability testing and on selecting and applying the appropriate usability testing methods.

Let's begin with a definition. Usability is the degree to which the design of a device or system may be used effectively and efficiently by a human. Some definitions of usability add user satisfaction as an additional factor for measuring usability, suggesting that usable design goes beyond merely enhancing operator performance; it must consider an operator's likes and dislikes as well. Allowing for differences in definition, however, the more important issue for engineers and designers is how to measure the usability of a system and how to put that knowledge to work in improving the design.

The term *usability testing* itself is misleading because it can be used generically to describe all types of usability evaluations or, specifically, to refer to a single technique. In its generic sense, usability testing refers to three broad categories of techniques and tools that measure system usefulness. The first category of usability evaluation is surveying, which relies on traditional, self-reported data collection methods to estimate product usefulness and acceptability. A second category of evaluation is usability inspections, where specialists scrutinize a design according to a systematic approach and judge its acceptability against certain criteria. The final category, experimental testing, includes any attempt to quantify operator performance using controlled data collection techniques. Today, all three categories of usability evaluation—surveying, inspection, and testing—are commonly referred to as usability testing.

Within each broad category of usability testing are a wide range of techniques and tools for collecting usability information. Naturally, each technique or tool has certain advantages and disadvantages, certain capabilities and certain limitations. Understanding when and how to use each method is important, and the bulk of this chapter is devoted to building such insight. Before discussing details of specific methods, however, we must spend a moment considering the proper preparation for any type of usability testing.

PREPARING FOR USABILITY TESTING

Conducting good usability evaluations begins with three prerequisites. First, all human activities in the system under development should be identified and described. From this complete accounting, a subset of these tasks then may be selected for usability evaluations. For example, when preparing to conduct usability tests for a new VCR design, evaluators might compile a list of operator tasks for the VCR, including perhaps inserting a tape, recording a TV program, or setting the VCR clock. From this overall list of tasks, a few key tasks then are selected for evaluation. Several popular methods for conducting task analyses can be found in industrial and organizational psychology textbooks.

Second, usability goals must be established before testing begins. During usability testing, an evaluator compares observed operator performance against expected or desired performance for a specific task. Differences between expected and measured performance may point to potential usability problems in the design. Absent usability goals, usability analyses lose their significance, and important questions become unanswerable. How much usability is enough? How much user satisfaction is enough? How much is too little? Without establishing usability goals, engineers and designers can never interpret the results of a usability test.

Furthermore, goals should be operationalized, or expressed in terms relevant to the domain of interest. For example, a usability goal for our new VCR under development might state that consumers will find the clock setting procedure to be much easier than on the previous VCR model. Although this goal sounds reasonable (especially for anyone who has struggled to set a VCR clock), it is too ambiguous to permit meaningful comparisons during usability testing. A better way to state this goal might be that 95% of consumers will be able to program the VCR clock successfully in less than 30 s and without use of the VCR remote or instruction manual. Operationalizing the goal in this way creates a quantifiable standard against which the design can be measured.

Finally, usability testing must be integrated into the larger development process so the findings of such testing can be used to improve the product. In other words, the results of a usability test should be available at a time and in a manner necessary to influence the design of the system. In practice this generally means conducting usability testing events early and often to maximize opportunities for identifying deficient designs and improving usability. If usability testing is accomplished only after most development has concluded, the effort holds little chance of improving the product or system.

With initial preparations complete, it's time to select the appropriate tools or techniques for the usability evaluation. The remainder of this chapter describes the common techniques available within each category, including an overview of each procedure, its advantages and disadvantages, and the most appropriate occasions in which to apply it. Additional references are also provided to guide the reader to learning more about the technique or tool.

SURVEYING TECHNIQUES

Perhaps the simplest way to evaluate the usability of a design is to expose prospective users to the new product and solicit their opinions on various aspects of the design. This basic approach is the common thread running through all surveying techniques, and, despite the comparatively uncontrolled nature of surveying, the results can be surprisingly helpful in detecting weakness in a design.

Questionnaires

A questionnaire is usually a form or booklet of questions provided to representative users soliciting feedback on specific issues related to a design under study. Administering questionnaires is an excellent method for quickly gaining an overall assessment of a design's strengths and flaws. Almost all usability evaluations can benefit from employing a questionnaire as part of an overall assessment. Questionnaires can be especially effective in judging the suitability of a commercial product design because eliciting a positive consumer response is an essential prerequisite to generating product sales. Questionnaires are generally less expensive than most other testing methods, depending on the scale of the questionnaire, and they can be administered usually without evaluator supervision (especially if distributed electronically).

Questionnaires differ from other survey techniques and from other usability tests in general because they are administered in written form and rely largely on the respondent to provide meaningful responses. Although questionnaires relieve the evaluator from much of the burden of data collection, they still require careful preparation on the part of the administrator. (See the earlier chapter in this volume for detailed guidance on questionnaire design.) Before administering a questionnaire, conduct a pilot test to gauge whether the instructions and content of the instrument are appropriate. Generally, a pilot test involves administering the questionnaire to a small group of participants to learn if the instrument itself contains any design flaws. Remember, in the pilot test the questionnaire itself is under scrutiny, not the product or design undergoing usability testing. Responses from the pilot test should never be included in the final usability test results.

Interviews

When time and resources permit, a personal interview with users can offer valuable insight into both how and why users reacted to a given design. Most interviews take the form of a discussion conducted between a moderator and a user who has been exposed to the design under study, although other interview formats are used occasionally (e.g., group discussions, panel discussions). Like questionnaires, interviews are a good addition to almost any usability evaluation. Often, interviews and questionnaires are combined to great advantage.

An interview can be very structured or free flowing, using either a question-and-answer format, open-ended discourse, or both. Obviously, the more structured the interview format, the easier it will be to compare responses across participants. Even in the case of an unstructured interview, the interviewer should prepare at least a few questions or topics of interest in advance to ensure meaningful results. Likewise, highly structured interviews should include at least some opportunity for participants to discuss their impressions freely. When interviews are so structured that all questions are set in advance, the experience resembles little more than an oral questionnaire.

Asking participants to discuss their experiences with a new design can reveal useful information, but like all sources of self-reported data, it has certain limitations. The quality of data varies with each participant's skill at self-observation. Participants may not notice important details of their experiences with the product or fail to attribute difficulties to the correct source. Second, a participant's recall of events may be incomplete or error prone. Finally, the interviewer must record, filter and organize the responses in a manner that permits comparison of responses across participants, potentially tainting responses with the interview's bias. The interviewer may interpret a participant's comments in light of previous feedback or according to the researcher's own preferences or expectations.

Direct Observation

Rather than relying on participants to recount their experiences, the evaluator may choose to observe the participant in action using the product under study. The researcher then records his or her observations. As with the interview technique, the degree of structure for this type of activity can vary widely, and the bias of the interviewer is still a factor. Compared to interviews or questionnaires, however, observation evaluations place considerably less responsibility on the participants for providing meaningful feedback and may reveal subtle usability problems that the participant themselves did not notice or report.

The typical direct observation technique begins by assigning participants a specific task or series of tasks to perform. The participant's instructions should be scripted, so that each participant is given exactly the same instructions. The test administrator (observer) monitors the participant's performance discretely to avoid influencing the participant during testing. As users perform the directed set of tasks, the test administrator watches for any erroneous or unintended actions, behaviors, and sequences and logs errors or other observations on a data collection form. The data collection form is a critical tool in this technique because it organizes the feedback by task, screen, or display module and minimizes the effect of observer bias. An example data collection form for an online catalog interface is shown in Fig. 18.1 and Fig. 18.2. Note that the steps for observation in the form may be as detailed or as general as desired for any evaluation.

Direct observation is a relatively quick, inexpensive, and capable means of identifying the critical flaws in a design. Because it evaluates every participant on the same design, the data collected for one participant can be compared with that of others, allowing discrimination between a design characteristic that was difficult for one participant versus one that prompted errors from many participants. These qualities make direct observation a strong candidate for use in iterative design processes.

Despite its speed and low cost, direct observation can still can be difficult to conduct, and it can yield inconsistent results if the observer is not well trained in the proper data collection procedures. For example, once a participant has been prepared to perform the desired task, the observer

Data Collection Sheet
TASK 1

User Task: Use the prototype interface to execute the purchase of a set of English Pub Glasses, and send them as a gift to Mr. John Doe, using the credit card number and address provided by the facilitator.

User Informati on
Name: _____ No.: _____

Table YYY, *continued*

Title/Rank/Position:
_____

Experience with the project or similar interfaces:
_____

_____

Exercise Information
Software/Build ID:_____ Date: _____

Location: _____ Start Time: _____

Observer: _____ Finish Time: _____

FIG. 18.1.   Example data collection sheet for direct observation—questionnaire.

should not provide the participant any additional information about how the product under study works or how to perform the procedure. Such unscripted direction risks contaminating the data. Observers also face the challenging duty of recording the participant's actions accurately and often in real time. The quicker the participant moves through the task, the more difficult the recording task becomes for the observer. Reviewing a videotape recording of each session can help the researcher capture missing data, but this procedure introduces new problems because it can be difficult to capture all the participant's activities with a camera, especially when the participant is interacting with a computer or other complex machinery.

INSPECTION TECHNIQUES

Standardization Reviews

A preventative measure for improving the usability of any product is to adhere to accepted human factors design guidelines from the beginning. Guidelines originate from many sources and address many different as-

Required Steps	Performed In Sequence?		Performed Correctly?		Comments & Observations
	Y	N	Y	N	
Enter "English Pub Glasses" in the Search field & Click "Go"					
From list of items returned, click on "English Pub Glasses – Set of 6"					
Click on "Purchase" icon for the set of 6 glasses.					
Confirm addition of the Pub Glasses to the shopping cart.					
Click on "Check Out Now" button.					
Enter credit card and billing address information: 1. CC type(menu) 2. CC number 3. Exp. Date 4. Name 5. Address 6. City, State, Zip 7. Country (menu)					
Click on "Submit" button					
Click on "Different Delivery Address" button					
Enter Delivery Address information: 1. Name 2. Address 3. City, State, Zip 4. Country (menu)					
Click on "Gift" checkbox					
Click on "Submit" button					
Click on "Enclose Gift Card" button					
Click on "Complete Transaction" button					

FIG. 18.2.   Example data collection sheet for direct observation—checklist.

pect of design, so choosing which standards to apply depends largely on the needs and expectations of the intended user population. For example, products destined for the military environment generally must conform to various Department of Defense standards (1989, 1996a, 1996b). Products designed for the commercial market usually follow standards established by ANSI or the ISO. Other useful standards include those of professional organizations (e.g., Human Factors and Ergonomics Society), industry affiliations, and major corporations (e.g., Microsoft, Apple Computer). Many applications will fall under several standards simultaneously.

Abiding by accepted usability standards offers numerous advantages to the designer and developer. First, standards promote external consistency. External consistency describes how closely one product or design compares to similar products or designs from other sources. For example, elevators from different manufacturers use controls that typically share many of the same design characteristics (e.g., button behavior, arrangement, size, labeling). The high external consistency of elevator controls enhances their usability. Why? Because external consistency encourages a transfer of learning. Users typically invest many hours building proficiency on complex systems, and they benefit from opportunities to transfer those learned skills to new systems. When designers adhere to widely accepted standards, they ensure that a wider audience of users can borrow from past experience in reaching full productivity sooner and with less effort.

Design guidelines also ensure internal consistency, or congruence among the different elements of a single design or product. Internal consistency means controls operate in a similar fashion, displays present information in a common format, and labels use comparable phraseology. People expect a product to present a unified appearance and exhibit predictable behaviors, yet when each part of a design follows different rules, the resulting product will seem more a Frankenstein patchwork than a carefully woven blend of features.

Increasingly, as the complexity of hardware and software systems grows, development responsibility is shared among large groups of engineers, designers, programmers, builders, and other professionals. Without guidelines and standards to ensure internal consistency, each team of contributors is free to develop their own conventions, resulting in a product that exhibits a disconnected personality, like loosely connected pieces rather than a coherent integrated whole. Standards and guidelines institutionalize the best design principles and ensure uniform application of those principles across all aspects of a product or system. For example, the procedure for closing a dialog box in a software program should remain the same for every such dialog box, regardless of when or where the box appears. If it varies arbitrarily from one circumstance to the next, the user can quickly grow frustrated with the product's internal inconsistency, and user performance may suffer.

Standards can be misapplied, however. When design guidelines are so rigid that no exceptions are recognized nor improvements incorporated over time, standardization becomes an impediment to good design rather than a guarantor of it. Once a standard is developed or adopted, reasonable allowances for exceptions or changes should be made. Likewise, standards must be capable of evolving and adapting to changing technology and needs. In the end, standards exist to serve the needs of designers, not the other way around.

Adopting design standards is a very practical and inexpensive way to improve quality. Little skilled expertise is required, so many members of the development team can apply the technique. Costs are negligible; in fact, standards may reduce overall development costs by encouraging reliance on proven designs and avoiding costly mistakes. In the end, compliance with design guidelines does not guarantee usability, but it usually helps eliminate major mistakes and promote consistency.

In addition to the guides mentioned earlier in this section, important standardization references include MIL-STD 1472E (DOD, 1996a) for design of military workstations, associated furniture, and the facilities in which they are placed. Designers of workstation displays should also consult ANSI's *National Standard for Human Factors Engineering of Visual Display Terminal Workstations* (1988). Finally, many software development teams or organizations rely on standardization documents called style guides. Essentially, style guides are in-house design standards sanctioned by management for use in all software development efforts. Style guide specifications usually address the appearance and behavior of individual interface components and provide rules for application and window design in software with a graphical or browser-based interface. Some well-recognized style guides are included at the end of this chapter.

Comparing Prototype Alternatives

Everyone likes to have choices. Engineers and designers are no different, and when budget and schedule permit, they prefer to pursue several design alternatives simultaneously, waiting to choose a single best design after an opportunity to study and compare the alternatives. Comparing alternatives is an excellent strategy for improving the usability of any design, especially when no clearly best solution is recognized, and it is a necessary prerequisite for many other forms of usability evaluations.

Not only can comparing prototypes improve usability by eliminating potentially poor designs, comparing prototypes during development promises significant reductions in development time. In fact, recent research suggests that the overall development process can be reduced by as much as 40% (Bailey, 1996) through prototyping techniques. If improved performance and reduced development time weren't reason enough to pursue this strategy, building prototypes for comparison testing yields a working model of the eventual product long before the real thing emerges in final form. This early preview version of the product can also serve as a valuable tool for the designer when describing the forthcoming design concept to management, trainers, programmers, prospective users, and anyone else involved in the project.

The level of fidelity of prototypes depends primarily on budget constraints. Fidelity can range from static (noninteractive) pictures of interface concepts to dynamic working models to production prototypes, where the actual system is partially built. Usually, the best investment return is found with prototypes in the middle to lower end of the fidelity scale. For high-performance systems such as military or emergency management interfaces, however, a more costly, higher-fidelity prototype may be the only way to identify critical safety issues and prevent casualties.

A popular means of low-fidelity prototyping is rapid prototyping, and it can be one of the most cost-effective means of usability testing. The hallmarks of a rapid prototype are low cost and quick turnaround. A rapid prototype includes only as much fidelity as is required to test the task or tasks under study. In some cases, static prototypes may be sufficient to support the type of testing desired; other times development of a dynamic interface mock-up may be necessary. The goal is to try many different ideas to generate feedback regarding what works best and what doesn't work at all.

When high human performance is critical, rapid prototyping is less viable. Usually more money must be invested in higher-fidelity prototypes. For example, display designs for a complex military command and control system likely would necessitate a richly detailed simulation to discern differences in design performance or to measure operator performance. In the extreme, a production prototype, where a working version of the actual system is built, may be required for an extremely complex system with a complicated set of operator control actions. Although it may be tempting to go straight for the production prototype every time, it's important to remember that the cost of prototyping goes up exponentially with increased complexity and fidelity.

Heuristic Evaluations

A specialized method for testing the usability of software interfaces, heuristic evaluations, works by employing small groups of evaluators to scrutinize a user interface design for compliance with specific usability principles (or heuristics). Like the standardization approach, a heuristics evaluation is a relatively low-cost means (sometimes called the discount engineering method) of identifying usability problems in a design without requiring access to prospective users. Unlike standardization reviews, however, heuristic evaluations look for the proper application of general design objectives rather than adherence to specific appearance and behavioral standards.

Heuristic evaluations are best conducted early in the design process, when usability problems can be identified before entering a production phase, or whenever budget limitations prevent more rigorous testing. Nielsen (1994) identified 53 heuristics that accounted for about 90% of typical usability problems found in a software user-interface designs. Using Nielson's heuristics, a good usability specialist can detect up to 60% of existing usability problems merely by inspecting the design carefully (Nielsen & Molich, 1990). Interestingly, Nielson (1994) found comparable evaluator performance regardless of whether the evaluator inspected a paper specification of the product or an actual prototype, suggesting that this technique may be employed reliably very early in the development process.

Heuristic evaluations produce the most reliable results when a team of evaluators examines the same design. How many evaluators participate in an evaluation depends on budget constraints and the availability of experienced evaluators, but Nielsen (1994) recommends no less than three and preferably five or more. As the criticality of a system's usability increases, so does the importance of having a larger group of evaluators because more evaluators generally means increased likelihood of identifying problems.

During a heuristics evaluation, evaluators should work independently to assess the design and then compare their results to identify common trends and unique observations. The actual process of inspection depends on a checklist guide of desired heuristics and their definitions. Evaluators familiarize themselves with the definition of each heuristic and then make note of any deviations from these ideals in the design under study. Nielsen and Molich (1990) offer perhaps the best-known checklist for heuristics evaluations, a summary of which has been included in this chapter (see Table 18.1). Consult Nielsen and Molich (1990) and Molich and Nielsen (1990) for a more complete explanation of heuristics checklists. Popular human interface standardization guides can be used in place of heuristics checklists, provided the guidance is expressed as general principles of usability and not specific interface traits. Apple Computer's *Human Interface Guidelines* (1987) and Sun Computer's SunSoft Design Guidelines are good examples of such documents (see Tables 18.2 and 18.3).

TABLE 18.1
Nielsen's Heuristic Checklist

1. Use Simple and natural dialogue: Dialogues should contain only relevant information.
2. Speak the user's language rather than in system-oriented terms.
3. Minimize the requirement for large memory load. Instructions for use of the system should be easily retrievable whenever appropriate.
4. Be Consistent. Consistent words, situations, or actions mean the same thing.
5. Provide Feedback: The system should always keep users informed about what is going on, through appropriate feedback within reasonable time.
6. Provide clear exits: Users need a clearly marked "emergency exit" to leave the unwanted state.
7. Provide Shortcuts/Accelerators for the User.
8. Provide clear error messages in plain language. Constructively suggest a solution.
9. Build Design to reduce probability or prevent error.
10. Provide Help and documentation. It is better if the system can be used without documentation, it may be necessary to provide help and documentation.

These heuristics, suggested by other authors, have been rewritten for the sake of brevity. The exact wording of these heuristics as printed here is therefore the responsibility of the present authors and does not necessarily correspond to the way the original authors would have edited their principles.

TABLE 18.2
Apple Computer Usability Guidelines

---

1. *Metaphors*: Use metaphors from the real world to take advantage of people's knowledge of the world.
2. *Direct manipulation*: Objects on screen remain visible while user performs physical actions on them, and the impact of these operations is immediately visible.
3. *See-and-point instead of remember-and-type*: Users may act by choosing between visible alternatives.
4. *Consistency*: Same thing looks the same, same actions are done the same way.
5. *WYSIWYG* (What You See Is What You Get): Do not hide features.
6. *User control*: Allow the user to initiate and control actions.
7. *Feedback*: Immediately show that a user's input has been received and is being acted upon. Inform users of expected delays.
8. *Forgiveness*: Make computer actions reversible and warn people before they lose data.
9. *Perceived stability*: Define a finite set of objects that do not go away (but may be dimmed).
10. *Aesthetic integrity*: Keep graphic design simple.
11. *Modelessness*: Allow people to do whatever they want whenever they want it.
12. *Accessibility*: Allow for users who differ from the "average."

These guidelines, suggested by other authors, have been rewritten for the sake of brevity. The exact wording of these guidelines as printed here is therefore the responsibility of the present authors and does not necessarily correspond to the way the original authors would have edited their principles.

Usability Inspection Case Study

To illustrate the use of the various inspection techniques described previously, we offer the following case study. Operators in a power-generating plant must monitor and control plant operations using a computer terminal. As engineers designed the user interface for the power plant computers, they considered two display alternatives for a plant monitoring screen. The alternatives, called Option A and Option B, differed in the way they permitted three different operators to individually select the variables they wished to monitor for a particular plant subsystem. The engineers decided to conduct usability inspection tests on these two alternatives, hoping to learn which one performed best.

The first display alternative, Option A (see Fig. 18.3), presented three scrolling lists in a window: One list showed the available subsystems in the plant, the second list displayed all the variables available for any selected subsystem, and the third list contained the variables that had been selected for monitoring on a particular subsystem. To use this display alternative, the three operators started by selecting the appropriate tab (Operator 1, 2, or 3) and then worked through the three lists sequentially until a subsystem and its corresponding variables had been selected for monitoring. Option B (see Fig. 18.4) used a pop-down menu for selecting the de-

TABLE 18.3
SunSoft Usability Guidelines

1. Basic functionality should be understandable within an hour.
2. System should use and understand the user's language.
3. Feedback should be provided for all actions.
4. Feedback should be timely and accurate.
5. UNIX concepts should be minimized (in general, minimize underlying system concepts).
6. User sensibilities should be considered.
7. Functions should be logically grouped and ordered.
8. Basic functionality should be clear.
9. Physical interaction with the system should feel natural.
10. System should be efficient to use.
11. Reasonable defaults should be provided.
12. Accelerators or shortcuts should be provided.
13. Users should not have to enter system-accessible information.
14. Everything the user needs should be accessible through the GUI (or, in general, through whatever interface style is chosen for the interface).
15. The user interface should be adaptable per user desires.
16. System should follow real-world conventions.
17. System should follow platform interface conventions.
18. System should be effectively integrated with the rest of the desktop.
19. Keyboard basic functions should be supported.
20. System should be designed to reduce or prevent errors.
21. Undo and redo should be supported.
22. Provide a good visual design.

These guidelines, suggested by other authors, have been rewritten for the sake of brevity. The exact wording of these guidelines as printed here is therefore the responsibility of the present authors and does not necessarily correspond to the way the original authors would have edited their principles.

sired operator and a list transfer process with arrow buttons, supporting the metaphor of transferring items from one list to the other.

Standards Inspection. The engineer's first step in analyzing the usability of the two display concepts was to perform a standard compliance test. The engineer noticed that Option A had three control push buttons located along the bottom of the window. A review of the Microsoft Windows conventions revealed that when push buttons are displayed horizontally in a window's action area, they should be placed at either the left or right margin of the window, as shown in Fig. 18.5. If there is an OK button, it should be placed first (even if it is not the default choice), followed by Cancel, and then the other buttons in the group, including Help, if appropriate. Consequently, the engineer concluded that Option A should be redesigned to right justify all the action buttons while reordering them in accordance with the Microsoft Windows standard of OK—Cancel Apply.

In the Option B display, the original designer chose to place the OK action button next to the list of variables chosen for monitoring. The de-

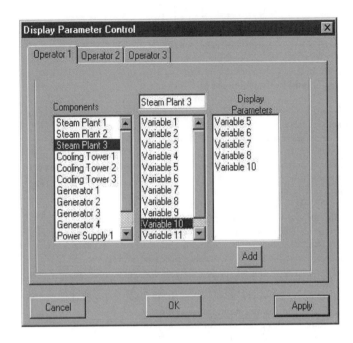

FIG. 18.3. Option A interface design.

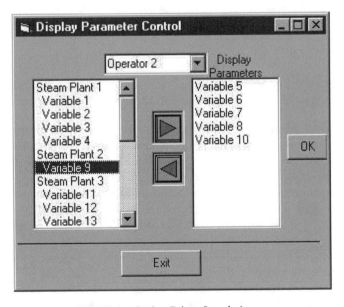

FIG. 18.4. Option B interface design.

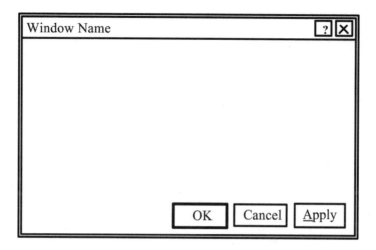

FIG. 18.5. A standard dialog box.

signer argued that the user's focus prior to pushing the OK button would be on that list region. Therefore, the approval action (OK button) was placed in close proximity to the variable list to minimize the operator's effort to locate the button. Despite the designer's logical argument, the engineer's standards inspection concluded that the OK button should be located at the bottom of the window. The designer overlooked the primary reason for standardization: to promote consistency across a system's entire design. Although the designer's choice for the location of the OK button may have improved operator performance when using that particular window, operator performance may have suffered elsewhere because of the designer's haphazard placement of such a common action button. Unpredictable deviation from accepted standards within a single design increases the likelihood of increasing operator errors, lengthening performance times, and lowering operator satisfaction.

Heuristics Evaluation. Following the standards inspection, the usability evaluator chose to accomplish a heuristics evaluation on each display alternative. Using Molich and Neilson's (1990) first four heuristics, the engineer identified several problems with both designs (see Table 18.4), which led to redesigns for both options (Fig. 18.6 and 18.7).

TESTING TECHNIQUES

To this point the usability evaluation methods described have reinforced or judged the quality of a design either by emphasizing usability standards before development or by performing subjective assessments during de-

TABLE 18.4
Heuristics Evaluation Results

Rule #1:	Dialogue should be simple and natural. Dialogues should not contain information that is irrelevant or rarely needed. Each additional unit of information in a dialogue competes with the relevant units of information and diminishes their relative visibility. All information should appear in a natural and logical order.
Finding:	In Option B's list transfer option, the user is required to scroll through a long list of information to find the golden nuggets desired for display. If the presentation of the information were filtered so that only relevant information were displayed, it could improve usability by reducing the need to scroll through a long list of unwanted items. Option B forces the operator to wade through superfluous information to find the essential item of interest, likely degrading the operator's performance significantly.
Rule #2:	Dialogue should speak the user's language. Dialogue should be expressed clearly in words, phrases, and concepts familiar to the user, rather than in system-oriented terms.
Finding:	The three operators of this system move around during their watch, and they commonly use each other's terminals. The operators refer to each workstation as Workstation 1, Workstation 2, and Workstation 3. Both interfaces use the engineering specification terms Operator 1, Operator 2, Operator 3. The labels on the interface designs should be changed to reflect the operators' terminology.
Rule #3:	The users' memory load should be minimized. The user should not have to remember information when moving from one part of a dialogue to another. Instructions for use of the system should be visible or easily retrievable whenever appropriate.
Finding:	Each interface contains menus of selectable items that minimize work load. Unfortunately, both interfaces demand that the locations each operator monitors be known and remembered. The interfaces could be enhanced if a geographic locator displayed the location of a selected operator position.
Rule #4:	Consistency should prevail throughout system. Users should not have to wonder whether different words, situations, or actions mean the same thing.
Finding:	Option B uses a pop-down window to select the operator console. Other windows within this same application, however, use tab windows (like Option A). To promote consistency within the interface, Option B should use tabs for selecting operator console as well.

velopment. Neither of these approaches, however, can offer the rigorous objectivity of empirical usability testing. Testing techniques place special emphasis on measuring a design's influence on human performance, an important consideration for many critical systems. Given the money, time, and expertise required to implement the techniques described in this section, a researcher can gain the clearest possible understanding of where a design succeeds or fails for its users.

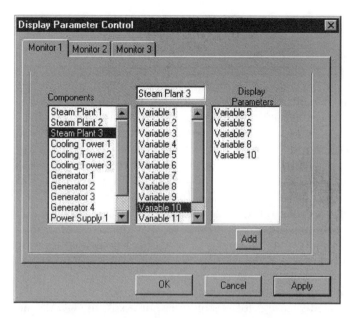

FIG. 18.6.   Option A redesigned according to heuristics evaluation.

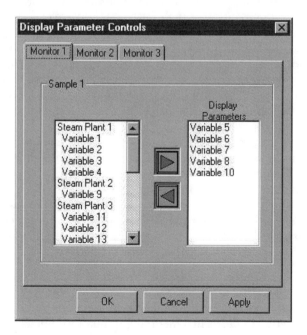

FIG. 18.7.   Option B redesigned according to heuristics evaluation.

Modeling and Simulation Techniques

Modeling and simulation methods offer an analytical approach to estimating a design's performance characteristics without the need to construct a fully working version of the design. In fact, simple empirical modeling approaches can be used to perform usability trade studies on several competing solutions before designers begin production of any design. Modeling methods vary by fidelity, but most work by comparing estimates of the time required for operators to perform certain discrete tasks using a particular interface design and drawing inferences on operator performance from the data. Usually, the timing data in these models is defined to the level of perceptual, cognitive, and motor response cycles, allowing a designer to appreciate the precise human performance characteristics of a design. Models following this technique include the GOMS (Goals, Operators, Methods, and Selection rules) family of models and model human processor models (Card, Moran, & Newell, 1983, 1986).

The basic GOMS family of models attempts to capture the cognitive behavior of a user by decomposing the operator's activity into subgoals and goal stacks. Because it focuses on operator goals, the GOMS model is an excellent tool for making error predictions for a particular design. The GOMS model assumes that the more methods and operations an operator must learn, the more chances the user has to make errors (Eberts, 1994). Besides error analysis, the GOMS models can be used to enhance ease of use and speed performance and increase ease of learning for users. Detailed information on applying the GOMS family of models, including the original Card, Moran, and Newell (1983) GOMS model, the Natural GOMS Language (NGOMSL), and the human information processing model can be found in Card et al. (1983, 1986), John and Kieras (1994, 1996), and Eberts (1994).

Human processor models are a good choice when user availability is limited and human performance time is a critical element in system performance. The modeling technique can yield impressive operator performance improvements. Using a GOMS model in one study, John and Kieras (1994) produced a 46% reduction in learning time and a 39% reduction in execution time for one design. Although not perfectly reliable at estimating task performance times, these tools still offer the potential for considerable usability improvements at relatively low cost.

Modeling Case Study

Returning to our earlier example of the power generating plant, the plant engineer decided to perform a modeling evaluation. The computer modeling technique selected by the engineer, Human Information Processor

Analysis (HIPA), compared the human performance impacts of the two computer terminal display designs under study, Option A and Option B (described in the inspection case study).

To begin the process, the evaluator produced a simple task analysis for each display option. The task analysis for Option A determined that operators searched through subsystems and selected items for monitoring. The engineer deconstructed this task into small steps (cycles), according to the type of human information processing required, and then he assigned small amounts of time to each step, according to tables provided in HIPA reference materials. The engineer identified three types of processing: perceptual cycles (t_p), motor cycles (t_m), and cognitive cycles (t_c). These discrete cycles can be combined in the HIPA analysis to estimate the overall performance time required by the operator to perform the required task with each display alternative (see Tables 18.5 and 18.6).

Once estimates of perceptual, cognitive, and motor response cycles were listed, the time estimates for these cycles were added. The average cycle times for a perceptual, cognitive, and motor cycle are 100 ms, 70 ms, and 70 ms, respectively. In the Option A screen, the engineer recorded 22 items in the component list, and each component had an average of five variables from which to select. In this design, the user would have to scroll to view items 11–22. Assuming equal probability of selecting all items, the user would scroll to view an average of 5.5 items. As a result, the operator would perform steps 1–2 and 16–22 only once, but the operator will perform steps 8–15 an average 5.5 times and steps 3–7 an average of 11 times. The engineer calculated the average time for an operator to complete the task using Option A.

In Option B, the engineer found 110 items in the component list, so steps 3–7 will be repeated (110 + 1)/2 times. The user will have to scroll to view items 11–110. Assuming equal probability of selecting all items, steps 8–15 will be repeated an average of (110 + 1 – 11)/2 times. The operator will perform steps 1–2 and 16–22 only once. The resulting estimate of the average time to complete the task can then be calculated for Option B as it was with Options A (see Tables 18.7 and 18.8).

As the engineer discovered from this HIPA analysis, operators using the Option A interface required considerably less time to complete the task than operators using the Option B interface. With such significantly different estimates for performance between Options A and B, the usability engineer felt safe in choosing the Option A interface as the baseline design. However, if the time estimates had been very similar, then the HIPA results would have been inconclusive, and the engineer would have considered developing a working prototype suitable for empirical testing.

TABLE 18.5
Option A Human Information Processing Analysis

---

1. (t_p)   Perceive list of system components by transferring from visual image store to working memory a verbal code of the label "components."
2. (t_c)   Decide to execute eye movements to list.
3. (t_m)   Execute eye movements to list item.
4. (t_c)   Decide whether item is last item on list. (If yes, skip to Step 8.)
5. (t_p)   Perceive item on list by transferring from visual image store to working memory of the verbal code of the item.
6. (t_c)   Match verbal code of item and component.
7. (t_c)   Decide if item matches the item stored in working memory. (If item is the desired item, skip to Step 16 or go back to Step 4.)
8. (t_p)   Perceive scroll bar or down arrow.
9. (t_c)   Decide to execute eye movements to down arrow.
10. (t_m)   Execute eye movements to down arrow.
11. (t_c)   Decide to move cursor to down arrow.
12. (t_m)   Move cursor to down arrow.
13. (t_p)   Perceive cursor over down arrow.
14. (t_c)   Decide to stop moving cursor.
15. (t_m)   Push cursor selection button; go to Step 5.
16. (t_c)   Decide to select item.
Repeat Steps 1–16 for the middle list. Once at Step 16, add the following steps:
17. (t_p)   Perceive the add button by transferring the verbal code of the button from visual image store to working memory.
18. (t_c)   Decide to move cursor.
19. (t_m)   Move cursor.
20. (t_p)   Perceive cursor over button.
21. (t_c)   Decide to stop moving cursor.
22. (t_m)   Select cursor selection button.
Repeat Steps 1–16 for the middle list. Once at Step 16, add the following steps:
23. (t_p)   Perceive the add button by transferring the verbal code of the button from visual image store to working memory.
24. (t_c)   Decide to move cursor.
25. (t_m)   Move cursor.
26. (t_p)   Perceive cursor over button.
27. (t_c)   Decide to stop moving cursor.
28. (t_m)   Select cursor selection button.

Think-Out-Loud Method

When resources aren't available to support more complex forms of usability testing, the think-out-loud method offers a simple yet structured process for increasing the quantity and relevance of the data collected. In this type of test, an operator is asked to use the design under study (typically a prototype) to perform tasks corresponding to the normal duties for the particular operator role in this system. For example, a cashier in a grocery store might be asked to check out several carts of groceries using a new bar

TABLE 18.6
Option A HIPA Results

Sum of Steps 1–2: 170 ms
Sum of Steps 8–15: 620 ms × 5.5 = 3,410 ms
Sum of Steps 3–7: 380 ms × 11 = 4,180 ms
Steps 1–2 and 3–7 are performed an average of (5 + 1)/2 for the middle variable list,
  therefore: 240 ms + 380 × ms = 1,380 ms
Sum of Steps 17–23: 550 ms
Total estimated time: 9,690 ms

TABLE 18.7
Option B Human Information Processing Analysis

---

1. (t_p) Perceive list of system components by transferring a verbal code of the label components from visual image store to working memory.
2. (t_c) Decide to execute eye movements to list.
3. (t_m) Execute eye movements to list item.
4. (t_c) Decide whether item is last item on list; if yes go to Step 8.
5. (t_p) Perceive item on list by transferring the verbal code of the item from visual image store to working memory.
6. (t_c) Match verbal code of item and component.
7. (t_c) Decide if item matches the item stored in working memory. If item desired item, go to Step 16 (select or go back to Step 4).
8. (t_p) Perceive scroll bar or down arrow.
9. (t_c) Decide to execute eye movements to down arrow.
10. (t_m) Execute eye movements to down arrow.
11. (t_c) Decide to move cursor to down arrow.
12. (t_m) Move cursor to down arrow.
13. (t_p) Perceive cursor over down arrow.
14. (t_c) Decide to stop moving cursor.
15. (t_m) Push cursor selection button, go to Step 5.
16. (t_c) Decide to select item.
17. (t_p) Perceive the add arrow button by transferring the verbal code of the button from visual image store to working memory.
18. (t_c) Decide to move cursor.
19. (t_m) Move cursor.
20. (t_p) Perceive cursor over arrow button.
21. (t_c) Decide to stop moving cursor.
22. (t_m) Select cursor selection button on arrow/add button.

---

code cash register. Rather than relying on posttest interviews or questionnaires to capture the participants' likes and dislikes with the design, the participants are asked to describe their impressions out loud as they perform each step of the scenario. Participants are encouraged to describe what they see, what information or controls they are seeking, and what outcome they wish to accomplish as they work through the scenario. The participant's vocalization of an ongoing thought process allows the evaluator to follow the user's train of thought and associate thoughts with actions.

TABLE 18.8
Option B HIPA Results

Sum of Steps 1–2: 170 ms
Sum of Steps 8–15: 620 ms × 55.5 = 34,410 ms
Sum of Steps 3–7: 380 ms × (110 + 1 – 11)/2 = 19,000 ms
Sum of Steps 17–23: 550 ms
Total estimated time: 54,130 ms

To lend the structure and discipline of a true empirical test to this process, evaluators must give substantial attention to test preparation. First, the tasks that participants will perform should be representative of those the users will perform with the future system. Furthermore, the scenario the test participants follow also should be representative of realistic, expected circumstances. Typically, tasks and scenarios are chosen to represent the most common or most critical tasks that users are likely to encounter. Second, the participants themselves should be members of the actual target user population (or at least very similar in most respects). Finally, all other factors in the test should be controlled to the greatest extent practicable. For example, evaluators should read the instructions to the participants from a script before each test trial. Equivalent data collection categories should be established and followed using data collection sheets designed specifically for the purpose (e.g., record points in the scenario where errors were made or where display information was misinterpreted). Finally, evaluators should keep test conditions constant while remaining vigilant to minimize distractions or interruptions in the test area.

As with the direct observation technique, the observer carries an enormous burden of capturing and organizing the volumes of data generated by the think-out-loud approach. It would be easy and even probable for an observer to miss a critical piece of information while busy recording a less-important item. Videotape recording of each trial can alleviate this problem, but it tends to increase costs and the time required to collect the data because each participant's trial must be reviewed in slower than real time (assuming the observer pauses the tape frequently to take notes).

If videotape recording is used, the observer should consider using separate, synchronized tapes to record both the screen picture and an over-the-shoulder or side-view picture of the participant. The dual camera technique ensures the observer will capture both the user's expressions and words as well as his or her interaction with the system under evaluation. If the interface being evaluated is a video display, it is desirable, if possible, to capture the input directly from the video feed. The dual camera approach is particularly important when evaluating computer displays because a view of the terminal is critical to understanding the participant's actions. When choosing to video tape a session, the evaluator should run pilot tests first to determine the best viewing angles, microphone locations, and lighting demands for the test setting.

Experimental Testing

Experiments are the most rigorous form of usability testing. As with any true experimental design, one or more variables of interest (e.g., background colors on a display) are manipulated in a controlled environment, while all other elements of the interface are held constant. Performing a thorough usability experiment is challenging and time-consuming, and specialized training or substantial experience is typically required of the designer. Although many good references describing the experimental design and analysis process are available (e.g., Box, Hunter, & Hunter, 1978; Linton & Gallo, 1975; Montgomery, 2000), someone preparing an experimental usability test for the first time should seek the assistance of an experienced person during the development of such experiments.

Deciding when to use experimental testing for a usability evaluation is also a challenge. Formal experimental testing is clearly the most complicated and costly method of usability evaluation, and often it is unnecessary. For example, although many of the test methods previously described in this chapter are suited to addressing a broad range of usability issues, the focus of an experimental test is necessarily limited to only those independent variables identified in the design. The narrow focus of most experiments means they produce results that, although revealing, tend to be difficult to generalize to larger usability questions. Consequently, experimental testing should be reserved for specific usability issues that must be resolved with high confidence while relying on objective data. In some cases, for example, experimental testing may be required for the hard evidence a designer may need to defend a design decision to skeptical system developers, management, or both. Further, although subjective methods are indeed useful, relying exclusively on subjective assessments may increase the risk of overlooking usability problems; user preferences can sometimes differ from user performance. The potential exists for users to report a preference for designs that unknowingly hamper performance, and only objective performance data is capable of persuading users of the superiority of a less-popular design.

Finally, only experimental testing can reliably estimate real-world user performance for a particular design and compare performance differences among several designs. For example, experimentation can report how long it will take users to complete certain tasks using several different designs, how often on average they will make errors on each design, and which types of errors are most likely to occur. Furthermore, if quantitative user performance goals are established at the outset of the design process, only experimental testing is powerful enough to validate whether a given design meets those goals.

Experimental Testing Case Study

To appreciate the power and complexity of a usability experiment, consider the following case study, where experimental testing was considered the best choice for a usability evaluation. The setting for this case study was a military battlefield headquarters, where personnel must be prepared during times of chemical warfare to operate computer terminals while wearing bulky protective clothing. The protective clothing covers the entire body, including rubber gloves for the hands and an oxygen mask for the face. In spite of the obvious performance handicap imposed by such clothing, the military personnel are expected to complete their assigned duties successfully using the computer terminals. Scientists studying this issue wanted to know how much an operator's performance on the computer terminals might suffer when required to wear the protective ensemble, and the scientists considered how the degradation could be minimized.

A research team working on the project proposed both hardware and software solutions that could compensate for the operators' reduced motor and perceptual capabilities while wearing protective clothing, and the team designed an experiment to evaluate the usability of each approach (Grounds & Steinberg, 1998). Their hardware solution focused on offering alternative input devices. In particular, the scientists compared operator performance using a mouse versus a trackball for typical cursor control tasks. Therefore, the hardware solution had two conditions for the experiment: mouse and trackball.

The scientists' software solution consisted of two new methods for selecting objects on the display: cursor proximity and autohighlighting. Cursor proximity was a feature built into the software that allowed an operator to select an object on the display simply by placing the cursor near the object rather than directly on it. The researchers believed the cursor proximity technique would demand less psychomotor precision from the operator than traditional selection techniques, thereby improving operator performance under adverse conditions. For the experiment, the researchers studied two activation ranges for the proximity effect (50 pixels and 200 pixels), as well as an off condition where the proximity feature was disabled. Therefore, the cursor proximity technique had three conditions: off (or 0 pixels), 50 pixels, and 200 pixels.

The second software solution, autohighlighting, worked in conjunction with the cursor proximity technique to place a distinctive green hexagon around a selectable object when the cursor moved within the preset range (either 50 or 200 pixels). For example, if cursor proximity were set to 50 pixels and autohighlighting were activated, a hexagon would appear around an object any time the cursor approached within 50 pixels of the object, thus alerting the operator when the object was selectable. If autohighlighting were disabled, the cursor proximity technique would still work, but the hexagon no longer appeared. Therefore, autohighlighting had two conditions: off and on. The scientists expected that when the highlighting technique was activated, it would enhance the usability of the computer terminals even further.

Finally, users could wear either normal or protective clothing while participating in the experiment trials. Therefore, normal and protective clothing were two conditions for clothing type. All these factors—hardware type, proximity setting, highlighting setting, and clothing type—were the independent variables in this experiment. The number of conditions possible for each independent variable were two for the hardware solution (mouse and trackball), three for the cursor proximity setting (off, 50, 200), two for the autohighlighting setting (on, off), and two for the clothing condition (normal clothing labeled MOPP0, protective clothing labeled MOPP4).

The usability experiment was designed as a 3 × 2 × 2 × 2 factorial: four independent variables, one with three levels and three with two levels, combined such that all unique conditions were represented (see Fig.

	INPUT DEVICE				
	MOUSE		TRKBALL		
AUTO HIGHLIGHTING:	ON	OFF	ON	OFF	
CURSOR PROXIMITY OFF	1a	4a	7a	10a	MOPP0
CURSOR PROXIMITY 50	2a	5a	8a	11a	MOPP0
CURSOR PROXIMITY 200	3a	6a	9a	12a	MOPP0
CURSOR PROXIMITY OFF	1b	4b	7b	10b	MOPP4
CURSOR PROXIMITY 50	2b	5b	8b	11b	MOPP4
CURSOR PROXIMITY 200	3b	6b	9b	12b	MOPP4

FIG. 18.8. The 3 × 2 × 2 × 2 experimental design for evaluating performance in MOPP4. Conditions 1a through 12a are performed in normal clothing (MOPP0). Conditions 1b through 12b are performed in chemical protective gear (MOPP4).

18.8). Each participant in the test performed each of the 24 task conditions, half performing the 12 MOPP0 tasks first and half performing the MOPP4 tasks first to eliminate any presentation-order bias. Under each experimental condition, participants performed a point-and-click task requiring them to select objects on a map display (e.g., put the pointer on an object and click the left button). As the participants performed their tasks, researchers collected the time participants required to complete the task and their errors. Seventeen participants completed the experiment.

The experiment produced meaningful objective data as expected, though the findings did not provide conclusive answers to the researchers' questions in every case (a common outcome of experimental evaluations). The most noteworthy result was a statistically significant interaction between the clothing and input device independent variables. Researchers learned that operators wearing protective clothing performed a particular task in significantly less time when using the mouse than when using the trackball (see Fig. 18.9). Other outcomes indicated that participants were fastest when cursor proximity was set to 50 pixels, were also faster with the mouse than with the trackball in general (regardless of clothing), and made more errors when autohighlighting was on than when it was off.

At first glance, the usability data seemed to support recommending operators use a mouse to increase their speed, regardless of clothing type. Unfortunately, other data from the same experiment indicated that operators committed more errors with the mouse than with the trackball, a finding that tempered the researchers' enthusiasm for either input device. The cursor proximity settings fare similarly. Although the data found that the 50-pixel setting prompted better operator performance than either

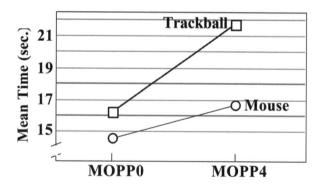

FIG. 18.9. Mean time to complete the point-and-click task for participants wearing normal clothing or protective clothing, while using a trackball or a mouse (p = .0183). The experimental data provided not only an indication of relative performance (which alternative was faster), but it also allowed the researchers to estimate actual operator performance when performing a certain task.

TABLE 18.9
Overview of Usability Evaluation Methods

Test Method	Description	Benefits	Disadvantages	When to Use
Standard compliance review	Validates using industry standard guidelines	Promotes consistent interface behaviors and interaction methods	Does not guarantee a well designed evaluation for a specific task	Low budget and the penalty for human errors have low impact.
Heuristic evaluation	Expert evaluation using a set of design principles	Low cost method of identifying usability problems on a design	May miss some significant usability problems	Low budget and no access to users
Questionnaires	User feedback obtained via paper feedback	Low cost and provides valuable feedback from users	Only direct questions; cannot explore user tangents	Low budget but have user accessibility
Interviews	Interview users to determine design issues	Provides valuable feedback from users; can ask open-ended questions	There is often a disassociation between what users prefer and what design provides the best human performance. Interviews may be tainted by this phenomenon.	High user accessibility but no prototyping capability available, or very early in development process
Observation evaluations	Evaluator observes operators performing a task on a prototype.	Allows evaluators to identify usability issues and solicit information from users to better understand the user's mental model.	Requires a prototype of some form	When budget has funds for a working prototype but not enough money for performance testing
Human performance modeling	Uses mathematical method to estimate operator performance on a design	Good method to assess performance trade-offs without building a prototype	Time-consuming and requires specialized expertise	No resources available to build prototypes, and human performance is high priority
Experimental testing	Users perform a task using a prototype and performance measures recorded.	Because there is often a disassociation between subjective preference and human performance, this type of testing is most accurate.	High cost, high effort, and time-consuming; the focus of the experimental test may be so narrow that it may not be cost effective.	When given enough resources and when the consequences of human error is high

the 200-pixel or off settings, it did not prove that 50 pixels produce the best possible operator performance. In fact, the optimum setting might be 5 pixels, 45 pixels, or any number other than 50 or 200, but the design of this experiment could only compare the specific variables tested. Consequently, the findings from this usability experiment taught the scientists important lessons about their designs, but it left other questions unanswered, requiring the scientists to perform additional statistical analysis on the data and perhaps conduct additional usability evaluations. This mixed result is typical for many experiments and demonstrates the importance of applying this usability technique only under specific circumstances.

SUMMARY

With enough resources, a usability testing program can grow as large as the rest of the development team, and some larger companies have invested significantly in state-of-the-art usability labs. Other companies hire outside experts to conduct their testing, while still others distribute in-house questionnaires on a shoestring budget. Many tools exist, and knowing which one is appropriate is important (see Table 18.9).

Regardless of which method is chosen, the key to building a successful usability testing program (and to producing good, usable designs) is consistency and determination. Stretching resources for a one-time, complex usability evaluation is less likely to improve design than a more modest usability test conducted consistently and professionally throughout the entire development effort. A little insight at many points in the process is better than a lot at one time, and establishing a consistent and comprehensive usability evaluation program can pay huge dividends in improving the quality of every product and design.

REFERENCES

American National Standards Institute. (1988). *National standard for human factors engineering of visual display terminal workstations*. Santa Monica, CA: The Human Factors Society.

Apple Computer, Inc. (1987). *Human interface guidelines: The Apple desktop interface*. Reading, MA: Addison-Wesley.

Bailey, R. W. (1982). *Human performance engineering: A guide for system designers*. New York: Prentice Hall.

Box, G. E. P., Hunter, J. S., & Hunter, W. G. (1978). *Statistics for experimenters: An introduction to design, data analysis, and model building*. New York: Wiley.

Card, S. K., Moran, T. P., & Newell, A. L. (1983). *The psychology of human computer interaction*. Hillsdale, NJ: Lawrence Erlbaum Associates.

Card, S. K., Moran, T. P., & Newell, A. L. (1986). The model human processor. In K. R. Boff, L. Kaufman, & J. P. Thomas (Eds.), *Handbook of perception and human performance, Vol. II* (pp. 45:1–45:35). New York: Wiley.

Defense Information Systems Agency. (1996). *Department of Defense human computer interface style guide. Department of Defense technical architecture framework for information management, Version 3.0*, Vol. 8. Washington, DC: Department of Defense.

Department of Defense. (1989). *Military handbook 761A. Human engineering guidelines for management information system*. Washington, DC: Department of Defense.

Department of Defense. (1996a). *Military standard 1472E. Department of Defense design criteria standard: Human engineering*. Huntsville, AL: U.S. Army Aviation and Missile Command.

Department of Defense. (1996b). *User interface specifications for the defense information infrastructure (DII), Version 2.0*. Washington, DC: Defense Information Systems Agency, Joint Interoperability and Engineering Organization.

Eberts, R. E. (1994). *User interface design*. New Jersey: Prentice-Hall.

Grounds, C., & Steinberg, R. K. (1998). Usability testing of input devices in MOPP IV for THAAD. *Proceedings of the 66th Military Operations Research Society Symposium* (pp. 272–273).

IEEE. (1992). *Recommended practice for graphical user interface drivability* (unapproved draft 2). Author.

John, B. E., & Kieras, D. E. (1994). *Using GOMS for user interface design and evaluation: Which technique?* [Online]. Available: http://www.cs.cmu.edu

John, B. E., & Kieras, D. E. (1996). *The GOMS family of analysis technigues: Tools for design and evaluation* [Online]. Available: http://www.cs.cmu.edu

Kobara, S. (1991). *Visual design with OSF/Motif*. Reading, MA: Addison-Wesley.

Linton, M., & Gallo, P. S. (1975). *The practical statistician: Simplified handbook of statistics*. Belmont, CA: Wadsworth.

Mayhew, D. J. (1992). *Principles and guidelines in software user interface design*. Englewood Cliffs, NJ: PTR Prentice Hall.

Mayhew, D. J. (1999). *The usability engineering life cycle*. San Francisco: Morgan Kaufmann.

Microsoft Corporation. (1995). *The Windows interface guidelines for software design*. Redmond, WA: Microsoft Press.

Molich, R., & Nielsen, J. (1990, March). Improving a human-computer dialogue. *Communications of the ACM, 33*(3), 338–348.

Montgomery, D. C. (2000). *Design and analysis of experiments (5th ed.)*. New York: Wiley.

Nielsen. J. (1994). *Usability engineering*. San Francisco: Morgan Kaufmann.

Nielsen, J., & Mack, R. L. (1994). *Usability inspection methods*. New York: Wiley.

Nielsen, J., & Molich, R. (1990). Heuristic evaluation of user interfaces. *Proceedings of ACM CHI'90 Conference*. 249–256.

Open Software Foundation. (1992). *OSF/Motif style guide. Release 1.2*. Englewood Cliffs, NJ: Prentice Hall.

Rubin, J. (1994). *Handbook of usability testing: How to plan, design, and conduct effective tests*. New York: Wiley.

Smith, S. L., & Mosier, J. N. (1986). *Guidelines for designing user interface software (ESDOTR086-278)*. Hanscom AFB, MA: USAF Electronic Systems Center.

BIBLIOGRAPHY

Apple Computer, Inc. (1997). *Apple web design guide* [Online]. Available: http://www.applenet.apple.com/hi/web/intro.html

Hom, J. (1998). *The usability methods toolbox* [Online]. Available: http://www.best.com/ ~jthom/usability/

Keeker, K. (1997). *Improving web site usability and appeal* [Online]. Available: http://www. microsoft.com/workshop/author/plan/improvingsiteusa.htm

Nielsen, J. (1996). *The alertbox: Current issues in web usability* [Online]. Available: http://useit.com/alertbox/

Raggett, D. (1997). *W3C HTML 3.2 Reference Specification* [Online]. Available: http://www.w3. org/TR/REC-html32.htmp

Richmond, A. (1994). *A basic HTML style guide* [Online]. Available: http://heasarc. gsfc.nasa.gov/Style.html

Tilton, J. E. (1997). *Composing good HTML* [Online]. Available: http://www.cs.cmu. edu/~tilt/cgh/

Chapter **19**

Operability Testing of Command, Control, Communications, Computers, and Intelligence (C4I) Systems

Lyn S. Canham
Air Force Operational Test & Evaluation Center

Command, control, communications, computers, and intelligence (C4I) systems collect, process, and disseminate information for the purposes of directing and assisting other operations or systems. The variety of C4I tasks include battle management; tactical planning, targeting, tasking, and execution operations; satellite command and control; unmanned aerial vehicle control; air traffic control; remote vehicle teleoperation; vehicle or service dispatch; and the many civilian emergency management and crisis response teams.

Although software applications have relieved C4I operators of many computational and repetitive tasks, those tasks that remain in human hands are perhaps the most challenging and critical (Bainbridge, 1982; Meister, 1996). Typically, these functions are deemed too critical to place beyond human control or too complex to automate with current technology. Unfortunately, the result of this allocation of tasks is too often a vaguely defined set of human cognitive tasks for operators, upon whose quick and accurate performance may rest the ultimate success of the system's mission (Lockhart, Strub, Hawley, & Tapia, 1993). Furthermore, as automation has become more intelligent, the growth in number of information sources and the resulting great quantity of information provided to the human authority requires automated decision support to ensure timely and accurate decisions (see Fig. 19.1).

The pivotal role of the human in complex, time-critical, and high-consequence C4I systems makes the testing of human factors an essential

433

FIG. 19.1. An Air Operations Center for command and control of airborne assets.

component of C4I system test and evaluation. Most of the HFTE methods discussed throughout this volume apply in one way or another to C4I systems. However, the special characteristics of C4I systems lead the tester to emphasize particular methods, as well as to extend HFTE methods in ways that differ from their employment in other test and evaluation settings. This chapter describes some of the special perspectives associated with C4I HFTE, discusses several approaches that have been successfully applied to assess C4I system operability, and leads the reader through the development and cost-benefit trade-offs of HFTE measures and methods for three example C4I systems.

C4I SYSTEM TEST AND EVALUATION PERSPECTIVES

Positive Human Control

The general concept of positive human control is that a human must retain executive authority over automated processes, especially where the consequence of an error is high (e.g., launching an attack or defensive strike). At the heart of this requirement lies the question of human trust in software automation. Typically, the developers of an automation system have faith and confidence in it (they also have many mental, emotional, and monetary resources invested in it), whereas the prospective users of the same automation system may have little trust and confidence in it. Frankly, this is often due to users having been long exposed to unreliable automation and poor performance. According to Masalonis and Parasuraman (1999), "users' trust in an automation tool can be increased by the automation's reliability, predictability, and performance" (p. 184). Too much trust, however, can be a bad thing. Parasuraman, Molloy, and Singh (1993) demonstrated a complacency effect that had a negative impact on human-system performance. Consistently reliable automation led to less monitoring of it by operator subjects, who missed occurrences of automation failures.

Meister (1996) states that the operators of highly automated C4I systems are primarily troubleshooters. As automation tools assume more capabilities and tasks, the role of the human operator is relegated to monitoring for situation anomalies, automation failures, or conditions that could trigger a change in course of action. The operator may take no action at all, or may furnish data or commands to the system, as required. Of course, the monitoring role can be more cognitively demanding than this description implies. It requires keeping track of a potentially large amount of information, understanding what the various system parameters mean, recognizing problems, and formulating a solution or corrective action. The operators need to know whether, when, and how to intervene to change system states, modify a course of action, and turn off, or otherwise override, the automation. Prior training, experience, and system documentation will contribute to this knowledge, but the automation should provide complete, relevant, and understandable status and corrective action information to keep the operator in the loop.

The human operator does not always maintain the continual awareness of automation performance and status required to know what to do and when to do it. An operator gets out of the loop (Endsley & Kiris, 1995) or becomes a victim of automation surprises (Sarter & Woods, 1994) due to a number of possible factors. Among those proposed are the lack of adequate system feedback, reduced interaction with the system (adopting a more passive processing role), the complacency effect described previously, or the loss of system operation proficiency over time (skill decay; Lockhart et al., 1993). Because awareness of system state precedes situation interpretation and the formulation of any control response, the loss of situation awareness (SA) translates to a loss of positive human control. The operator out-of-the-loop effect might manifest as operator errors of omission or delayed decision time.

Given the importance of operator cognitive processes in these systems, how can positive human control be measured? In cases where the operator is consistently required to execute a given task (exert control) for the system to perform its mission, positive human control is inferred from the operator performing that task when needed and in accordance with mission requirements. The speed and accuracy with which the operator performs those required functions can then be taken as measures of positive control.

Thus, human performance measures, such as response time and errors, can provide reasonable estimates of the degree or quality of positive human control, not just notations of the presence or absence of it. The effectiveness of the C4I human-system interface is reflected in human performance, and, in turn, human performance directly impacts system performance. Hence, the examination of human performance can greatly aid in the characterization of overall system performance (see the SITE approach in chapter 3 of this book). Unfortunately, this fact seems to be neglected all too often during test and evaluation. It is not uncommon for C4I system performance metrics (e.g., message speed) to be tested and reported exclusive of operator performance times, even though the system cannot function without operator inputs.

However, because positive human control is exercised through the largely cognitive means of monitoring, problem solving, and decision making, actual operator intervention (and thus measurement opportunities) may be rare. In some cases, it may be necessary to determine where and when human control should have been exercised and the degree to which any observed operator behavior corresponds with the desired behavior. Under some test conditions, incidents requiring operator intervention can be scripted into test scenarios to elicit the behaviors for measurement. Time taken to detect and to resolve anomalies are examples of such measures. Meister (1996) also provides a list of possible operator errors for covert troubleshooting activities, such as failure to detect an anomaly, failure to recognize a change in the situation, misinterpretation of status information, and pursuit of an incorrect solution to a problem in the presence of contrary evidence.

Decision Support

A decision support or decision aiding system (DSS) is special software automation that works with the C4I operator to effect a timely and accurate decision. A model of human decision making that is well suited for understanding C4I DSS is Klein's naturalistic decision (or recognition-primed decision) model (Kaempf, Klein, Thordsen, & Wolf, 1996; Klein, 1993). This model predicts that a human will perceive environmental features and quickly recognize the perceived features as a typical situation from past experience (feature matching). In turn, this memory will activate situation-specific expectations, goals, and actions, and the appropriate course of action will immediately follow. Klein's model, in tandem with Endsley's

(1995) situation awareness model, successfully explains a growing body of data from a variety of naturalistic settings that relate to complex tactical decisions made under time stress. Accordingly, these theories inform much of the decision support research base that should guide DSS design and current HFTE methodologies.

Many studies have demonstrated the negative impact on system performance of various types of human-automation relationships. The introduction of more autonomous automation, which can also preempt or modulate human authority, has heightened the attention and knowledge demands on operators of these systems (Sarter & Woods, 1997). As stated in Bainbridge (1982), "automation can make the difficult parts of the human operator's task more difficult" (p. 132). Humans working with decision aids can be less productive because of insufficient information processing, termed *automation bias* by Mosier and Skitka (1999); less informed and thus less likely to detect and correct both system and automation failures; and less knowledgeable and thus less likely to anticipate system behavior (Sarter & Woods, 1994). If an operator has a good mental model of the automated aid, he or she can form the critical expectations on which human override decisions are based. The converse is also true. If a decision aid works similarly to the human (i.e., is based on an operator model), then its actions can be more readily understood; it will possess the property of inspectability (Thurman, Brann, & Mitchell, 1999). Therefore, an automated aid is more likely to be used, and used effectively, if the DSS design is based on human performance requirements, as opposed to system technical capabilities (Rasmussen, 1983).

Again, it is difficult to obtain test measures that reliably reflect the human decision-making processes of information seeking, problem detection, hypothesis testing, activation or modification of mental models, or course of action selection. Objective measures that have been applied to evaluate the human-DSS interface include decision time, display dwell time, and percent correct decisions. Some C4I systems specify the minimum time in which a critical mission decision should be made. Such a test criterion translates into a measure of the average time required for the human decision maker, in conjunction with the supporting decision aid, to perform the critical mission decision task, whether this is situation assessment, weapon engagement authorization, or course of action selection (Canham, 1997). The underlying assumption is no different from general reaction time measurement assumptions (e.g., Sternberg, 1969), namely, that a longer decision time represents greater difficulty in processing decision information and thus implicates inadequate DSS display design.

Decision-time measures should be qualified with other measures, such as decision accuracy (or appropriateness), information overload (e.g., workload measures), or information insufficiency (e.g., SA measures), in recognition of the fact that a decision may be made in timely fashion yet not have been the most appropriate decision. However, accuracy measures, such as percent correct decisions, can only be employed in cases where there is no ambiguity about the correctness of operator decisions. Additional qualifying information can come from subjective measures of DSS design quality collected from operators. Some rating dimensions that have proved useful in past testing are amount (or rate) of information presented to the operator; amount (or rate) of false information; amount (or rate) of irrelevant information; accuracy of information; comprehensibility of information; completeness of information; timeliness of information; and, most important, the resulting operator confidence in the information provided.

Although the DSS perspectives described previously have been studied in the context of individual decision making, special research consideration has more recently been given to supporting team decision making. A team of C4I operators will make or contribute to team decisions in the process of navigating mission surprises or ad hoc mission conditions to achieve desired mission outcomes. The construct that best describes the effective functioning of a team of decision makers is the degree to which the team members have shared understanding or mental models of the situation and goals toward which they are working. One of the benefits conferred by such shared understanding is the ability of any team member to accurately predict other team members' performances in a given situation, resulting in the desire of team members to work together to more effectively realize the common mission performance goals.

One indicator of the degree to which mental models are shared among team members may be the types of communications they exchange (Fowlkes, Lane, Salas, Franz, & Oser, 1994; Jentsch, Sellin-Wolters, Bowers, & Salas, 1995). In other words, participants of a good team will promote the timely sharing of mission-relevant information among their members. Using a crew coordination behavior classification scheme, Jentsch et al. (1995) demonstrated that frequencies of situation awareness and leadership statements and adherence to protocol communications among aircrew members reliably predicted faster problem-identification times in laboratory flight scenarios. Using the targeted acceptable responses to generated events or tasks (TARGETs) behavioral checklist methodology, Fowlkes et al. (1994) successfully discriminated team performance (defined as exhibiting behaviors pre-judged as critical for mission success) between crews given aircrew coordination training (ACT) and crews that did not receive ACT training. Upon examination of observed versus expected coordination behaviors, the proportion of hits was significantly greater for the ACT-trained crews. These research findings clearly suggest crew decision-making skills and behaviors that could be improved through targeted efforts in crew training programs.

Situation Awareness

As Billings (1995) and others have noted, situation awareness is a theoretical abstraction that is often casually used to define the very phenomenon the term was originally coined (by fighter pilots) to describe. It is important not to lose sight of the explanatory cognitive concepts that underlie SA and to employ the SA construct usefully to bound research and test of this operational issue. Others have stressed the importance of the distinction between SA *process* and *state*. The SA perspective that is adopted has implications for the way SA is measured. The term *situation awareness* itself is a noun that naturally implies concern with the product of complex, interrelated processes. Information about the C4I operator's state of mission-relevant knowledge at a point in time can form, analogously to workload, the basis for system (display) design decisions. However, some SA researchers (e.g., Orasanu, 1995; Sarter & Woods, 1991) believe study of the process used to build SA, usually called situation assessment, will produce more effective design guidance to improve operator SA.

Like other covert mental processes whose characteristics impact human-system performance, it is difficult to measure SA and its cognitive components. Although task performance measures are generally favored for their many advantages, including objectivity, nonintrusiveness, and facilitated linkage to system performance indices, they are not sufficiently diagnostic regarding potential variations in cognitive phenomena, such as workload and SA (Endsley, 1995). Although performance can often be successfully predicted through the mediating concepts of SA and workload, it is also well-known that humans can use workarounds and other cognitive strengths or strategies to compensate for system-induced cognitive deficits, especially when there is some performance latitude (Endsley, 1996).

As most testers realize, a robust measurement approach employs multiple, convergent techniques to yield a body of data that instills increased confidence. Meister (1995) reaffirmed the need to use subjective or self-report measures of SA in conjunction with objective measures. One approach to objective measurement of SA is based on queries of mission-critical operator knowledge during explicit pauses in a mission simulation. The dependent measure is percent of mission variable values reported correctly. This objective SA accuracy measure is frequently collected and compared across alternative system designs and is ultimately compared to system performance. The most well-known of these methods is Endsley's (1996) SAGAT.

Modification of SAGAT is required for HFTE employment in operational settings that preclude the freezing of displays during a simulation.

Called the real-time SA probe technique (RT-SAP; Jones & Endsley, in press), the intent of RT-SAP is to obtain information about the SA state through queries of operators performing a mission task but use the time required by the operator to verbally respond to the SA probe as the direct measure of SA. The displays containing evolving mission data still being in view of the operator necessitates the response time measure. Therefore, accuracy is a potentially compromised measure, but it is still collected, as correct SA response times are of most interest. Endsley (1996) has proposed that the time to respond to queries about mission-relevant aspects of SA reflects the amount of knowledge the operator has about this information. Alternatively, it could reflect workload (spare attention capacity), as measured by traditional secondary task techniques, or some other mental process.

As with operator decision making, theories of SA were initially concerned with explaining the SA of individuals. Yet SA appears to have a significant impact on team decision making and performance (Cannon-Bowers, Tannenbaum, Salas, & Volpe, 1995). Salas, Prince, Baker, and Shrestha (1995) believe a good model of team SA will eventually derive from a better definition of individual SA, as well as the growing body of research on how teams work, or how team members coordinate together and share information. In their review of a variety of studies, two fundamentally common elements of team SA emerged: individuals' extraction of important environmental states and the sharing of this information through crew communication. The definition that emerged of the state of team SA is "the shared understanding of a situation among team members at one point in time" (Salas et al., 1995, p. 131).

Because team SA is believed to entail shared mental models among members, process-tracing techniques developed to study the construction of a mental model of the situation could be used to assess the process to create shared mental models. Alternatively, the nonintrusive observation and behavioral classification of team communications may be a fruitful approach to examining crew SA assessment processes (Fowlkes et al., 1994; Jentsch et al., 1995).

Workload

Operator workload in C4I systems is often highly variable and unpredictable because of the event-driven nature of these systems. C4I task loads often arise from external sources, such as higher headquarters, customers who need information services, unexpected changes in the weather, threats, enemy maneuvers, a system's own performance, or the readiness status of units under a system's command.

It is often observed that operators of C4I systems do not act as passive processors of tasks put before them but rather as active managers of the sequencing and pacing of those tasks, especially during periods of high workload. Workload management is an attempt by the operator to cope with high workload demands by shedding or deferring tasks of lesser importance and urgency in favor of tasks that command immediate priority and attention. In doing so, the experienced operator attempts to balance her or his workload demands over time.

Human factors test and evaluation of C4I systems must be especially sensitive to workload management strategies that may be adopted by the C4I operator. Summary workload measures taken once a work shift are of less use than those examining ongoing changes in operator behavior. Further, operators will not always perform tasks in the expected sequence. Instead, tasks may be performed out of sequence or differently from the standard procedure, using workarounds the operator may have learned from experience in that position. In more extreme cases, the C4I operator may even be forced to hand off responsibilities to other operators, facilities, or resources. Although difficult to predict, it is quite useful to observe and record the task-related points at which an operator or a team decides to off-load tasks to automation. For example, depending on mission complexity and difficulty, the operational modes of many C4I systems range from fully manual, through varying degrees of manual-automation collaboration, to fully automated operations. The HF tester may create a scripted test scenario that triggers each operational mode in a representative sequence to capture workload data.

Time- and event-referenced observations of task deferral, task workarounds, and task shedding provide valuable insights into the behaviors of skilled C4I operators, and they can also be taken as indirect measures of operator workload. Moreover, the potential for workload management behavior underscores the need to provide repeated opportunities, through critical incident review or debriefing techniques, for operators to explain to testers the reasoning behind their actions.

Software Usability

The heavy emphasis on operator-software interaction in C4I systems makes the HFTE of C4I system operability largely a test of the usability of the operator-software interface. The usability evaluation community has evolved an informal, fairly common definition of software usability, similar to one provided by Corbett, Macleod, and Kelly (1993): Usability is the extent to which the operator-software interface can be used with efficiency and satisfaction by specific users to effectively achieve specific goals in specific environments.

The development process of iterative design and evaluation (Hewett, 1989), often termed *build a little, test a little*, has been frequently and successfully applied to the operator-software interface. The approach may have gained extra impetus as a methodology that facilitates user involvement in the design cycle. The use of this strategy is an invitation to the HFTE specialist to become involved with the design team early, to be integrally involved with the multiple disciplines represented on the design team, and to stay involved through all phases of C4I system development. In doing so, two of the practitioner's key challenges are to successfully adapt HFTE methods to fit the software development phase and to find appropriate measurement venues to collect the most useful and meaningful usability data.

Verbal protocol analysis techniques are useful for directing or assessing conceptual and early design stages of the operator-software interface. The methods include cognitive walk-throughs (e.g., Jeffries, Miller, Wharton, & Uyeda, 1991) and think-alouds (e.g., Virzi, Sorce, & Herbert, 1993), and they typically place representative system users (both novice and expert) in front of storyboards or early prototypes of the user interface to elicit the users' approaches to performing tasks and accomplishing goals. The users verbalize how they would do something, what they choose to do, and why, as they do it. This more open-ended approach guides the evolving design in potentially unanticipated directions that are consistent with the users' conceptualizations of system purpose and structure.

Usability evaluations of more mature, though still preoperational, software interfaces often employ expert inspection techniques. Some of these methods check for user interface standard and guideline compliance or usability deficiencies before the software design is locked in. The most common inspection method is a usability checklist consisting of characteristics drawn from one of the available design guidelines or standards (e.g., Open Software Foundation, 1993; Ravden & Johnson, 1989; Smith & Mosier, 1986). Thus, for example, a check may be made to confirm that the left mouse button is exclusively used to select items that are being pointed to on the screen. Adherence to these design features ensures a minimal degree of uniformity and good design practice and may well be necessary for good software usability. At the very least, the method generates important usability deficiencies to be fixed by the developer at an early stage. However, design guideline compliance alone is not sufficient to guarantee that an operator will be able to accomplish the operational C4I tasks using the software interface.

The close coupling of software usability with C4I system operability means that, ultimately, software usability must be evaluated through observations of operator performance. Lab or field observations of representative operators performing realistic sequences of tasks with a test version of the software interface comprise usability testing. Usability data may be collected under test conditions or test scenarios that thoroughly exercise the expected user interface tasks. Diagnostic information about the location of usability deficiencies in the test user interface can be inferred from measures of task completion time, time spent searching for information, time spent resolving a problem, and task path and sequence errors, among others, which indicate the costs in time and accuracy of pursuing user goals using the system under test.

Usability questionnaires have often been applied successfully to assess the C4I operator-software interface. These instruments are typically administered to representative users upon completion of usability test scenarios (for more diagnostic information) or upon completion of usability testing (for summary level data). Questionnaires that ask the user to identify specific software usability deficiencies are generally preferable to those that simply ask the user to rate general characteristics of the software. However, because of the complexity of C4I operations, software usability deficiencies alone may be misleading unless accompanied by meaningful operator expert opinion concerning the impact of the specific usability deficiencies on the associated operational tasks. In this regard, the software usability evaluator (SUE) questionnaire (Taylor & Weisgerber, 1997) is attractive because the respondent is asked to provide, for each usability attribute tailored to the system, both an example of the reported software usability deficiency (from his or her own experience) and a rating of the operational impact of this deficiency.

C4I SYSTEM TEST AND EVALUATION APPROACHES

In this section, the reader will be guided through three examples of C4I system test and evaluation planning efforts. Although all three HFTE case studies are Air Force C4I systems, all of the information presented applies to other service, government, civilian, and industry HFTE. The case studies follow the chronological C4I system development time line, and all are current test programs.

HFTE Early Involvement

As most of us know, when the HFTE practitioner can get in on the ground floor of C4I system development, good things can happen. The development of most military and commercial C4I systems now follows the integrated product development (IPD) model, also dubbed *evolutionary acquisition*. Typically, software-intensive C4I systems proceed through iterative or incremental design and testing phases, where software build requirements are updated and system design is further refined. In this process, integrated product teams (IPTs) are formed with allocated responsibilities for design, engineering, and production of specific system functions. IPTs are integrated because their membership usually includes representatives from the government system program office (SPO), the contractor developer, the using command (military user), and the government tester. The practice of involving the HF specialist from one or more of these organizations in at least one IPT has now become quite commonplace. The early, iterative, and integrated involvement of the HFTE practitioner was probably spurred by the increased focus of software and C2 IPTs on system-user involvement in the user interface design process.

A good process is in place to include HF considerations in design and to perform regular HF evaluation for the Battle Management (BM) and C4I portions of the Airborne Laser (ABL) system. Briefly, the ABL system is designed to shoot down theater ballistic missiles during the boost phase of flight by disabling them with directed laser energy. The ABL BMC4I program is currently in a program definition and risk reduction (PDRR) phase. The ABL BMC4I is using an organizationally integrated team (like an IPT) called the Crew Systems Working Group (CSWG). The CSWG, cochaired by the SPO and the user, meets periodically to manage, assign, discuss, and resolve crew-system design issues and tasks. Contract HF specialists form the backbone of the CSWG and determine many of its concerns, activities, and tasks, including mission requirement-function decomposition; crew-automation function allocation; preliminary workload analysis; crew station equipment design reviews; early display concept prototyping; development of the ABL *Human-Computer Interface (HCI) Style Guide*; enlistment of each segment IPT's participation in *HCI Style Guide* compliance reviews of segment displays and controls; expert HF review of displays and controls; and display and control design evaluations direction and analysis.

Soon after the CSWG was initiated, a satellite group of the CSWG was formed, consisting of a variety of representative future users of the ABL BMC4I system. This group is called the Crew Systems Evaluation Team (CSET). The overarching CSWG identified user subject-matter experts (SMEs) who closely matched the expected skills and positions of the ABL mission crew members to the extent these position descriptions were known at this early date. The contract HF members of the CSWG are in charge of the early design assessment activities of the CSET. The CSET members review and input to the operator-software interface design within each current software spiral (i.e., development increment) of each major software build as soon as a concept or design is ready for evaluation. An analogous process is in place for user reviews of hardware design. Activities have included CSET SME reviews of developer and user function analysis, crew-automation function allocations, and early workload analysis, as well as iterative review and feedback on crew station equipment, design layouts for hardware panels, and prototype displays and controls. The HF specialists then work to incorporate the CSET inputs, along with their own HF analyses, into that iteration of the design build.

The ABL contractor has proposed a systematic process to evaluate both the functionality and usability of each operator-system interface component of the ABL system. Once the supporting operator interface for two or more roles is sufficiently developed, task sharing and teamwork will be evaluated. These assessments will be performed during simulated mission scenarios; will be based on mission task decision times and errors, audio recordings of participants' verbalizations as they work through the scenarios, and video recordings; and will identify control and display problems. An extensive debriefing phase, which will include interviews and questionnaires, is planned to follow the scenario-based data collection.

In the current PDRR phase, the ABL contractor will have accomplished the initial levels of mission function, task, phase, and mode analyses, which will be used to support a five-step workload ratings analysis of each operator mission task. The CSWG has requested each CSET SME provide Cooper-Harper workload rating estimates of all operator functions (from the function analyses) for the ABL position he or she represents. The HF specialists will then perform a series of analyses on this data set. First, they will sort CSET SME workload ratings into each expected tactical mission phase. Then they will review the CSET workload ratings to estimate the weakest decision-making link for each tactical phase. Next, they will then perform a series of workload rating paired comparisons to detect the heaviest workload areas. Last, they plan to perform a series of workload rating paired comparisons to analyze interference and interdependency among operator mission task phases. It is intended that this analysis will form the requirements for design of operator-, task- and mission phase-dependent decision aids.

Operator-software interaction data have been, and will continue to be, collected from an ABL human-in-the-loop virtual simulation that is tied to periodic major theater mission C4I exercises employing the Theater Air Command and Control Simulation Facility (TACCSF). The former prototype displays (developed with TACCSF resources) are gradually being replaced with the actual ABL displays, and eventually when decision aids are developed, they will be inserted. It is expected that this data collection effort will become more systematic and performance based as the program matures. System and human performance data (e.g., decision times, task completion times, and success rates) collected in the exercise environment will help establish the operational validity of the ABL system.

OT&E of one preproduction ABL aircraft will occur late in the last development phase prior to initial production, a phase called engineering, manufacturing, and development (EMD). The EMD phase for the ABL

TABLE 19.1
Human Factors Methods and Measures for
Airborne Laser Operational Test and Evaluation

ABL HF OT&E Measure	*Methodology*
Crew ratings of mission planning tools	Survey, observation
Pilot ratings of ABL cockpit design	Survey, observation
Mission crew ratings of battle management (BM) tools	Survey, observation
Pilot ratings of workload (flight)	Modified Cooper-Harper
Mission crew ratings of workload (BM)	NASA TLX or CSS
Pilot SA (flight simulator)	SAGAT
Aircrew SA/crew coordination	Communication, observation
Mission position SA (simulator)	SAGAT or RT-SAP
Mission crew SA/coordination	Communication, observation
Mean crew laser firing consent time	Observation
Percent of crew auto laser firing overrides	Observation
Other potential BM decision times, errors	Observation
Crew ratings of postmission operations	Survey, observation
Crew ratings of task performance in gear	Survey, observation
Crew ratings of training programs	Surveys
Pilot ratings of flight simulator	Survey, observation
Mission crew ratings of BM simulator	Survey, observation

program is unusually short, thus imposing several resource and time constraints on testing. For example, OT&E will be performed in conjunction with developmental testing and evaluation (DT&E) and will depend in large part on the DT&E program. At this time, OT&E-unique plans call for operationally representative flight tests and one large-scale emulated deployment scenario. Pilot and operator task-specific performance, workload, and situation awareness data, along with crew member ratings of all crew-system interfaces, should be collected by HF evaluators on both the PDRR and EMD air vehicles. The PDRR data would not be collected on operationally representative equipment, but there would be some legacy to the EMD aircraft and future operations. In a case such as this, with appropriate caveats in place, OT&E would use DT&E data to project program risk areas and potential future operational system impacts. HF measures and methods planned for the dedicated OT&E portion of the combined testing program are shown in Table 19.1.

HFTE in Advanced System Development

The multiservice National Missile Defense (NMD) program has been in various phases of concept, architecture, and engineering development since 1984 (under different program names). The current architecture is only now approaching an administration decision point that will determine whether this system of systems will be developed for deployment by 2005 or if it will maintain a lower level of funding and slower technical development. The involvement of HF practitioners in the design phases, shadowed by HFTE in test planning, has been underway for almost as long.

NMD is designed to protect the United States against a limited intercontinental ballistic missile attack. It consists of several radar systems and C4I nodes working in conjunction to direct the impact of U.S. ground-based interceptors with enemy reentry vehicles outside the atmosphere. At the heart (or brain) of the NMD system is an immensely complex and highly automated C4I infrastructure and process, overseen by a human decision maker who primarily monitors and overrides functions. U.S. Space Command, as the NMD user, specified a key performance parameter for positive human in control (HIC) over NMD engagement operations but has had difficulty writing testable requirement parameters for HIC.

System user requirements are an important source of HFTE issues and form the basis for many HFTE measures, but there are many other information sources for the maturing HFTE concept. A rich legacy of command center and decision support HF experimentation in the NMD program dates back to the late 1980s. For example, in an experiment designed to investigate the most effective role for NMD centralized decision makers, MacMillan and Entin (1991) examined the effects of data-oriented versus decision-oriented display design on operator decision time and workload. Decision times for four critical threat assessment tasks were significantly decreased, and display feature utility and workload ratings were more favorable for the decision-oriented displays, compared with the data-oriented displays. This, and a number of other controlled decision-making experiments, have fed into NMD system development over the years. A new generation of HF experiments continue to this day.

In contrast to laboratory simulations and controlled experiments, U.S. Space Command hosts an annual NMD distributed-C2 simulation exercise that has been a significant assessment event for HFTE. Typically, a huge corpus of data is gathered at these war games, through complex observation schemes, automated recordings, audio and video recordings, postscenario and postgame surveys, and debriefing sessions called hot washes. The war-game analysts have had their work cut out for them, trying to make sense of the enormous amount of diverse and confusing data. Perhaps surprisingly, decision making, workload, SA, and display design issues have not usually been the focus of the C2 simulation analyses. Space Command wants information to help them converge on a C2 architecture, concept of operations, and overall rules of engagement for NMD. However, because the war games assemble representative operators to staff prototype C2 centers with developmental versions of the battle management software, they have been useful events for understanding prospective system operations, crew staffing, and crew procedures. For a time, an Army Research Institute HF analysis team used the war game events to elicit detailed NMD mission information requirements from the players in multiple centers and echelon levels. Multidimensional scaling of information similarity judgments was used to derive and validate information requirements for the design of potential NMD displays and decision support tools (Kobler, Knapp, Ensing, & Godfrey, 1994).

Based on years of participation in experiments, design reviews, war games, and other test events, we developed HFTE measures to provide a comprehensive and converging evaluation of the NMD HIC parameter. This experience, along with a SITE test planning structure, led to the development of a set of measures, examples of which are shown in Table 19.2. Constructing a multiservice test plan, however, is an exercise in compromise (and patience). Each service has its own perspectives and priorities, resulting in different terminologies and procedures for planning, conducting, and reporting tests. An important lesson to learn is that usually the truly critical test information is preserved, in spite of what may at first seem like an unacceptable change in test structure or emphasis.

HFTE in Near-Operational or Fielded Systems

The mission of the Space-Based Infrared System (SBIRS) is to perform detection and early warning of potential threats to national and U.S. and allied theater forces. Operators are required to assess infrared event tracks as being either valid threats or invalid noise data. SBIRS is an excellent example of a highly dynamic acquisition and test program. SBIRS has been near operational for almost 2 years, triggering a host of funding, manning, technical, and training problems peculiar to these circumstances. This section will briefly describe the SBIRS HF initial operational test and evaluation (IOT&E) planning effort and the test planning issues associated with ongoing, continual changes in system hardware and software maturity, procedures, and training and operational concepts (see Fig. 19.2).

HFTE planners spent several years in an early involvement role as part of the preparation for IOT&E, conducting two operational assessments (OAs), one in 1996 and one in 1998. An OA is an early assessment of a system under development for the government by the independent government OT&E agency. Evaluators attempt to project expected system effectiveness and suitability, based on thorough reviews of system design documentation, design reviews, lab tests, mock-ups, demonstrations, and other early indicators of system performance. It has been more difficult in recent years to obtain traditional HF design documentation for review because the government instituted streamlined acquisition, a design process

TABLE 19.2
Example Test Measures for National Missile Defense Program

HIC Requirements

Operators with decision support will be able to validate a threat event within (X) time-unit of launch detection notification.

Commanders with decision support will be able to characterize an attack within (X) time-unit of launch detection notification.

Commanders with decision support will be able to authorize defensive engagement within (X) time-unit of threat assessment notification.

Operator Performance Measures	Measure Elements
C/DSS* response time to assess situation	Time of launch notice, time of C2 node assessment report
Proportion correct situation assessments	Number of assessments that match range, truth, or simulation inputs; number of situation assessments
C/DSS response time to select COA	Time of launch notice, time of C2 node COA selection
C/DSS response time to release defensive engagement authorization	Time of threat assessment notice, time of C2 node release of direction

Operator Situation Awareness Measures	Measure Elements
Proportion of situation information values correctly reported in real time	Number of crew-reported probe values that match info requirements
Time to respond to SA probe	Time probe presented; time operator initiates verbal response
Commander/operator ratings of SA	Ratings of information accessibility, relevance, quantity, appropriate detail, and so on

Operator Workload Measures	Measure Elements
Number and type of operator task deferrals and workarounds	For key HIC tasks: the number and types of observed task deferrals and workarounds
Commander/operator ratings of workload	Periodic ratings of workload and fatigue, by position, crew, shift, and HIC task/scenario

Human Factors Design Measures	Measure Elements
Operator/Maintainer ratings	Ratings of workstations, layout, documentation, crew procedures, communications, maintenance accessibility, decision support system, etc.
Software Usability Evaluation (SUE)	For key HIC tasks: number of user-interface deficiencies and their mission impact ratings

*C/DSS = Commander with decision support system

FIG. 19.2.   Space-Based Infrared System operations center.

that gives the developer plenty of latitude to employ best commercial practices in developing and fielding a system. The SBIRS developers did produce a human factors engineering plan and attempted, as funding waxed and waned over development cycles, to maintain a structured HFE program. During both OAs, HFTE assessed the following SBIRS program elements: the HFE program, contractor trade studies, operator function allocation, operations concept and operator tasks, operator workload analyses, early indications of operator performance (compared with operational requirements), evolving display and control design, and the HF test plan. The two OAs differed primarily in terms of the maturity of the design processes and the types of events examined.

In both OAs human factors information was provided to the SBIRS SPO, contractor, and user. For example, in the first OA, HFTE personnel recommended improvements in the operational relevance of the display design and operator review process. These suggestions were implemented by the contractor, who subsequently led operators through representative task threads employing the development displays, providing a more meaningful context for the operators' display design feedback.

Significant issues noted in the OAs were carried as watch items after the OAs. The government tracked these watch items to closure, forcing a series of actions, resolutions, and agreements among all concerned SBIRS parties. In particular, HFTE planners followed the issue of human performance developmental testing. Although human performance measurement is always useful, it took on additional importance in the SBIRS program because the requirements documentation contained manual task time and error rate requirement thresholds to test. It thus became critical to assess the progress of the system design toward meeting these test criteria, especially while the design could still be modified to permit achievement of the manual performance requirement thresholds. Ultimately, the contractor performed an evaluation during segment testing and followed with a test during system-level DT&E to demonstrate human-system performance against the manual requirements.

The conduct of OAs by the operational tester does more than contribute to system development risk reduction. The information gleaned about system design and program resources leads to the formulation of the OT&E plan, refinement of the plan, or both. The HFTE portion of the SBIRS IOT&E plan has undergone many refinements prior to, during, and since the OAs, as system development took new turns or revealed newly pertinent details. With changes in system design, the associated user operations concept and procedures have also taken course corrections.

The SBIRS human factors IOT&E plan currently calls for dedicated HFTE scenarios to be conducted relatively late in the 3-month IOT&E time period. HFTE data collection is timed to follow on the heels of the mission-ready certification training evaluation trials of all five SBIRS crews. Unique scenarios are being developed for the planned HF test effort because the system-level IOT&E scenarios are not compatible with SBIRS operator-system interaction in a day-to-day mode. The factors included in the SBIRS HFTE scenarios are designed to exercise the range of mission conditions to which operator task time, error rate, workload, and situation awareness measures are sensitive. It is expected that the HF test will be more realistic and representative of the spectrum of expected future operations than any other part of the SBIRS IOT&E. The scenarios, integrated over all operator positions and functions, will span 7 days and will simulate events that trigger a range of operator actions across the SBIRS Satellite Operations Center (SOC). Scenario events for different SOC functional areas will sometimes occur almost simultaneously. HFTE measures will be taken from several identified operator positions, including the position at the crossroads of all this activity, the crew commander. All pertinent human-in-the-loop data will be automatically or manually recorded at key test points in the scenarios, over a 24-hr period, for the specified operator positions, using all five SBIRS crews. To preserve the operational realism of the HFTE scenarios, the presentation orders of the levels of the scenario factors to be experienced by all operators cannot be randomized independently for each operator. As just one example, the half-dozen different types of threat events sensed by the SBIRS infrared sensors (and then validated by the operators) cannot be randomly intermixed in equal numbers to form test scenarios for different subsets of mission operators, as this test procedure would markedly violate SBIRS operational event expectations and event-processing procedures. Thus, the HFTE scenario design does not translate into a valid experimental design, for example, the hierarchical within-subjects (or mixed) design to which it most closely approximates.

Some of the HFTE measures and associated methodologies planned for the postponed SBIRS IOT&E have been touched on in the previous discussion. These include automated recording of operator event validation times (evaluated by a number of factors, such as event type and threat size), manual calculation from logs of the frequencies and types of operator event validation errors, and other near-real-time and posttest data, as shown in Table 19.3. Except for the measures of event validation times and errors (that

TABLE 19.3
Example Operational Test Measures
for Space-Based Infrared System

SBIRS IOT&E Measure	*Methodology*
Operator strategic event validation time	Automated recording
Operator theater event validation time	Automated recording
Manual strategic false report rate	Operator logs
Manual theater false report rate	Operator logs
Intelligence Situation Analyst assessment	Questionnaires, observations
Operator ratings of HF mission impacts	EOT* questionnaire
Operator and crew situation awareness	Observations, survey, RT-SAP**
Operator ratings of workload and fatigue	CSS every 2 hr
Software usability evaluation	EOT* questionnaire

*End of Test
**Real-time Situation Awareness Probe

map directly to the user's requirements), the measures shown in Table 19.3 apply across the three primary functional areas of SBIRS, which are mission event processing, satellite system control and status operations, and ground system control and status operations. Objective and subjective data will be broken down by relevant test conditions and summarized with descriptive statistics: means and standard deviations for the ratio scale data and medians and ranges for the ordinal-interval scale data.

Event validation time for each of the two primary missions (strategic and theater) is the elapsed time between the display of the event track message and the operator's selection of the yes button in the mission display. SBIRS contractor software automatically logs the occurrences of these software actions, and the SBIRS test team will transfer this data to an analysis program to compute the elapsed times and descriptive statistics. Test team observers, working in shifts to cover the 24-hr operations during the test, will be stationed at specific operator positions (two or three mission positions, one satellite operator position, and one ground controller position) with two types of observation worksheets. On one the test observers will record relevant SOC activity and crew communication context for the HF measures (e.g., performance times, workload, display and control issues) and will also log any event validation errors made by the mission crew members.

The other form is a crew SA communication observation log, similar to the Fowlkes et al. (1994) TARGETS methodology described in the previous discussion of team decision making. SBIRS crew members operate using controlled procedures and a rigid communications nomenclature. The observation methodology will capitalize on this fact. Expected communication behaviors, along with what they signify, will be generated in advance for each scenario event and for mission, satellite C2, and ground operations, with the assistance of SMEs from each of these functional areas. Communications will be selected that signify one of the three SA levels (perception, interpretation, and projection; Endsley, 1995). The scenario event triggers, the associated expected communications, the titles of the crew positions expected to share the communications, and the SA level will be preprinted on the observation form, such that the busy observers will only have to note whether the observed crew communication matched the expected communication and then record a hit. Number of hits out of total scenario events will be used to compute proportions of crew communication hits for each of several scenario variables of interest. It will not be possible to form an absolute judgment of the level of SBIRS crew SA, but any differences linked to threat scenario size or event type, for example, will be informative.

Alternative real-time SA methods are also being explored as potential SBIRS test measures. The SBIRS preoperational environment is being used as a validation test bed for the new RT-SAP methodology (Jones & Endsley, in press) described earlier. In agreeing to the validation study, the SBIRS user requested that the research focus on the ground-system control operator because this position is believed to have serious workload and SA problems. Two pairs of ground and satellite operators (the expected operational crew configuration) will participate in three challenging satellite and ground system anomaly-resolution scenarios. Anomaly detection and resolution times will be collected during all three scenarios. Four SAGAT freezes containing six queries each will occur in the SAGAT scenario condition. The same queries in the form of RT probes will be distributed throughout the RT-SAP condition. A third control condition will not contain any SA queries to assess whether the SA measures were intrusive. This design permits the collection of more RT probe data per scenario, hopefully eliminating any sensitivity advantage for SAGAT due to a larger overall number of queries. The outcome of the validation study will determine whether RT-SAP will be used to collect individual and crew SA data in the SBIRS IOT&E.

Other IOT&E measures will include the operator-software interface assessment tool called SUE (described previously) and HF questionnaires covering SOC equipment layout, workstation design (e.g., monitors, controls, work space, seating, storage), communications equipment, working environment (e.g., noise, lighting, climate), crew procedures and coordination, and the effectiveness of specific SBIRS displays.

Once IOT&E is completed, the HFTE results will be assessed as a whole to support a judgment of the overall acceptability of the SBIRS human-system interface design. The summary assessment and detailed reports of each measure will be included in a dedicated HFTE report annex, whereas the HFTE findings judged to critically impact other effectiveness and suitability test areas will be summarized with the reports of those results.

SUMMARY

The test issues and data collection concerns described in this chapter cover the broad range of operability testing considerations associated with C4I systems. Yet it must be remembered that the evaluation of C4I operability can never stray far from consideration of the operational tasks and mission outcomes desired for the C4I system as a whole (Andre & Charlton, 1994). Rather than a parochial consideration of software usability, human factors standards, or even accomplishment of operator tasks, as goals unto themselves, the test and evaluation of C4I systems must be based on the somewhat more removed goal of ultimate mission accomplishment. It has been our experience that to be effective, the human factors tester must become a team player and direct his or her energies to the design and testing issues of the development team and the larger goal of fielding an effective C4I system.

REFERENCES

Andre, T. S., & Charlton, S. G. (1994). Strategy to task: Human factors operational test and evaluation at the task level. *Proceedings of the Human Factors and Ergonomics Society 38th Annual Meeting*, 1085–1089.

Bainbridge, L. (1982). Ironies of automation. In G. Johannsen & J. E. Rijnsdorp (Eds.), *Analysis, design and evaluation of man-machine systems* (pp. 129–135). New York: Pergamon.

Billings, C. E. (1995). Situation awareness measurement and analysis: A commentary. In D. J. Garland & M. R. Endsley (Eds.), *Experimental analysis and measurement of situation awareness* (pp. 1–5). Daytona Beach, FL: Embry-Riddle Aeronautical University Press.

Canham, L. S. (1997). Issues in human factors operational test and evaluation of command and control decision support systems. *Proceedings of the Evaluation of Decision Support Systems Symposium* (pp. 1–11). Bedford, MA: International Test and Evaluation Association.

Cannon-Bowers, J. A., Tannenbaum, S. I., Salas, E., & Volpe, C. E. (1995). Defining competencies and establishing team training requirements. In R. Guzzo & E. Salas (Eds.), *Team effectiveness and decision making in organizations* (pp. 333–380). San Francisco: Jossey-Bass.

Corbett, M., Macleod, M., & Kelly, M. (1993). Quantitative usability evaluation: The ESPRIT MUSIC project. In M. J. Smith & G. Salvendy (Eds.), *Human computer interaction: Applications and case studies* (pp. 313–318). Amsterdam: Elsevier.

Endsley, M. R. (1995). Toward a theory of situation awareness in dynamic systems. *Human Factors, 37*, 32–64.

Endsley, M. R. (1996). Situation awareness measurement in test and evaluation. In T. G. O'Brien & S. G. Charlton (Eds.), *Handbook of human factors testing and evaluation* (pp. 159–180). Mahwah, NJ: Lawrence Erlbaum Associates.

Endsley, M. R., & Kiris, E. O. (1995). The out-of-the-loop performance problem and level of control in automation. *Human Factors, 37*, 381–394.

Fowlkes, J. E., Lane, N. E., Salas, E., Franz, T., & Oser, R. (1994). Improving the measurement of team performance: The TARGETS methodology. *Military Psychology, 6*, 47–61.

Hewett, T. T. (1989). Toward a generic strategy for empirical evaluation of interactive computing systems. *Proceedings of the Human Factors Society 33rd Annual Meeting*, 259–263.

Jeffries, R., Miller, J. R., Wharton, C., & Uyeda, K. M. (1991). User interface evaluation in the real world: A comparison of four techniques. *Proceedings of the Computer-Human Interaction (CHI) Conference* (pp. 119–124). New York: ACM Press.

Jentsch, F. G., Sellin-Wolters, S., Bowers, C. A., & Salas, E. (1995). Crew coordination behaviors as predictors of problem detection and decision making times. *Proceedings of the Human Factors and Ergonomics Society 39th Annual Meeting*, 1350–1353.

Jones, D. G., & Endsley, M. R. (in press). Examining the validity of real-time probes as a metric of situation awareness. *SA Technologies Tech Report*.

Kaempf, G. L., Klein, G. A., Thordsen, M. L., & Wolf, S. (1996). Decision making in complex naval command-and-control environments, *Human Factors, 38*, 220–231.

Klein, G. A. (1993). A recognition-primed decision (RPD) model of rapid decision making. In G. A. Klein, J. Orasanu, R. Calderwood, & C. E. Zsambok (Eds.), *Decision making in action: Models and methods* (pp. 138–147). Norwood, NJ: Ablex.

Kobler, V. P., Knapp, B. G., Ensing, A. R., & Godfrey, S. L. (1994). A strategy for risk mitigation in BM/C3: Human-in-control subsystem. *Proceedings of the American Defense Preparedness Association Meeting No. 492*, 453–464.

Lockhart, J. M., Strub, M. H., Hawley, J. K., & Tapia, L. A. (1993). Automation and supervisory control: A perspective on human performance, training, and performance aiding. *Proceedings of the Human Factors and Ergonomics Society 37th Annual Meeting*, 1211–1215.

MacMillan, J., & Entin, E. B. (1991). Decision-oriented display design and evaluation. *Proceedings of the 1991 Symposium on Command and Control Research*, 8–15.

Masalonis, A. J., & Parasuraman, R. (1999). Trust as a construct for evaluation of automated aids: Past and future theory and research. *Proceedings of the Human Factors and Ergonomics Society 43rd Annual Meeting*, 184–188.

Meister, D. (1995). Experimental analysis and measurement of situation awareness: A commentary. In D. J. Garland & M. R. Endsley (Eds.), *Experimental analysis and measurement of situation awareness* (pp. 43–47). Daytona Beach, FL: Embry-Riddle Aeronautical University Press.

Meister, D. (1996). Human factors test and evaluation in the twenty-first century. In T. G. O'Brien & S. G. Charlton (Eds.), *Handbook of human factors testing and evaluation* (pp. 313–322). Mahwah, NJ: Lawrence Erlbaum Associates.

Mosier, K. L., & Skitka, L. J. (1999). Automation use and automation bias. *Proceedings of the Human Factors and Ergonomics Society 43rd Annual Meeting*, 344–348.

Open Software Foundation. (1993). *OSF/Motif style guide, revision 1.2.* Englewood Cliffs, NJ: Prentice-Hall.

Orasanu, J. (1995). Evaluating team situation awareness through communication. In D. J. Garland & M. R. Endsley (Eds.), *Experimental analysis and measurement of situation awareness* (pp. 283–288). Daytona Beach, FL: Embry-Riddle Aeronautical University Press.

Parasuraman, R., Molloy, R., & Singh, I. L. (1993). Performance consequences of automation-induced "complacency." *International Journal of Aviation Psychology*, *3*, 1–23.

Rasmussen, J. (1983). Skills, rules, and knowledge: Signals, signs, and symbols, and other distinctions in human performance models. *IEEE Transactions on Systems, Man, and Cybernetics*, *13*, 257–266.

Ravden, S. J., & Johnson, G. I. (1989) *Evaluating the usability of human computer interfaces: A practical approach.* Chichester, England: Ellis Horwood.

Salas, E., Prince, C., Baker, D. P., & Shrestha, L. (1995). Situation awareness in team performance: Implications for measurement and training. *Human Factors*, *37*, 123–136.

Sarter, N. B., & Woods, D. D. (1991). Situation awareness: A critical but ill-defined phenomenon. *International Journal of Aviation Psychology*, *1*, 45–57.

Sarter, N. B., & Woods, D. D. (1994). Pilot interaction with cockpit automation: II. An experimental study of pilots' model and awareness of the Flight Management System. *International Journal of Aviation Psychology*, *4*, 1–28.

Sarter, N. B., & Woods, D. D. (1997). Team play with a powerful and independent agent: Operational experiences and automation surprises on the Airbus A-320. *Human Factors*, *39*, 553–569.

Smith, S. L., & Mosier, J. N. (1986). *Guidelines for designing user interface software* (ESD-TR-86-278). Hanscom Air Force Base, MA: U.S. Air Force Electronic Systems Division.

Sternberg, S. (1969). The discovery of processing stages: Extensions of Donders' method. *Acta Psychologica*, *30*, 276–315.

Taylor B. H., & Weisgerber, S. A. (1997). SUE: A usability evaluation tool for operational software. *Proceedings of the Human Factors Society 41st Annual Meeting*, 1107–1110.

Thurman, D. A., Brann, D. M., & Mitchell, C. M. (1999). Operations automation: Definition, examples, and a human-centered approach. *Proceedings of the Human Factors and Ergonomics Society 43rd Annual Meeting*, 194–198.

Virzi, R., Sorce, J., & Herbert, L. B. (1993). A comparison of three usability evaluation methods: Heuristic, think-aloud, and performance testing. *Proceedings of the Human Factors and Ergonomics Society 36th Annual Meeting*, 309–313.

Chapter 20

Ergonomics, Anthropometry, and Human Engineering

Aernout Oudenhuijzen
Peter Essens
TNO Human Factors

Thomas B. Malone
Carlow International

This chapter examines methods and techniques for evaluating a system's human engineering and ergonomics design. There is a noticeable trend in ergonomics from late testing toward concurrent ergonomics. As part of the product design process, ergonomics specialists had to rely on indepth knowledge of ergonomics. They could not quite foresee the detailed interaction between the user and the product. They could only test this interaction using a prototype or a mock-up; if done, this was often at the end of the product development phase. To manage costs, modifications were only implemented if those modifications were viewed as critical to the product's success. Today, ergonomics specialists can simulate the physical relations between the product and its users. This has become available recently, following developments in the field of human modeling and virtual environment (VE) technology.

In the first part of this chapter, we consider anthropometric accommodation in work spaces: how well a system fits the intended user population. Anthropometric assessments of work spaces, systems, and components concern the physical human interface. Essential for these assessments is the availability of knowledge about the variation in the physical sizes of the people that interact with the system. Anthropometrics is about the sizes (*metrics*) of the human (*anthropos*). The anthropometric assessments could include highly complex systems, such as a manned space vehicle, or it could include simple equipment, such as a hammer. Besides knowledge on sizes, knowledge about a person's job is also required. One cannot assess the anthropometry of a work space without knowledge of the requirements of the tasks in terms of the interaction with the system.

Assessments are usually performed in developmental test settings. Occasionally, anthropometric evaluations are performed during final operational tests or with final products. This may be necessary to check whether the anthropometric fit conforms to requirements. We will present methods for evaluating work space design against established ergonomic standards and criteria.

ANTHROPOMETRY IN TEST AND EVALUATION

In the first edition of this book, we related an incident in which a U.S. Army general asked a human factors evaluator whether the new M1E1 Abrams Main Battle Tank, a 60-ton metal monster packing all the latest weaponry and fighting gear, "fits." Seemingly an absurd remark at the time, his point was well taken. In the past, design engineers were trained to design systems strictly toward technical functionality (i.e., speed, braking, and other technical performance attributes). They designed systems relying on their own ideas as to how the humans would eventually have to sit, reach, and move about at workstations. If the user was lucky, he or she happened to have the right anthropometric characteristics (e.g., stature and reach).

It was not long before manufacturers recognized that designing to technical requirements alone created too many constraints on manning. Through World War II, for example, pilots had to be selected based on stature, among other things, to fit the cockpit of some aircraft. Soon, however, research psychologists began to work closely with the design community to both design and test for a larger, more robust population. Today, we see a much more concerted effort to design around the user. In the case of the M1E1 Main Battle Tank, earlier experiments with various M1E1 crew station configurations, seating designs, and computer-assisted fire control subsystems, among others, were put to the test using anthropometric measurement techniques. Prototypes were "wrung out" in field trials under the scrutiny of HF specialists. Human performance was measured using anthropometrically representative people. Problems that were found to occur with relative fit were fixed through design changes, so that by the time the general had asked his question, there was no doubt that the armored system did indeed fit.

The notion of anthropometry testing and evaluation is quite simple: It helps us to determine whether the fit of an item (particularly, items in development) is adequate with respect to its intended user population. In addition, anthropometry helps us determine whether our design will accommodate people wearing specialized clothing, including thick gloves, snow boots, and others.

This section describes some of the basic principles of anthropometric testing. Although the basic principles have changed little since its original concepts were described by Morgan et al. in *Human Engineering Guide to Equipment Design* (1963), anthropometric data have been updated to reflect our "growing" population. The scope of this section includes a discussion on why and when an HF specialist would want to perform anthropometric tests and provides some basic test procedures for conducting the more traditional anthropometric evaluation practices using manual methods. There are some relatively inexpensive software tools that can be used by a novice anthropometrist. The reader is warned, however, that interpreting anthropometric test results may be tricky, and one should consult an experienced anthropometrist before drawing conclusions.

Background

Anthropometry and the closely related field of biomechanics deal with the measurement of the physical features and functions of the human body. These measurements may include linear dimensions, weight, volume, range of movements, muscle strength, dexterity, and more. Both research and practical experience have shown that people's performance and general acceptability of facilities, for example, those physical systems, including apparatus and workplaces, with which we interact on a daily basis, are affected by how well they fit us (Morgan, Cook, Chapanis, & Lund, 1963; McCormick, 1976). More to the point, failure to provide a few more inches in a system design, which might be critical to the operator, could jeopardize performance, operator safety, and machine reliability (VanCott & Kinkade, 1972). Although anthropometric tests may be conducted during operational testing, anthropometric measurements are traditionally gathered during developmental testing to determine whether the design adequately accommodates a certain portion of the intended user population.

Data Sources

Rarely is a HF specialist called on to actually measure a sample of the population along anthropometric dimensions and then apply those data toward an evaluation. Typically, data on static and dynamic body dimensions are already available through various sources. The real question is, which data source is the most appropriate for our particular test? One such source is MIL-STD-1472. This database, however, is based on a sampling of U.S. military personnel and does not cover the anthropometric profiles of the general population, nor does it characterize the anthropometrics of armies of other countries. Notably, other armies use databases based on civilian populations because it reflects the recruiting population better than military databases.

Anthropologists in Europe use several anthropometric databases. These databases include MIL-STD-1472D, NASA's *Anthropometry Source Book*, and databases consisting of measurement among the European civilian population. In the Netherlands, for example, it is essential that Dutch data be used because the Dutch are currently among the tallest populations of the Western world (young Dutch males were measured at an average of 1,848 mm in 1999).

Other new sources are currently being developed. One such source is the Caesar database. This database will differ from the more conventional databases consisting of static body dimensions, such as stature, shoulder width, and forward-reach length. The CAESAR database will consist of three-dimensional shapes of 4,000 American, 2,000 Dutch, and 2,000 Italian civilians. This database consists of the tallest and smallest populations living in the western world. Numerous body dimensions can be determined using the database. Sources of anthropometric data, including human static and dynamic dimensions, can be found among the following sources:

> *Human Engineering Design Criteria for Military Systems, Equipment, and Facilities* (MIL-STD-1472D, 1983).
>
> *Man-Systems Integration Manual* (NAS87; NASA, 1987).
>
> *Military Handbook Anthropometry of U.S. Military Personnel* (DOD-HNBK-743,).
>
> *Anthropometry Source Book* (NAS78; NASA, 1978).
>
> *Antropometrische steekproef onder HAVO/VWO schoolverlaters* (Daanen, H. A. M., Oudenhuijzen A. J. K., & Werkhoven P.J., 1997).
>
> *DINED, Dutch anthropometric data* (J. F. M. Molenbroek & J. M. Dirken, 1986).

Of these references, the two-volume *Source Book* (NAS78) contains the most complete list of populations of adults from around the world and has become one of the foundation sources for contemporary anthropometry (Badler, Phillips, & Webber, 1993).

Functional Versus Dynamic Bodily Dimensions

When testing for anthropometric fit, two kinds of body dimensions can be considered: structural dimensions (also called static dimensions, or the body at rest) and functional dimensions, also referred to as dynamic anthropometry (McCormick, 1976; VanCott & Kinkade, 1972). The distinction between structural and functional dimensions is less strict when you consider that the body in motion is only a special case of the body at rest. Consequently, VanCott and Kinkade argued that there is really only one anthropometry. Nonetheless, it is sometimes necessary to describe these differences for test and evaluation purposes.

Structural dimensions are body dimensions of subjects in a static (fixed, standardized) position. Structural dimensions are useful when evaluating drawings and diagrams of systems, especially for single-dimension parameters, such as sitting height, head clearance in vehicles, or head circumference, to determine the fit of newly developed headgear. However, in most real-life circumstances people are more often in motion than not. Thus, when evaluating how well the person fits the equipment, or vice versa, the HF specialist should also assess functional body dimensions.

Functional body dimensions are body positions that result from motion, such as arm reach, leg extension, and so on. Here, the central postulate is that in performing physical functions, the separate body members normally do not act independently, rather they move in concert (McCormick, 1976). When, for example, a driver extends an arm to activate a switch on the dashboard, the head may move forward, the trunk may rotate partially, and the back may bend somewhat. Thus, when anthropometric fit is evaluated, the human factors test specialist should consider both structural and functional body dimensions.

Basic Principles in the Evaluation Process

There are two approaches to anthropometry in the evaluation process: the ideal approach and the practical approach. In the ideal approach, the work space or the equipment should accommodate all intended users. This is in general not technically feasible. For instance, the intended population of users may include both small Asian females ($P = 5$ for stature; 1,380 mm) and tall Dutch males ($P = 95$ for stature; 1,984 mm). A practical approach is what we call the peeling process: the lower and upper testing limits are established in a discussion between the users and anthropologists. The discussions are about how many rejects are acceptable. For example, to require the anthropometric design of a fighter aircraft's crew station to accommodate more than the expected population, for example, 95%, may require extraordinary design features and might prolong system development beyond a reasonable delivery date. Conversely, to constrain the design to accommodate only a small percentage of the available population may result in too small a pool of qualified pilots from which to choose.

The anthropometric portion of ergonomics testing can be accomplished in four steps:

1. Determine the target user population: males, females, Caucasians, non-Caucasians, somatotype.

2. Determine the expected usage (life of type) of the product. For example, will the ship be used for the next 35 years; will the fighter last until 2020; or, is the car being replaced in the next three years? Based on the results, one can calculate the expected growth of the population due to the secular trend of acceleration. It is possible that a certain work space will not accommodate the expected population in the near future if the secular trend of acceleration is not taken into account.

3. Determine the critical anthropometric parameters. For example, stature is essential because it is the dominant parameter for standing work spaces. However, stature is not directly usable for a seated work space; here, sitting height, buttock knee length, and popliteal height are the controlling parameters. However, sitting height is not a steering parameter for the design of gloves; here, metacarpal width and other data are the steering parameters.

4. Determine the percentile range as a test limitation, that is, use percentiles as a representative measurement medium. The anthropometric evaluation should consider design and operation for extremely small or extremely large individuals. The evaluation should consider design and operation for adjustability. Several ranges are used for determination of the test limits. In some cases, the system is tested for accommodation of 95% of the target user population, starting at the $P = 2.5$ for the lower and the $P = 97.5$ for the upper limit. In other cases, 90% is acceptable (between the $P = 5$ and the $P = 95$).

Percentiles. Percentiles offer a more realistic concept of the range of dimensions to be accommodated than does the average anthropometric value, or the full range from the lowest to the highest value encountered in the normal distribution. Anthropometry data expressed as percentiles indicates where a person stands in relation to the rest of the population. For example, if you were told that your score in standing height was 90, this alone wouldn't tell you much. If we tell you that a score of 90 represented the 97th percentile of all those measured in height, this would indicate that 97% of all the people who were rated had lower scores than you.

There are two basic rules to follow when using percentiles. The first is to use the critical parameters for the work space being designed. The second rule is to use percentiles one by one. Often a design is intended to fit from the $P = 5$ to the $P = 95$ for stature. The intent is that the design will fit 90% of the target population. However, this is often not true. For a seated work space, dimensions other than stature are critical; one should use the percentile ranges of the design-critical parameters for the seated work space: sitting height, upper and lower leg length. Using stature one combines the sitting height and leg length in one percentile. However, the combined $P = 95$ for sitting height and leg length results in a taller person than our $P = 95$ person. If done in this fashion, one would fail to accommodate the intended 90% of the population.

Evaluating the Extreme Case. When designing a product one of the first questions asked is: Who will use the item? There's no way we can accommodate all potential users. Consequently, the question becomes, what is a reasonable anthropometric range with which to judge the fit of our product?

Frequently, system specifications will state the anthropometric range of users for which the system must be designed. Often, this is a simple statement such as, "The automobile driver's seat must accommodate the full range along structural and functional dimensions of 5th percentile females through 95th percentile males from among the population of North Americans." Only a comparatively few would be either too large or too small to fit the physical design of the system.

A 90% coverage would be acceptable for most commercial and military products considering the costs associated with developing systems, return on investment, and operational implications. It is because of such factors that the U.S. military has adopted a standard user specification—5th percentile female through 95th percentile male—for systems in which females are expected to interact, which is to say, most systems. With an anthropometric range specified, the job of the HF specialist would be to apply available anthropometric data to determine whether the design accommodates this range.

Returning to the example of designing a new automobile's interior to accommodate the range of users from 5th percentile females to 95th percentile males of a population of drivers, the critical factor is maximum visibility over the dashboard. This has to be applied to the dimension sitting height, erect. This latter dimension was chosen so that the designer can attain a design for minimum visibility over the dashboard. We would first measure the seat adjusted fully down (if this is an option). This is accomplished by measuring from the seat reference point (SRP) to the ceiling to determine maximum allowable head clearance; then the same is done with the seat adjusted fully up. With the seat adjusted fully up, we would also measure from the SRP to the top of the dash to determine minimum clearance for visibility (so that a 5th percentile female can see over the dashboard). We would compare these and other measures with our anthropometric data source. Measures falling outside of this range would lead us to conclude that the design does not accommodate the full range of intended users. (Procedures for conducting anthropometric evaluations are presented later in this chapter.)

Adjustability.   What is the first thing people do when they get into an automobile? They adjust the seat in its forward-rearward and back-angle positions, according to the best fit for operating the foot pedals and steering wheel. If it feels good, they buy it. But what about others in the family who may be taller, shorter, or more robust? In its forward-most position, the seat should accommodate females of at least the 5th percentile (U.S. female) in leg length. This would be 99.6 cm. Adjusted fully back, the seat should accommodate 95th percentile male (Dutch) individuals, or those who are 131.7 cm. The same is true for extended functional reach, popliteal height, and other dimensions associated with an automobile's seating.

MEASUREMENT METHODS FOR WORK SPACES

There are several methods to measure workspaces: traditional methods, digital tests, checklist tests, and combined tests. Traditional methods use measuring branches, calipers, anthropometers, scales, and so on. The traditional method can also be effectively supported using a digital measuring device. An example of such a device is the FARO arm. Other digital tests use computer aided design (CAD) techniques and digital human modeling systems. (The reader is referred to the following chapter for additional information on these techniques.)

Traditional Methods

Morgan, Cook, Chapanis, and Lund (1963) suggested certain procedures for conducting an anthropometric evaluation of operational or prototypical equipment, starting with the measurement of relevant bodily features (i.e., sitting height, standing height, and weight), for at least 10 test subjects. The conventional (i.e., nonautomated) method for evaluating work space anthropometry entails certain steps found in TECOM's *Test Operations Procedures* (U.S. Army Test and Evaluation Command, 1983a, 1983b). The following is a compilation of those procedures.

Criteria.   Criteria are summarized for standing operations and seated operations. For example, for standing operations, work benches and other work surfaces should be 915 mm (± 15 mm) above the floor. For seated operations, lateral work space at least 760 mm wide and 400 mm deep should be provided whenever practicable. Desktops and writing tables should be 740 to 790 (up to 900 mm for tall Dutch) mm above the floor.

Instrumentation. Instrumentation required for anthropometric measurements includes an anthropometer, straight measuring branches, curved measuring branches, sliding calipers, spreading calipers, weighing scale, goniometer (to measure angles), and tape measure. It is also possible to use a digital measuring device (see "Digital Methods" later in this chapter). An example of such a device is the FARO arm. The only difference is that the FARO arm produces digital results. The digital results can be used in a CAD system to reproduce a CAD model of the work space.

Test Setup. The test specialist should select body dimensions to be considered, select test participants classified as 5th and 95th percentile (or others specified) along the appropriate dimensions, and have them perform required activities with the item under representative operation/ maintenance and clothing conditions. Clothing consistent with the most extreme weather conditions under which the test item may be expected to operate or be used should be worn, for example, overalls and parkas for extremely cold climates. The fullest range of operator activities should be exercised, including such tasks as adjusting seats, mirrors, and seat belts. The natural position should be considered in seating requirements but should not be used as a justification of inadequate clearance.

From the task checklist, a checklist should be constructed of those tasks in which body parts play a functional role. From this list, tasks that cover measurement criteria should be checked off. The remaining tasks should be observed and any difficulties recorded.

Data Collection. Data should be recorded for dimensions of all work spaces to be assessed and for dimensions of the most representative test participants with 5th and 95th percentile measurements available for the test. For each participant, a list should be made of those difficulties encountered with work space in the performance of assigned tasks. With any difficulty encountered, measurements should be made of the specific work space involved and the dimensions of the affected body parts of the 5th and 95th percentile personnel involved, including extrapolations.

Data Reduction and Analysis. Data obtained from measuring work space should be compared with the criteria mentioned earlier, or other specified design and operational criteria. For dimensions falling outside prescribed criteria, a separate list should be developed to contain these discrepancies.

Evaluating Work Space Anthropometry. The purpose of an anthropometric evaluation is to draw conclusions about the total population based on task criticality. Conclusions should be based on both the frequency of occurrence and severity of inconvenience or inoperability. This also applies to special use situations, such as those for the physically impaired.

For example, if measurements reveal that only 10% of the intended population will be inconvenienced by having to stretch beyond a reasonable limit to activate a switch, but that the switch involved controls a critical safety feature of the system, then the conclusions should consider the criticality of the switch involved. Thus, designers might want to solve this particular problem so that no user is inconvenienced. Another consideration in the evaluation is to present anthropometric findings and draw conclusions for each dimension measured. The results of the anthropometric evaluation should state in each dimension measured the percentage of operators inconvenienced so that the designers can estimate which of the problems are serious enough to merit remedial action.

Digital Methods

There are several digital methods available for work space measurement. Often these methods are called computer-assisted methods or automated methods. However, the digital methods used nowadays may be far from automated. That is, they may be supported by a computer but still rely heavily on manual measurement.

There are two computer-assisted methods. The first method is based on the use of digital human modeling systems (HMSs). The second method uses VE techniques. HMSs are used for anthropometric accommodation in work spaces, VE for visual judgment of work spaces.

Digital Human Modeling Systems. Continuing advances in high-speed computers and software has made available HMSs. HMSs are used more often during the development of work spaces, tools, and so on. HMSs can also be used for ergonomic testing. They have several advantages and make anthropometric test and evaluation infinitely more convenient and practical. Today several HMSs are commercially available: Boeing's Human Modeling System (BHMS), Safeworks, RAMSIS and Jack, and MADYMO (Mathematical Human Model) to mention a few. All of these HMSs have one thing in common: they can produce manikins in various sizes and proportions. Additionally, they have their own specific distinguishable features. HMSs are essentially visualization tools providing an interactive, graphical man-model. The HMSs are designed to interface with, or even reside within, CAD computer-aided engineering or (CAE) tools. The advantage of HMSs is that they enable computer-assisted human-in-the-loop assessments of system concepts at various stages of the design process. This human-in-the-loop assessment reduces the need to build full-scale mock-ups. Today's HMSs' basic functions support three specific physical ergonomic tests:

Fit test: to test if the target population fits the workspace.

Reach test: to test if the target population can reach controls and displays.

Vision test: to test if the target population can view adequately controls and displays and to ensure that their view is not obstructed.

All these aspects relate to anthropometry. Other functions are also supported: comfort analysis, maintenance studies, injury analysis, motion analyses, and thermal tests. Also, other human models are being developed for human performance and cognitive studies.

It is almost impossible to discuss the detailed possibilities of all mentioned human models because of their inherent differences. Some HMSs are very easy to use; some require extensive training. The easy HMSs support mostly very basic functions; the more difficult support complex functions, such as injury or comfort analyses. Some tools are developed for a specific application; some are used for ergonomics in a more general sense.

McDonnell Douglas originally developed BHMS, for anthropometric cockpit design and certification activities. However, it is usable for design activities of ship bridges, is fairly easy to use, and does not require an extensive amount of training. Its weakness is a systematic error in arm reach.

Safework, developed by Genicom in Canada, is a HMSs for general ergonomic studies. It has a high anthropometric accuracy. The basic functions are easy to use; the more detailed functions, for instance comfort analyses, require more training.

RAMSIS is developed for and by the German car industry. Tecmath currently markets RAMSIS. RAMSIS has certain functions specific for the car industry. These functions require some training to be used effectively. Other functions are usable for general ergonomic studies.

Jack is in some aspects different from BHMS, Safework, and RAMSIS. It is more tuned toward studies for assembly lines, Mean Time Maintenance (MTM) studies, and so on. The anthropometry is less developed compared with other HMSs. The system is very easy to use without a need for extensive training courses.

All mentioned HMSs have a model of the human skeleton and an enfleshment. The HMSs also have joint models, often including a 17- to 22-segment torso. In addition, all HMSs have a complete hand model. However, for all their automated features, HMSs require many of the manual measurement inputs, selection of anthropometric baseline criteria, test constraints, and so on, provided by the traditional methods described earlier. The only difference is the instrumentarium; the work space is not available in hardware but in software. Thus, the instrumentarium is a CAD tool instead of a caliper, a measuring branch, or a scale.

The HMSs' anthropometric databases provide the user the capability to import a variety of empirically measured body dimensions. Some HMSs offer databases from which to choose (e.g., NASA astronaut trainees, Army soldiers, and data from the Society of Automotive Engineers). The HMSs allow the user to define individual body structures (i.e., length, width, and depth) and set joint limits for each degree of freedom. Interactive access is provided through a spreadsheet like interface. The HMSs are continually being improved.

An example of an analysis for a bus driver's cabin is provided in Fig. 20.1. The BHMS was used in this study.

Virtual Environment Techniques. When designing something as complex as a new frigate, you do not go from the drawing board to the fabrication yard in one step. On the contrary, it is normal practice to build mock-ups of key areas, such as the bridge or engine room, to see what things will look like in three dimensions. After all, it would be a pity to find out, too late, that the view from the bridge is obscured by jamming stations or that the engine room is too cramped for people to stand in. Building full-sized mock-ups and scale models is an expensive and time-consuming business. Further, every time you change a design, you must build a new mock-up. That is why designers have turned to new design tools, such as VE.

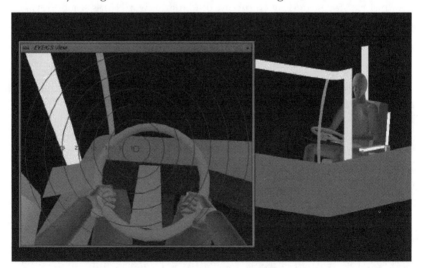

FIG. 20.1. A virtual bus driver at the wheel, showing the driver's perspective of the instrument panel.

FIG. 20.2. Getting to know the design with a virtual reality helmet combined with real railing and wall.

Human factors engineers have come up with a way to mix fact and fiction by using realistic props in a design environment specially created on the computer with the help of virtual reality (see Fig. 20.2). Crew members can check the ergonomics of proposed designs by wearing a virtual reality helmet, which gives them a lifelike view of what the new facilities will look like. They are encouraged to walk around the proposed facilities and to check things out from different angles. To give the demonstration an extra degree of realism, a few physical barriers, such as railings and walls, are erected at the test site.

This mix of design fact and fiction has proven to be very successful in ship design. Several ship bridges have been designed and evaluated using VE techniques. Among these were the ATOMOS II standardized ship control center, the bridge for the Air Defence and Command Frigate for the Royal Netherlands Navy, and for international ship builders, such as Boskalis.

During the design process of the bridge and bridge wings of the Air Defence and Command Frigate (see Fig. 20.3), the use of virtual reality techniques has shortened the design cycle, improved the participation and

FIG. 20.3.   The Air Defence and Command Frigate as seen in the virtual world.

communication, and reduced both the risk and the cost of the design process. Current developments at TNO (Human Factors) are focused on digital human models and on improved interaction methods (e.g. virtual hand control) in virtual environments.

REFERENCES

Badler, N. I., Phillips, C., & Webber, B. (1993). *Simulating humans: Computer graphics animation and control.* New York: Oxford University Press.

Daanen, H. A. M., Oudenhuijzen A. J. K., & Werkhoven P. J. (1997). *Antropometrische steekproef onder HAVO/VWO schoolverlaters* [Anthropometric survey Antro '95] (Report TNO-TM-97-007, in Dutch). Soesterberg, The Netherlands: TNO Human Factors.

Gagge, A. P., Burton, A., & Bazett, H. (1941). A practical system of units for the description of the heat exchange of man with his environment. *Science, 94,* 428–430.

DOD-HNDBK-743. *Military Handbook Anthropometry of U.S. Military Personnel.*

McCormick, E. J. (1976). *Human factors in engineering and design.* New York: McGraw-Hill.

Military Standard, MIL-STD-1472D. (1983). *Human engineering design criteria for military systems, equipment, and facilities.* Washington, DC.

Military Standard, MIL-STD-1474 (1984). *Noise limits for army materiel.* Washington, DC.

Molenbroek, J. F. M., & Dirken, J. M. (1986). DINED Dutch anthropometric data. Faculty of Industrial Design, Technical University, Delft.

Morgan, C. T., Cook, J., Chapanis, A., & Lund, M. (1963). *Human engineering guide to equipment design.* New York: McGraw-Hill.

National Air and Space Administration. (1978). *Anthropometry source book*, NAS78. Washington, DC.

National Air and Space Administration. (1987). *Man-systems integration manual*, NAS87. Washington, DC.

U.S. Army Test and Evaluation Command. (1983a). *Human factors engineering data guide for evaluation*. Aberdeen Proving Ground, MD.

U.S. Army Test and Evaluation Command. (1983b). *Test operations procedure 1-2-610*. Aberdeen Proving Ground, MD.

Van Cott, H. P., & Kinkade, R. (1972). *Human engineering guide to equipment design*. Washington, DC: American Institutes for Research.

Chapter 21

Training Systems Evaluation

Jerry M. Childs
TRW

Herbert H. Bell
Air Force Research Laboratory

Training is the process of designing and delivering a managed set of experiences so that people gain the knowledge, skills, and attitudes (KSAs) that will prepare them to successfully perform their jobs in what Vincente (1999) has described as complex sociotechnical systems. These increasingly complex, information-intensive systems are leading to a shift in the nature of the workforce. Drucker (1994) has characterized this shift as the replacement of the industrial worker with the knowledge worker. Instead of simply refining techniques, it now seems that we must continually acquire new knowledge and skills to keep up with more diverse and complicated systems.

The demand for continual increases in our knowledge and skills means that training is a big business, for both the public and private sectors of our economy. The United States spends billions of dollars annually for training and training-related activities. The following examples illustrate the magnitude of these annual investments from the military and private sectors:

- Over $200 million for formal training courses in the Air Force (Air Education and Training Command, 2001)
- Between $20-$30 billion for training within the Department of Defense (Department of Defense, 2001)
- Over $40 billion for formal courses in the private sector (American Society for Training & Development, 1997)

- Over $120 billion for on-the-job training in the private sector (U.S. Department of Labor, 1999)

These examples show that the development and delivery of training represents a significant investment. The assumption is that such investments will result in better performance of the job-specific behaviors necessary to meet system objectives (Meister, 1989). Whether or not better performance actually occurs depends on many factors. These factors include the abilities and aptitudes of the individuals being trained, the environment in which they perform their tasks, and the nature of their training experiences.

This chapter is primarily concerned with evaluating the effectiveness and efficiency of training. Given the dollars invested in training, it is reasonable to ask questions regarding the value of training programs and how they can be improved. This chapter introduces the procedures and tools used to answer such questions. It presents an overview of training as a means of enhancing human performance and discusses the design of training systems using the instructional systems development (ISD) model. It then describes the fundamentals of the evaluation processes, provides examples of training evaluation tools and procedures, and discusses some of the issues involved in training system evaluation.

WHAT IS A TRAINING SYSTEM?

A training system is a purposeful set of hardware, software, courseware, personnel, procedures, and facilities designed to impart, sustain, or improve skilled operational performance (Childs, Ford, McBride, & Oakes, 1992). Training systems are generally designed using the ISD process. Figure 21.1 depicts a generic ISD model that is representative of those employed for many training systems applications.

Andrews and Goodson (1980) identified over 60 different ISD models. Although these models differ in emphasis and level of detail, they each represent a systems engineering approach to training development. These models identify the different activities involved in developing a training program, allocate these activities to various phases or subsystems, and specify the interactions between the various phases. Each activity within a phase is further subdivided into steps to be addressed for a specific system development effort. (See Air Force Manual 36-2234, 1993; Army Regulation 350-7, 1982; Dick & Carey, 1990; or Gagne, Briggs, & Wager, 1988; for more detailed treatments of the ISD process.)

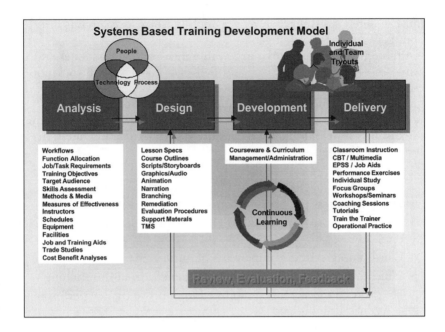

FIG. 21.1. The instructional systems development model.

Effective training system evaluations require an understanding of the entire process of training systems analysis, design, production, implementation, and management. If only the outcomes of training are assessed, no opportunity exists for increasing efficiencies or improving resource use, an expensive and wasteful situation. Therefore, the activities and components of each phase must be considered in designing the evaluation. To conduct an effective evaluation, the training evaluators must understand the phases of the ISD process and what each phase accomplishes. This is necessary to determine what aspects of the training system's performance should be evaluated, what tools should be employed, and how they should be applied. Training evaluation must be considered as an integral part of the ISD process rather than a discrete phase or step.

THE NEED FOR EVALUATION

A training system consists of a set of resources and activities directed toward improving human and organizational performance. The goal of a training system may be to improve management skills as a result of a 4-hr workshop on diversity in the workplace, or it may be to produce a pilot who can safely and effectively fly a high-performance aircraft after 18 months of intensive training.

Although there are clearly large differences in the personnel and resource costs required to design and deliver a management workshop as opposed to training a pilot, it is hoped that both training systems were designed and implemented to meet specific organizational needs. It is reasonable, therefore, to ask questions about how well each of these training systems meets these organization needs. Typical evaluation questions include the following:

- Is the training system producing the desired product?
- Has job-related performance improved?
- What are the strengths and weaknesses of the current training system?
- Can training system components be improved?
- Which individuals are most likely to benefit from the training?
- How should the system be modified to improve trainee performance or reduce training costs?
- Were the modifications to the training system successful?

All too frequently, these questions are answered based on personal experience, casual observation, and informal feedback. However, today's competition for funding and other resources within both public and private organizations means that these organizations are demanding more accountability from their service activities. Training system evaluations can provide the systematic data necessary to help assess the value of a training system in light of organizational objectives. Simply stated, we evaluate training systems to make decisions regarding effectiveness or efficiency (Goldstein, 1986).

Well-designed training evaluations can provide data that can guide performance improvement at both the individual and organizational level. For example, training evaluations can address one or more of the following:

Personnel KSA Assessment

- Mastery of course/lesson objectives
- Certification of operators
- Acquisition and retention of multiple skills
- Acquisition of a broader or more detailed knowledge base
- Reduction in performance variability for production jobs

- Improvement in rate of progression through the system
- Reduction in need for recurrent training

Training System

- Component effectiveness, efficiency, currency
- Administration and management training and reporting
- Feedback to training system personnel
- Facilitation of training system decisions
- Reinforcement of the learning process

Operational Environment

- Personnel selection and assignment
- Improved job performance standards
- Validation of allocation of training and operational resources
- Verification that training KSA transfer positively to the job

For training systems evaluation, the first goal is to determine if the system characteristics (e.g., a device, instructor guide, learner throughput, or training module) meet the desired performance criteria. For instance, a high remediation rate for a given course suggests that various components of the course (e.g., content or format) should be examined for possible revision. Similarly, poor transfer from a training device to job performance with the actual equipment may indicate that critical physical or functional cues that are present in the actual job setting may be lacking in the training device.

A second training systems evaluation goal is to determine student performance (e.g., rates of progression through the system, time on modules, number of objectives mastered) within the system. This can help both to identify the need for design revisions and to identify students who require remediation or recurrent training. Other evaluation goals might include measurement of support and logistical variables in operating and maintaining the training system.

PLANNING THE TRAINING EVALUATION

An effective training evaluation plan provides data for upgrading the training system, for identifying operational elements that require modification, and for developing new training systems. The evaluation plan should receive the same attention and resources that have been devoted to the planning and development of the training system itself.

In principle, training system evaluations are similar to other program evaluations because they seek to provide information on program costs, program operation, or program results. Because of the similarity between training system evaluation and other types of program evaluations, training evaluators would benefit from the more general evaluation literature (e.g., Rossi & Freeman, 1989). For example, key stakeholders often have a great deal of uncertainty regarding the value of any program that requires extensive resources. A well-designed evaluation can reduce that uncertainty by providing data regarding the strengths and weaknesses of the program. That data can assist decision makers in determining a program's future—whether or not it should be continued, expanded, modified, or terminated. Because these types of decisions must made, it is essential that they be based on data regarding goals, resources, processes, and outputs that have been collected in a systematic manner.

Training evaluations determine the extent to which the training system affects worker performance and whether or not that effect is consistent with the organization's goals. Evaluation determines the extent to which actual performance matches desired performance. This determination is a quantitative description of the match between actual and desired performance. If actual and desired performances are essentially identical (i.e., actual = desired ± an operationally defined tolerance limit), no intervention is necessary and the learner progresses to the next stage of training or the training system remains unmodified. If desired performance is not met, then some type of intervention (e.g., learner remediation or redesign of a training module) is needed to reduce the difference between the actual and desired performance.

One of the first steps in developing a training evaluation plan is the identification of the training requirements for the system. A training requirements matrix, such as the one shown in Table 21.1, is a useful method of documenting the desired training system characteristics and for developing criteria to certify that these characteristics are applied to training system design. Such tools function not only to support training systems designers but also to assist evaluators in ascertaining the overall integrity of the training system. In Table 21.1, the parameter specifies the desired characteristic of the system. The requirements describe the general functionality that is necessary to address the parameter, and the criteria describe or quantify the level of that functionality. Parameters, requirements, and criteria are system specific; the examples presented in the table are for a military training system.

Once identified, the training system requirements enable the evaluator to pose the specific evaluation questions that, in turn, will drive the selec-

TABLE 21.1
Training Requirements Matrix

Parameter	Requirements	Evaluation Criteria
Guaranteed proficiency level of learner	Time/accuracy measures on graduates	Perform task in +/– ____ minutes with < 1% error
Integrated training technologies	CBT, DVI, CDI, EPSS	Measures of processing speed; memory, graphics, video, audio, networking, fidelity to operational equipment
Personnel reductions	Decrease to control cost	20% reduction
Proficiency-based progression	All phase criteria mastered prior to advancement to next level	Training objectives identified/quantified; mastery levels defined; use of technology for training delivery; 100% mastery (with remediation) of objectives by all students
Media selection model	Characteristics and fidelity of media linked to knowledge, skills, and attitudes	Model must show traceability of media mix to knowledge, skills, and attitudes
Training management system	Scheduling, courseware QA, record keeping, facilities, instructors, throughput, progress evaluation, remediation, resource management	Descriptions or quantitative statements of each requirement
Training device configuration/ content	Visual system CGI 150° field of vision Freeze capability 400 initial conditions	Descriptions or quantitative statements of each requirement
Student throughput	Must keep job 100% staffed to job requirements	2,000 graduates per year
Surge in throughput	Accomodate increase or decrease in students across all sites	5% surge per month 10% surge per quarter
Training support costs	Reduce with no compromise to student proficiency	10% reduction
Recurrent/refresher training	Self-test to diagnose need; equipment or lab on demand for delivery	12 CBT self-evaluation modules; 2,000 ft² lab to accomodate 30 students

tion of data collection and evaluation methods. Some potential evaluation questions include the following:

- What resources are to be used in developing the training, what is the available budget, and what constraints are acting on these resources?

- What is the primary training product or service to be delivered and how will this product/service benefit the student graduates in performing their jobs/missions?
- Based on considerations of lessons learned from predecessor efforts, how can the training product/service be improved?
- How will product/service effectiveness be assessed?
- Are clear and effective user-system interfaces being designed?
- How will learners' attention and interest be maintained?
- How will learners' skill mastery by determined?
- What is the role of instructors in learning?
- Are instructors skilled in the use of devices and other training resources?
- How and when will learners use the skills acquired in training?
- What is the level of workplace morale surrounding the training?
- How can effective training impact productivity and profitability?

As we have shown, evaluation is a critical component of a training system and should be integrated throughout the life cycle of the training. Training evaluations provide data to show how the training contributes to organizational objectives, to determine whether or not the training program should be continued, and to identify areas of the training program that need improvement. From a human resource perspective, evaluation is necessary to determine the long-term impact of training on the individual. A well-designed evaluation provides information that helps identify whether or not the participants perceived the training as worthwhile, increased their confidence in their knowledge and skills, and enabled them to perform their job more effectively. From a business perspective, evaluation data helps to justify training in terms of both job satisfaction and increased productivity.

LEVELS OF EVALUATION

One of the most influential models of training evaluation over the past 30 years is Kirkpatrick's four-level model (Kirkpatrick, 1976, 1983, 1998). Kirkpatrick suggests that training evaluations should address four general areas: reaction, learning, behavior, and results. An adaptation of the model and its components is presented in Table 21.2. Although we believe that Kirkpatrick has only scratched the surface regarding the variety of evaluation methods and data that can be used to collect data in these four areas, we believe his model highlights the various levels that should be addressed as part of training system evaluations.

TABLE 21.2
Four Levels of Training System Evaluation

Level and Type	What Is Measured and Evaluated	Measurement Method
1: Learner and instructor reaction	Satisfaction (The Smile factor) Course materials ratings Content delivery effectiveness	End of training Evaluation or critique
2: Demonstration of learning	New knowledge, skills, and attitudes attainment Objectives mastery	Final Examination Performance Exercise Pre- and Posttests
3: Application of learning on the job	Use of knowledge, skills, and attitudes on job Training transfer Individual or team improvement	Job performance outcomes
4: Operational outcomes	Return on training investment Organizational or corporate benefits	Cost-benefits analysis Business outcomes

Level 1

The first level of training evaluation focuses on the participants reaction or critique of the training. This level of evaluation identifies what the participants thought of the instructors, program, materials, and delivery. Most often, this level of evaluation provides a customer satisfaction index by focusing on how well the participants liked the training.

Level 2

This level identifies the degree to which the participants learned the knowledge, attitudes, or skills that were presented during the training. This level of evaluation helps us to determine whether or not there is any reason to expect behavioral change once the participants return to the workplace and, in conjunction with Level 1 data, the factors that may or may not have facilitated learning.

Level 3

This level evaluates whether or not the participants are using the KSA they learned in the workplace. This is the level of evaluation that we typically refer to as transfer of training. It is important to realize that positive data from Level 1 and Level 2 evaluations do not guarantee that the training will result in improved performance in the actual workplace.

Level 4

Evaluations at this level are used to determine whether the training has an impact on the organization. This level of evaluation seeks to measure the benefits of training for the organization. For example, has the improved performance by workers reduced the amount of rework enough to justify the cost of the training?

Level 1 Complacency

Kirkpatrick's model can be used to identify training system components and measurement methods for evaluation. For each level, measures of effectiveness and criteria must be identified. Note that in Table 21.2, training evaluation benefits, relative to operational requirements, increase as we move from Level 1 to Level 4. At the same time, evaluation becomes more difficult (and easier to overlook) as we move from the training environment (Levels 1 and 2) into the job settings (Levels 3 and 4). So, like almost everything else in life, the highest payoffs are achieved with the greatest expenditures of time, care, and rigor. Procedures for obtaining training effectiveness data for Levels 2, 3, and 4 are discussed later in this chapter.

In reality, evaluation beyond Kirkpatrick's Level 1 is rarely done. In their survey of management training practices in U.S. companies, Saari, Johnson, McLaughlin, and Zimmerle (1988) found the following:

> In general, there is limited evidence of systematic evaluations of management training by U.S. companies. To the extent that evaluations are conducted, the primary method used is evaluation forms administered after program participation. . . . this study indicates that evaluations of management training, other than reactions of participants following program attendance are not evident in U.S. companies. (p. 741)

The American Society for Training and Development (1997) reported that only about one third of the 100 largest companies surveyed conduct Level 2 evaluations (i.e., those that document that learners have mastered all program objectives). Figure 21.2 shows an exponential decrease in the incidence of training evaluation as we move from Level 1 to Level 4 data. Bell and Waag (1998) found a similar pattern in their review of the effectiveness of military flight simulators for training specific combat skills. They found that Level 1 reactions were by far the most often used method of evaluation.

The limitation of Level 1 affective reactions is that there may be little relation between affective measures of training value and either learning

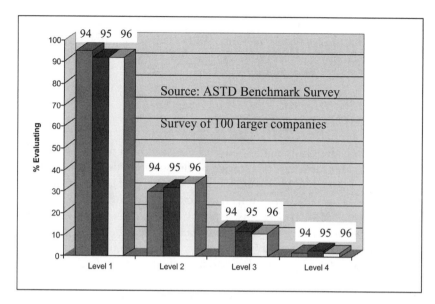

FIG. 21.2. The frequency of each of the four levels of training evaluation.

or transfer (Alliger, Tannenbaum, Bennett, Traver, & Shotland, 1998). Although Level 1 evaluations cannot provide direct information regarding learning and transfer, they do not have to be limited to simply asking whether or not the participants liked the training. It is possible to ask questions about what specific KSAs are most likely to benefit from the training or how the participants believe they would use the KSAs. Alliger et al. (1998) found that such utility-based Level 1 evaluations were more predictive of the amount learned than simple affective judgments.

An evaluation conducted by Houck, Thomas, and Bell (1991) illustrates a Level 1 evaluation that measured the perceived value or utility of the training for specific job-related skills. Houck et al. used the reactions of mission-ready fighter pilots to evaluate a simulator-based air combat training system. As part of the evaluation, pilots rated the training value of both the simulator program and their existing in-flight training for specific air combat tasks. The results indicated that pilots perceived simulator training and in-flight training as having different utilities. Simulator training was viewed as providing better training for tasks involving large numbers of aircraft, extensive use of electronic systems, beyond-visual-range employment, and work with air weapons directors. On the other hand, unit level, in-flight training was perceived as having more utility than simulator training for air combat and basic fighter maneuvering.

Although Level 1 evaluations cannot provide definitive proof regarding the effectiveness of a training program, it is difficult to imagine how a training program that is perceived as being poorly presented or having little relevance can survive. As Bell and Waag (1998) have pointed out, participant satisfaction is probably a necessary but certainly not sufficient condition for a successful training program.

MEASUREMENT ISSUES

A distinction is generally made between measurement and evaluation (or assessment). Measures are qualitative or quantitative descriptions of student or training system performance. Measures might include test scores, training device visual and motion characteristics, authoring package specifications, number of classrooms, student training hours, or number of computer-based training development hours per finished hour of instruction. To reduce interpretation subjectivity in the evaluation process, attempts should be made to develop and use robust quantitative measures of student and system performance.

If the performance measures are imprecise or ill defined, the evaluation requires a greater degree of interpretation to determine the match between actual and desired performance. Further, too much subjective interpretation in the evaluation process generally reduces validity and reliability, leaving our concerns about training effectiveness and efficiency essentially unaddressed. The degree of subjectivity in the evaluation process is also related to the definition of precisely what aspects of the training system or learner performance are to be measured, as well as when and how they are measured. Evaluation uses performance measures to make a judgment or decision regarding the degree to which actual and desired performance match. Because some performance measures require interpretation or judgment to render useful evaluations, the distinction between measurement and evaluation can be confounded.

Once the evaluation issues to be addressed in the plan have been identified, the next step is to determine the measurement metrics and methods to be employed in conducting the evaluation. An example from American and European industry may illustrate the importance of measure selection. One of the most commonly used measures of training effectiveness is the number of students who participate in training over a given unit of time (the training factor). However, learner throughput measures indicate only the number of students who have attended a course. These measures are misleading in that they fail to reflect the proficiency attained by the students as a result of their involvement in the course. The training factor is an example of a measure that has content validity (measuring what is intended), can be reliable (consistently measured within and across measurement intervals), and can be objectively determined (no judgment or interpretation involved in its designation), yet its predictive validity relative to job performance is largely unknown. Simply being in the class does not ensure that students acquire the proficiency that transfers positively to the operational setting.

Muckler (1977) has suggested that we can seldom avoid subjectivity in the evaluation process. The key is to structure that subjectivity in terms that are understandable and commonly shared. For example, 5-point ratings issued for Communication Effectiveness may mean very little unless we provide an anchored definition of that variable in the context of the operational requirements and constraints on which the rating is rendered. After these definitions are developed and validated by operational personnel, we must then ensure that raters use the definition in arriving at their ratings. Valid and reliable measurement of human-centered systems is a time intensive process that requires careful planning and execution. Good measurement doesn't just happen; it's purposefully planned and carried out.

One common misconception is that automating the measurement process renders it more objective. Although automated performance measurement generally improves sampling reliability, the data gathered may be based on very subjective measures of effectiveness (Roscoe & Childs, 1980). Questionnaires and surveys appear to be overused relative to other methods for training assessment purposes, and this tendency appears to be increasing, at least in industry. This predilection to collect attitudinal data is partially due to our focus on employee satisfaction and team building, and our assumption that the Smile factor (tapped by questionnaires) validly reflects both satisfaction and performance. Yet discussed earlier, affective reactions are not necessarily good predictors of learning (although organizations will have a hard time supporting training that trainees don't like). This confusion between training utility and trainee satisfaction is unfortunate from a training evaluation viewpoint. We cheat the learners and ourselves when we fail to verify that they can perform their job functions as a result of their participation in training.

Tests of training effectiveness and learner proficiency should match content to the objectives via item analyses and other test construction tools. Test items should not be open to interpretation because this generally reduces both validity and reliability. The items should relate to concepts and issues that are commonly understood by the learning audience. Finally, the items should match the performance and conditions to be assessed to those of the operational settings. A test construction checklist adapted from Berk (1986) is provided in Table 21.3.

For training effectiveness, learner KSA must meet the performance objectives required by the operational setting as a result of training. Thus, we must have the capability to measure progress toward mastery (i.e., achieving the objectives by demonstration of KSA in the operational setting).

TABLE 21.3
Test Construction Checklist

☐ State what is to be measured (system and learner performance components) in observable and, if possible, quantitative terms.
☐ Ensure that the skill to be measured is the same as the skill specified in the training objective.
☐ State operational equipment and simulation needs for measurements.
☐ Ensure that equipment and simulation used for testing will permit the skills to be demonstrated.
☐ Ensure that outcomes and critical processes leading to outcomes are measured.
☐ Determine the best sources of measurement for system and learner speed and accuracy.
☐ Determine an operationally valid weighting scheme for all measures based on actual operational data.
☐ Ensure that the composite scores are meaningful relative to the actual jobs and missions.
☐ Ensure that performance and written tests are assessing the same general knowledge, skills, and attitudes sets.
☐ Ensure that all learners are exposed to comparable test conditions.
☐ Ensure that items are representative of operational conditions in difficulty and comprehensiveness.
☐ Ensure that aspects of observing, recording, and scoring are objective.
☐ Verify that a subject-matter expert is capable of correctly responding to the items.
☐ Consider individual tryouts of items to determine the minimum acceptable accuracy.
☐ Ensure that test instructions specify how learners will be evaluated.
☐ Ensure that the instructions for completing the metric are clear and understandable and can be applied logically to every item in the metric.
☐ Ensure that items address all training objectives.
☐ Ensure that items enable learners to demonstrate the extent to which objectives are mastered.
☐ Ensure that adequate space is available on the metric for writing descriptive information and for item responses.
☐ Ensure that tricky, obvious, and irrelevant questions have been screened.
☐ Check that metric items are generally independent (i.e., collecting different information).
☐ Ensure that similar items are grouped by function, location, or other logical classification.
☐ Ensure that the metric can be scored easily.
☐ Ensure that feedback is provided for both correct and incorrect responses.

Training metrics provide us with this measurement capability. Like any training system component, valid and reliable metrics require careful planning, development, and continuous evaluation and improvement. Metrics are linked to the objectives of the training system but are developed with learner expectations and experience under full consideration.

EVALUATION PHASES

The ultimate measure of training effectiveness is the demonstrated proficiency of learners performing their missions and jobs in the operational setting following their training. The goal of evaluation is to use training measures to verify that the necessary KSA are attained. Three phases of evaluation can be specified, as shown in the Training Evaluation Model in Fig. 21.3. These are formative, summative, and operational evaluations. Formative evaluation is conducted throughout the training development and delivery process. Summative evaluation is conducted at the completion of training, at the beginning of the job for which training has been conducted, or at both times. Operational evaluation is conducted as part of the job setting to verify that the KSA are present. Note that part of the operational evaluation process consists of continuous evaluation to promote continuous learning.

Formative Evaluation

Formative evaluation is applied to the training system as it develops. Effectiveness of components is addressed by sampling the following aspects of the system:

Design/Development
- Training standards applied
- Instructional strategies linked to knowledge and skills
- Learning objectives defined and organized

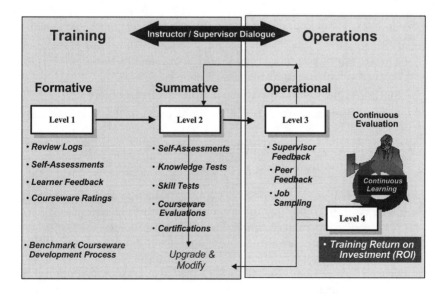

FIG. 21.3. The three phases of training evaluation.

Delivery
- Methods and media that are keyed to objectives
- Instructor effectiveness
- Capability for web-based training delivery
- Learner interactivity (for computer-based or web-based training)
- Coaching and mentoring techniques

Learning Outcomes
- Use of training measures and metrics
- Remediation capability
- Transfer potential to applied setting

Formative evaluations emphasize levels of effectiveness and efficiency as a result of the interaction between the learner and the training system. Consequently, instructional materials should be developed and evaluated relative to a training test and evaluation plan that includes strategies, schedules, and resources for validating all training system components. One way to validate courseware effectiveness is via the conduct of individual tryouts (ITO) and team tryouts (TTO).

The purpose of the ITO is to examine weaknesses in the instructional materials and in the delivery process with a representative sample of surrogate learners. The ITO serves as the test of instructional integrity in that it evaluates the extent to which the following occur:

- Courses, lessons, and tests are structured to facilitate the learning process.
- The length of courses, lessons, and tests matches the objectives and level of detail required for KSA mastery.
- Methods and media are effective and linked to objectives.
- Learner interactions with the material provide the necessary feedback.
- Sequence and pacing of delivery holds learner interest.
- Learners are assimilating the intended KSA.

Figure 21.4 shows a generic process for conducting an ITO. Each step should be tailored to the specific context of the training system and to operational goals and expectations. The ITO may include questionnaires and focus groups, courseware test item analyses to pinpoint needed exam item revisions, or performance analyses to indicate needed revisions in practical, hands-on exercises. The ITO may also employ videotaped observations of learners interacting with the materials to facilitate discussions of potential improvements.

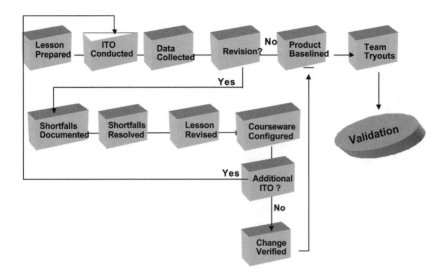

FIG. 21.4. Validating courseware effectiveness by conducting ITOs.

TTOs follow the same general strategy as ITOs in assessing instructional integrity. In addition, they serve to provide a system-level evaluation by examining the interrelationships of components in affecting learning effectiveness and efficiency. TTOs assess the following:

- Lesson compatibility with learning objectives and learner expectations
- User orientation of materials and supporting resources
- Operational effectiveness of all components
- Training management system devices and other equipment (e.g., software, facilities, delivery systems)
- Instructors/facilitators
- Complexity of content
- Pacing and sequence of instruction
- Operation of computer-based training (CBT) and other automated media
- Communication and coordination among team members

Most important, the TTO establishes the suitability of courseware for use by learning teams. TTOs, unlike ITOs, help to identify and resolve deficiencies in the training process by focusing on collective performance. Emphasis should be on identifying and improving the training system components that prevent KSA mastery. Informal interviews can be conducted between instructors and learners to get feedback and to clarify observational data.

Summative Evaluation

Summative or outcome evaluation determines the fit between the training system and the initial operational performance requirements. Specifically, this type of evaluation determines the degree to which graduates can perform the full range of requirements in their mission or job setting. It confirms the degree to which learner proficiency is achieved and produces results that can be used to optimize training system effectiveness and efficiency. Summative evaluation addresses the same factors as formative evaluation but focuses more on full-scale test and evaluation of all training components as they operate collectively under representative operational requirements and constraints. Summative evaluation also addresses the allocation of training resources and empirically assesses the performance of groups and teams under these requirements or constraints.

Summative evaluation for a pilot would include a flight checkride conducted by one or more qualified instructors. A software applications course might include a scenario-based evaluation that requires learners to effectively apply the functionality of the program.

A typical ISD metric used to provide summative evaluation consists of specifying the tasks, conditions, and standards for the variables of interest. The task specifies the performance to be evaluated, the condition states the circumstances or situation within which the task is performed, and the standards specify, typically in quantitative terms, the level of desired performance for the task. For an excellent treatment of both formative and summative evaluation factors applied to an aircrew training system, see Spears (1986).

Operational Evaluation

Operational evaluations assess training effectiveness and efficiency during real-world operations. Like summative evaluations, they determine whether training is meeting operational requirements, but they also extend the evaluation into the actual job setting. Operational evaluations seek to ascertain whether graduates are meeting or exceeding their job requirements, whether specific training components are facilitating job performance, whether job standards require modifications relative to training objectives, and whether training can be accomplished less expensively or better.

Learner performance may be evaluated during any or all three of the evaluation phases. Several methods are available to the training analyst for evaluating learner KSA. These include observation or recording of performance during training, job performance observation or recording after training, supervisor evaluations on the job, structured interviews with learners and instructors, questionnaires and surveys, computer-based training test scores, instructor-based training test scores, lesson completion scores or times, and hands-on exercise scores.

For online learning systems, frequent self-assessments accompanied by remediation facilitate learning. A learning management system should be included to track student performance, document areas of improvements, and record levels of proficiency or certification achieved by learners and graduates.

Follow-up operational measures are often hard to obtain because of resources and cost (Sanders, 1989). Priorities frequently are shifted from training issues to operational performance, and training is relegated to a subordinate role. However, because one of the most important outcomes of evaluation is to establish the connection between training content and mission or job performance (Brethower, 1989), we have to find the resources to provide feedback from operational setting to the training system. Otherwise, we risk perpetuating training systems that do not fully meet our operation requirements.

Field studies and observations can be used to determine the need for training system modifications. Supporting data can be gathered on system performance, effectiveness of training and performance support materials, or student performance in training or job settings. For example, a large telecommunications firm might use field observations, termed *service sampling*, to identify the need for improvements in providing customer service. Data are gathered via structured checklists and surveys, team members are coached in customer service improvement areas, and positive feedback is given on desired performance. One of the benefits of field observations of this type is that system and human performance can be sampled within an operational environment that is at or near full capacity.

An example from the Department of Energy might include hazardous material control on the job and relate observed KSA to those acquired in the course. A dental education application might evaluate the accuracy of hygienist procedures on practice patients. A course on business ethics might be operationally evaluated by a computer-based test delivered to a random sample of students at a designated time following training completion.

Case methods use focused work-based materials, either real or fictional, to enable learners to practice and make decisions about critical areas addressed by instructional design. Case methods generally are more efficient than field studies. This is because events that are extraneous to the learning activity can be eliminated, and critical issues can be designed into the methods. Field studies typically are aimed more at the training system, whereas case methods are learner tools.

Two of the pioneers of instructional systems design have advocated teaching multiple learning outcomes together and tying these outcomes to operational measures of effectiveness. According to Gagne and Merrill (1990), if the outcomes consist of teaching teams to increase their output over time and to provide customer-oriented service, the training content should stress how overall service is improved by these productivity increases. This is supported by our consulting experience with military and commercial organizations. Not only do learners need to know why KSA are useful, but they must also know how they can use their proficiency to improve their performance. Although the application of learning theory stresses the importance of these practices, they are often overlooked in conventional training efforts.

ANALYSIS ISSUES

Media Assessment

One of the more important activities in the evaluation process is to determine whether learning media are optimally applied to training objectives. The general goal of the media allocation process is to apply the medium with all, but no more than, the fidelity necessary to provide the KSA for operational proficiency. Examples for the military include avoiding the use of expensive visual- and motion-based simulators to teach basic procedures. For commercial application, we should not tie up expensive operational equipment with regularly scheduled work cycles to conduct a basic equipment operation course for a few students.

There is a growing trend to load high volumes of information and training content onto the Web. Learning may be compromised because content is often not interactive or engaging. Poorly designed, sequenced, or paced content is not improved by delivering it over the Internet. Table 21.4 presents an example of a media allocation matrix that was applied to a distance learning system. For each cell of the matrix, the designers assessed the utility of the training medium with respect to the KSA to be imparted. Where a given medium was determined to apply to the KSA, consideration was then given to the cost and logistics for its use relative to other applicable media. Where multiple media were applied, their mix was determined over the scheduled training time using a basic building block approach. That is, simpler skills were taught using lower-cost media. These skills were reinforced and further developed in conjunction with

TABLE 21.4
Media Allocation Matrix

	Media									
Characteristics	Instructor	Overhead	Board	Video	CBT/IVD	Multimedia	Peer	Print	Simulator	Group Exercise
Instructional pace										
Student control										
Multiple examples										
Track time										
Track errors										
Practice										
Segmentation										
Complex analysis										
Complex content										
Stable content										
Realism										
Graphics										
Animation										
Text required										
Audio										

(Continued)

TABLE 21.4
(Continued)

Characteristics	Media										
	Instructor	Overhead	Board	Video	CBT/IVD	Multimedia	Peer	Print	Simulator	Group Exercise
Narration										
Build-up exercises										
Motion required										
Hands-on equipment										
Tactile cues										
Kinesthetic cues										
Immediate feedback										
Complex development										
Content motivating										
Proficiency based										
Variable learning levels										
Contingencies										
Discriminations										
Generalizations										
Information exchange										
Perceptual-motor										
Problem solving										

more complex skills by introducing progressively more complex and costly media.

Trade-Off Assessments

For more complex and resource-intensive training systems, trade-off studies are often conducted to determine the most cost-effective alternatives. Various names are given to these studies that follow a systematic set of procedures for arriving at the best training solutions. First, a set of criteria to be applied to the training system is chosen and agreed on. Criteria may apply to (1) the operational system (e.g., fidelity of motion, work-flow and people interface, microprocessor speed or capacity, communication, or audiovisual cue presentation); (2) the cost associated with developing and maintaining the system (e.g., research and development, analysis of training requirements, courseware and media production, hardware and software engineering, training management, and other such costs); or (3) its ease of use, its safety, reliability, expandability, flexibility, maintainability, and other "ilities" associated with its general long-term operation.

Table 21.5 shows a criteria set that was recently applied to courseware authoring and training management system assessments. These criteria were developed and evaluated for the purpose of moving Federal Express air operations from wide area networks (WAN) to the Web for 6,000 pilots and maintenance technicians (Hughes & Childs, 1999). An anchored 5-point Likert scale was used to rate various products against these criteria.

After criteria are selected, each criterion is weighted according to its relevance and criticality to the system as a whole. A frequent weighting procedure is to subdivide the weights such that they sum to 100%. Next, measures are designated for each criterion. For example, if voice communications effectiveness is a criterion, the frequency of work-related dialogue among team members could serve as a measure. If safety is a criterion, the number of incidents and accidents associated with the use of the

TABLE 21.5
Evaluation Criteria for a Courseware Authoring
and Training Management System

Training effectiveness	Upper-level programming
Ability to migrate with technology	Level of interactivity
Ability to interface with current hardware and software	Collaboration with other tools
Speed of modifying courseware	HTML instructional levels
Courseware configuration management	Built-in benchmarking
Use of centralized graphics libraries	Integrated instructional design
Multimedia support	Replicates instrumentation
Ease of authoring	

system over a designated time interval could serve as the measure. If media production costs serve as a criterion, labor and material associated with generating the finished product could be estimated or reconstructed from existing cost accounting files.

Measures for each criterion are empirically or analytically gathered. A data collection methodology should be selected that is economical and effective. If the systems being evaluated are under development, it is necessary to derive measures from predecessor systems having characteristics that are as similar as possible to the new systems. Alternatively, ratings by subject-matter experts may be obtained if the rating categories are structured with operational definitions and all raters follow those definitions in issuing their ratings.

Analysis of these data should reveal trends indicating strengths and shortfalls in training system. These trends provide a basis for recommendations that will improve the suitability and cost-effectiveness of the training system. As trite as it may sound, these recommendations should be based on the trade study and other criterion-referenced data, not political and subjective factors that often intervene in real-world settings. Table 21.6 presents a hypothetical matrix of ratings issued by subject-matter experts concerning eight criteria applied to three training systems. For each criterion, the systems were assigned a rating of 1 (*least desirable*) to 3 (*most desirable*). Note that for some criteria (e.g., reliability and availability), two systems were rated equally. Also note that availability was assigned a high weighting factor for this example. For each criterion, the rating is multiplied by the weight to derive a composite value. These values are then summed for each system to obtain an overall score. For this example, System 2 is the overall choice as a result of an overall score of 250.

TABLE 21.6
Sample Trade-Off Analysis Ratings

| Criteria | Weight | Alternative 1 | | Alternative 2 | | Alternative 3 | |
		Score	Value	Score	Value	Score	Value
Cost	15	3	45	2	30	1	15
Risk	15	2	30	3	45	1	15
Fidelity	15	1	15	3	45	2	30
Expandability	10	3	30	2	20	2	20
Reliability	5	1	5	1	5	3	15
Processing	5	2	10	2	10	3	15
Availability	30	2	60	3	90	3	90
Maintainability	5	3	15	1	5	1	5
Total			210		250		205

PERSONNEL AND ORGANIZATIONAL ISSUES

The selection and training of personnel is critical to the success of training systems evaluation. The role of the instructor is transitioning from one of principally a lecturer and teacher to that of coach and evaluator of student and training system performance. The following list presents some of the skills and knowledge areas that training evaluation personnel should possess: instructional design, development, delivery, and management; system, student, and team performance evaluation; test construction; data recording and analysis; descriptive, inferential, and correlation statistical methods; quality improvement processes; training systems; media allocation and use; team coaching and mentoring; instructional support features and job aids; adult learning theory and applications; practice and reinforcement; oral and written communication skills.

CONDUCTING TRAINING EVALUATIONS
IN OPERATIONAL SETTINGS

When we conduct a training program, the principal underlying assumption is that the participants learn something that will improve their performance. Otherwise, there would be no point in expending the time, energy, and money necessary to design and deliver the training. Therefore, let us assume that our basic problem in measuring training effectiveness is that of measuring the type and degree of performance change that can be attributed to the training program. The fundamental question is how do we know when a person (or a team) has learned something?

In its simplest form, the answer is that we must be able to measure changes in job-related knowledge, skills, or attitudes. Simply, there is no other way to measure learning except to measure changes in behavior. For example, after completing a training program, the individual demonstrates learning by performing a task at a higher level of skill than previously, or by answering questions about a particular subject, or by making different statements or taking different actions regarding a particular matter. Thus, the central problem is the measurement of change. Ideally, this measurement would be the result of carefully controlled experimental manipulations.

Training evaluations can be accomplished using a variety of approaches. For example, Caro (1977) identified 10 different approaches for estimating the training effectiveness of flight simulators. Bell and Waag (1998) grouped these approaches into three major categories: utility evaluations, learning, and transfer. Utility evaluations, which correspond to Kirkpatrick's Level 1, have already been discussed. Learning and transfer, which correspond to Kirkpatrick's Level 2 and Level 3, are discussed in this section using what we shall loosely term *the experimental approach*. This approach attempts to identify valid cause and effect relationships between the characteristics of the training program—the independent variable(s)—and the resulting performance—dependent variable(s)—by directly manipulating one or more characteristics of the training program itself. The distinguishing feature of this approach is the systematic design of the evaluation such that one can be confident that the change in the trainee's performance depends on the characteristics of the training program.

This section describes some of the basic designs used to assess training effectiveness. However, there are significant differences in the strength of our confidence in the conclusions that can be drawn from these different designs. These differences are caused by different designs not being equally effective in eliminating a number of threats to the validity of our conclusions. These threats involve alternative explanations (e.g., intervening events, maturation, or instrumentation changes) for performance differences that cannot be ruled out based on the design of the experiment. These threats are well described in Campbell and Stanley (1966). Additional detail is available in a variety of sources (Bracht & Glass, 1968; Cook & Campbell, 1976; Kerlinger, 1973). Because these threats provide alternative explanations for the observed change in performance, they weaken our ability to conclude that the training was responsible for the performance change. Obviously, the ideal solution would be to use the most rigorous design and thereby eliminate the greatest number of validity threats. Unfortunately, this is often impossible to achieve when evaluating training devices in operational settings (Boldovici & Bessemer, 1994).

Although we cannot often obtain the degree of control that we would like to achieve, we should still attempt to apply as much rigor as possible to our evaluations and recognize both their strengths and weaknesses. The systematic collection of evaluation data is necessary to support "effective training decisions related to the selection, adoption, value, and modification of various instructional activities" (Goldstein, 1986, p. 111).

Experimental and Nonexperimental Designs

True experiments are characterized by two features: control over the administration of the independent variable(s) and random assignment of participants to the various experimental conditions. In an experiment, the experimenter provides two or more groups of participants with different experiences, representing two or more levels of the independent variable, and measures how those differences impact subsequent performance on one or more dependent variables. To be reasonably confident that any statistically significant differences in performance were the result of the experimental manipulation, the experimenter must rule out the possibility that some other intervening event or preexisting differences between the various groups caused the observed differences in performance. A true experiment minimizes these possibilities by controlling as many other random or extraneous variables as possible and by randomly determining the level of the independent variable that each participant receives. If this is done, then it is reasonable to assume that the biggest difference in the experiences of the groups is that produced by the independent variable and that the groups were statistically equal at the beginning of the experiment.

Preexperimental Designs

The three most common types of preexperimental designs are shown in Table 21.7. Each of these designs is limited because they cannot effectively rule out alternative explanations for observed results. The classic discussion of the threats to the internal and external validity of various types of research designs is available in Campbell and Stanley (1966).

One-Shot Case Study. Although Campbell and Stanley (1966) identified a number of limitations with this design, it is still widely used. As shown in Table 21.7, the one-shot case study involves providing training to a group of individuals and then measuring their performance. This design is fatally flawed because no performance baseline is established prior

TABLE 21.7
Preexperimental Designs for Training Evaluation

	One-Shot Case Study		
Group 1		T	O
		Training	Posttest
	One Group Pre- and Posttest Design		
Group 1	O	T	O
	Pretest	Training	Posttest
	Static Group Comparison		
Group 1	O		O
	Test		Test
Group 2	O	T	O
	Pretest	Training	Posttest

to the introduction of the training program. Therefore, it is impossible to know whether or not there has been any change in performance because there is nothing to compare performance against. In addition, factors other than training could account for the behavior measured after training. For example, trainees could have heard the company was planning to open a new plant and that bonuses were being offered to top performers to entice existing company personnel to transfer. Given that it is impossible to determine whether or not there has been any change in performance, it is difficult to imagine how one can justify the use of this design for any serious effort to empirically evaluate training effectiveness. At best, this design is limited to providing Level 1 type reaction data.

One-Group Pretest-Posttest Design.   Although this design allows us to determine whether or not performance has improved, it still does not allow us to logically rule out the possibility of alternative explanations. For example, it is still possible that any change in performance could be the result of other variables, such as a change in company policy. It is also possible that the participant's performance increased as a result of learning that was associated with the pretest rather than as a direct result of the training program.

Static-Group Comparison.   This design corrects some of the problems associated with the previous designs. Assuming that both groups are tested at approximately the same time, each group would be equally likely to be influenced by extraneous events, such as a change in company policy. In addition, because neither group received a pretest, learning as a result of pretesting can be ruled out. However, because the groups were not formed randomly, it is impossible to assume that they were equal in terms of their preexisting KSAs. Therefore, it is possible that there were preexisting differences between the control group that did not receive training and the experimental group that did receive training. Because of the possibility of preexisting differences, we cannot determine whether or not the training was responsible for any difference in performance between the two groups.

A recent evaluation by Huddlestone, Harris, and Tinworth (1999) illustrates a variation of the basic static-group comparison. The purpose of this study was to assess the effectiveness of a new ground-based aircrew team training device (ATTD). The purpose of this ATTD was to help prepare two aircrews, each composed of a pilot and navigator, to operate as a team to defeat enemy aircraft. Prior to the introduction of this new ATTD, many of the fundamental aspects of team tactics were taught in the aircraft because of limitations in the previous ground-based training device. Because the students were learning advanced procedures and maneuvers and, at the same time, attempting to integrate those into a complex sequence of in-flight behaviors, there were a number of ineffective sorties in which the students did not achieve satisfactory levels of performance. The hope was that the introduction of the ATTD would reduce the number of ineffective sorties.

The basic research design is shown in Table 21.8. Baseline data indicating the percentage of student failures for each of six in-aircraft training sorties were collected on a group of students who were trained prior to the introduction of the ATTD into the curriculum. Following the introduction of the ATTD, similar data were collected on another class of students who received ATTD training in addition to the in-flight training received by the no-ATTD group. The data indicate that the percentage of failures for the ATTD group was significantly lower than that of the no-ATTD group. In addition, the percentage of unsatisfactory ratings for behavioral processes (e.g., tactical leadership, tactical awareness, communication/coordination) was also significantly lower for the ATTD group.

The Huddlestone et al. (1999) study represents one of the most common problems encountered in training evaluations: the inability of the evaluator to run an experimental and control group concurrently. Typical reasons for this problem include insufficient resources (e.g., facilities, instructors, devices), a limited training population, and political/legal concerns involved with giving different groups of trainees different learning experiences. As a result, even though preexperimental designs are inherently weak, they are frequently used because of real-world limitations in-

TABLE 21.8
Research Design Employed by Huddlestone et al.

	Time 0		Time 1	
	Training	Measures of Effectiveness	Training	Measures of Effectiveness
Group 1: No ATTD	Learn and demonstrate proficiency in aircraft	Instructor assessment of student performance in aircraft		
Group 2: ATTD			Learn and practice team skills in ATTD Learn and demonstrate proficiency in aircraft	Instructor assessment of student performance in aircraft

Note. Groups 1 and 2 were not randomly formed and represent different training programs occurring at different times.

volving time and resources. Therefore, training system evaluators must often use preexperimental designs as part of their evaluation. In many cases, the quality of these evaluations can be improved by obtaining baseline data documenting performance prior to a training intervention as employed by Huddlestone et al. Readers who are unable to conduct true experiments to assess learning and transfer should consult Cook and Campbell (1976) for examples of quasi-experimental designs that overcome some of the problems associated with preexperimental designs. In addition, the quality of evaluations conducted using preexperimental design can be significantly improved by incorporating case study methodologies and design considerations, such as those described by Yin (1984).

True Experiments

Each of the preexperimental designs presented in the previous section involved the presentation of training and the measurement of performance following the presentation of training. However, as we discussed, any evaluation based on these designs contains a number of potentially confounding variables that make it impossible to conclude that the observed results were due to training rather than one or more confounding factors. The overwhelming power of a true experiment is its ability to simultaneously rule out these other confounding explanations for the results.

Table 21.9 shows the basic structure of a true experiment. The defining characteristics of a true experiment are two or more conditions and the random assignment of participants. This design is very efficient and powerful. Even without a pretest it is logical to believe that the experimental and control groups are equal, within chance limits, at the time of the posttest on everything except their training experience. Therefore, any difference in their posttest performance is most likely due to one group receiving training that the other group did not receive. Other factors such

TABLE 21.9
True Experimental Designs for Training Evaluation

	Two-Group Randomized Posttest Only Design			
Random	Group 1			O
assignment	Group 2	O	X	O
		Pretest	Training	Posttest

	Two-Group Randomized Pretest/Posttest Control Group Design			
Random as-	Group 1	O		O
signment	Group 2	O	X	O
		Pretest	Training	Posttest

as age, education, experience, and intelligence level that might explain a difference in the test scores of the two groups should be randomly spread out within the two groups. Therefore, we can have much greater confidence that any difference between the two groups is primarily the result of training when we evaluate training using a true experiment.

Training effectiveness evaluations are frequently conducted to compare different training media. An example of such an evaluation is a comparison of two aircrew training devices conducted by Nullmeyer and Rockway (1984). The purpose of this evaluation was to determine whether or not a C-130 Weapon System Trainer (WST) provided significantly better training than a C-130 Operational Flight Trainer (OFT). The difference between these two training devices was that the WST had a wide field-of-view visual system providing computer-generated out-the-window imagery, whereas the OFT was an instrument flight trainer. In this study, three separate experiments were conducted to assess training effectiveness for three different populations of trainees: students undergoing initial qualification training to become copilots, copilots upgrading to aircraft commanders, and mission-ready aircraft commanders enrolled in an instructor course. Each study used a control group and an experimental group. The primary dependent variable in each experiment was the number of flights in the actual aircraft, following ground-based training in either the OFT or the WST, needed to demonstrate that they met minimum requirements for the training events shown in Table 21.10. The data indicated that for each of the training events, pilots who received WST training reached the criterion level of performance significantly quicker than students who were trained in the OFT.

Taylor et al. (1999) used a two group randomized design to determine the effectiveness of a personal computer aviation training device (PCATD). The objective of this experiment was to determine whether a PCATD was effective for training private pilots to accomplish the procedures and maneuvers necessary for instrument flight. Table 21.11 illustrates the basic design used in this experiment. Students in the experi-

TABLE 21.10
Training Tasks Measured During C-130
Weapons System Trainer Evaluation

Course		
Copilot	*Aircraft Commander*	*Instructor*
Approach	Formation airdrop	Windmill taxi start
Landing	Lead airdrop	Three engine takeoff
Engine out	Recoveries	
	Assault landing	

TABLE 21.11
Basic Design Employed by Taylor et al.

	Group	Measures of Effectiveness
Randomly assign students to groups	*Control Group* Learn and demonstrate proficiency in aircraft	Number of trials in aircraft required to meet criteria Time required to complete each flight lesson Total aircraft time required to complete course
	Experimental Group Learn and demonstrate proficiency in PCATD Learn and demonstrate proficiency in aircraft	

mental group were introduced to each procedure and maneuver using the PCATD and were required to demonstrate task proficiency in the PCATD before moving to the aircraft. Students in the control group were taught each procedure and maneuver using only the aircraft. Both groups of students were required to meet the same performance criteria in the aircraft to complete that specific lesson. For both the experimental and control group, the instructors recorded the number of trials needed to reach criterion and the time to complete each lesson. In addition, check pilots who were unaware of whether or not the student was assigned to the experimental or the control group conducted the instrument flight checks in the aircraft.

The measures of training effectiveness used in this experiment involved the number of trials and the amount of time. The first, percent transfer, was defined as the following:

$$((Ac - Ae)/Ac)100 = \text{Percent Transfer}$$

where Ac is the time or trials required in the aircraft for the control group and Ae is the time or trials required in the aircraft for the experimental group. The second measure, transfer effectiveness ratio, is defined as the following:

$$(Ac - Ae)/Te = \text{Transfer Effectiveness Ratio}$$

where Te is the time or trials required in the PCATD for the experimental group.

The results of this experiment indicate that PCATDs are effective for training instrument flight tasks. In general, the experimental group was superior to the control group on trials to criterion, time to complete flight lessons, and time to complete course. Taylor et al. (1999) also discuss the importance of matching device use to training needs and the potential impact of training strategies on cost-effectiveness.

Unfortunately, it is often not feasible to employ true experimental designs when assessing training effectiveness in applied settings. For example, it is frequently impossible to obtain a sufficient number of trainees to establish both an experimental and a control group. Similarly, specific trainees may be designated to receive training at specific times based on production and manning constraints, thereby making random assignment impossible. Hence, although true experimental designs provide us with the most powerful vehicle for determining whether or not the training program produced changes in job-related performance, they are frequently impossible to actually implement.

Although true experiments provide us with much greater confidence regarding the likelihood that the difference between the experimental and control group is due to the difference in training, no experimental design can ever completely rule out the possibility that these differences were caused other factors. This is especially true in applied research, in which we are often faced with a variety of real-world problems that may limit the power of our research designs. These problems frequently include small sample sizes, the need to assign trainees as groups rather than individuals to the various training conditions, and trainees who represent only a limited subset of the full range of KSAs that may be encountered in an actual training program. Therefore, training researchers should employ a number of techniques to assess the value of a training program. These techniques may include utility evaluations by participants and major stakeholders, program evaluation measures, and quasi-experimental and experimental methods. It is hoped that the combination of such techniques will allow the researcher to triangulate or converge on the relative effectiveness of the training program.

LEARNING AND TRANSFER

Each of the designs discussed previously included measurement of performance after training was completed. Kirkpatrick (1998) correctly notes that there are really two separate but related issues involved in measuring performance in response to training that he labeled learning and behavior. Evaluations focused on learning seek to determine the degree to which trainees have acquired the KSAs that the training was designed to convey. These are specified as part of the ISD process, and both the content and the structure of the training system are geared toward enabling the acquisition of those KSAs. If trainees do not acquire the KSAs that were identified during the ISD process, then the training system has failed to accomplish its mission and requires modification.

This level of training effectiveness evaluation is very similar to academic testing and follows the same general approach. It should be based on a careful analysis of the different types of learning and instructional objectives involved in the training program (Gagne, Briggs, & Wager, 1988). The test items and instructional objectives should be explicitly linked following Mager's (1962) guidance to name the desired terminal behavior, describe the conditions under which the behavior will be demonstrated, and specify the criteria for acceptable performance.

Behavior focuses on what happens when trainees return to their jobs. Because trainees simply have acquired specific KSAs does not mean that those KSAs will improve job performance. The objective of this evaluation level is to determine if the training impacts performance in the real job setting. Successful transfer reflects not only the principles of good instructional design and delivery but also the accuracy of the task analysis that was used as the foundation for the training. Successful transfer also requires that trainees have the opportunity to demonstrate the KSAs they acquired in training as part of their job and that the organization rewards the job-specific behaviors that are associated with those KSAs. As Goldstein (1986) and others have pointed out, transfer sometimes fails to occur because workers simply do not have the opportunity to demonstrate their new KSAs in the work environment or because the actual job environment does not reward workers for using those specific KSAs.

QUALITY ASSURANCE

The costs of poor training often result in quality deficits, unnecessary rework, errors of omission or commission, and unsafe, inefficient, or ineffective operations, all of which reduce productivity and employee morale. Such practices also increase employee turnover in the commercial sector, thereby raising operating costs and reducing profit. It has been estimated that poor quality requires 25% of corporate resources for correction in the manufacturing industry and even more (30%–40%) in service firms (Peters, 1988).

Quality assurance (QA) is a process that supports the training system in meeting operational requirements by providing student graduates with the KSA needed for the mission or job. Training quality assurance can be of two principal types: internal or external. Quality should be driven by client expectations, which in turn define the services, products, and other performance outcomes to influence operational success. Because these outcomes are influenced by factors outside the organization, primarily client factors but also potentially those of vendors, suppliers, contractors, competitors, and other organizations, external QA generally should be the starting point for implementing the training QA process. Internal quality will evolve with the application of QA principles to external sources.

External QA factors as applied to training systems include training effectiveness (i.e., providing KSA) in preparing graduates for the operational setting, graduate proficiency levels relative to standards, required training system revisions, and standardization of procedures across organizations and sites.

Internal training systems QA factors include test scores, hands-on performance evaluations, student and instructor attitudes and motivation, remediation requirements, instructor training (train-the-trainer), media allocation, and efficiency (optimal use of training resources).

Every employee should be trained and actively involved in the quality improvement process. In industry, performance-based measurement should be carried out by teams, to assess themselves and other system parameters. Quality measurement should be performed early and often in the operational development cycle. Operational problems should be checked relative to those encountered in training. If the operational problem is determined to be related to a training deficiency, its nature and location in the training process should be pinpointed and immediately corrected.

SUMMARY

To conduct effective training systems evaluation, we must understand and address system components and their interrelationships. This chapter has described training as one form of human performance improvement and has focused on evaluation tools relative to all components of the training system—the hardware, software, courseware, procedures, and personnel. Because training systems vary in scope, charter, and application, the reader is cautioned to apply only those aspects of the evaluation process that meet the needs of the organization.

ACKNOWLEDGMENTS

The views expressed in this chapter are those of the authors and do not reflect the official position of the U.S. Air Force or the U.S. Department of Defense.

REFERENCES

Air Education and Training Command. (2001). Air Education and Training Command, http://www.aetc.af.mil.USAirForce.

Air Force Manual 36-2234. (1993). *Instructional system development*. Washington, DC: U.S. Air Force.

Alliger, G. M., Tannenbaum, S. I., Bennett, W., Jr., Traver, H., & Shotland, A. (1998). *A meta-analysis of the relations among training criteria* (AFRL-HE-BR-TR-1998-0130). Brooks, Air Force Base, TX: Air Force Research Laboratory.

American Society for Training and Development. (1997). Available: www.ASTD.org

Andrews, D. H., & Goodson, L. A. (1980). A comparative analysis of models of instructional design. *Journal of Instructional Development*, *3*, 2–16.

Army Regulation 350-7. (1982). *Systems approach to training*. U.S. Army.

Bell, H. H., & Waag, W. L. (1998). Evaluating the effectiveness of flight simulators for training combat skill: A review. *The International Journal of Aviation Psychology*, *8*, 223–242.

Berk, R. A. (1986). *Performance assessment: Methods and applications*. Baltimore: Johns Hopkins Press.

Boldovici, J. A., & Bessemer, D. W. (1994). *Training research with distributed interactive simulations: Lessons learned from simulator networking* (ARI Tech. Rep. No 106). Alexandria, VA: U.S. Army Institute for the Behavioral and Social Sciences.

Bracht, G. H., & Glass, G. V. (1968). The external validity of experiments. *American Educational Research Journal*, *5*, 437–474.

Brethower, D. M. (1989). Evaluating the merit and worth of sales training: Asking the right questions. In R. Brinkerhoff (Ed.), *Evaluating training programs in business and industry*. San Francisco: Jossey-Bass.

Campbell, D. T., & Stanley, J. C. (1966). *Experimental and quasi-experimental designs for research*. Chicago: Rand McNally.

Caro, P. (1977). *Some factors influencing Air Force simulator training effectiveness* (Rep. No. Hum-RRO-TR-77-2). Alexandria, VA: Human Resources Research Organization.

Childs, J. M., Ford, L., McBride, T. D., & Oakes, M. (1992). Team building and training. In R. Bakerjian (Ed.), *Tool and manufacturing engineers handbook, design for manufacturability* (Vol. 6). Dearborn, MI: Society of Manufacturing Engineers.

Cook, T. D., & Campbell, D. T. (1976). *Quasi-experimentation. Design and analysis issues for field settings*. Chicago: Rand McNally.

Dick, W., & Carey, L. (1990). *The systematic design of instruction*. Tallahassee, FL: Harper Collins.

Department of Defense. (2001). Defense Link News, *http://www.defenselink.mil*, Department of Defense.

Drucker, P. E. (1994). The age of social transformation. *Atlantic Monthly*, *274*(5), 53–80.

Gagne, R. M., Briggs, L. J., & Wager, W. W. (1988). *Principles of instructional design*. Fort Worth, TX: Holt, Rinehart, and Winston.

Gagne, R. M., & Merrill, M. D. (1990). Integrative goals for instructional design. *Educational Technology Research and Development*, *38*(1), 22–30.

Goldstein, I. L. (1986). *Training in organizations: Needs assessment, development, and evaluation*. CA, Brooks/Cole Publishing Company, 2nd ed.

Houck, M. R., Thomas, G. S., & Bell, H. H. (1991). *Training evaluation of the F-15 advanced air combat simulation* (AL-TP-1991-0047). Brooks Air Force Base, TX: Armstrong Laboratory.

Huddlestone, J., Harris, D., & Tinworth, M. (1999). Air combat training—The effectiveness of multi-player simulation. In *Proceedings of the Interservice/Industry Training Simulation and Education Conference*. Arlington, VA: National Defense Industrial Association.

Hughes, K., & Childs, J. M. (1999). *Federal Express Internet-based training study: Phase II—Evaluation of authoring and student administration alternatives* (Tech Report, TRW/A99-0010). Albuquerque, New Mexico: TRW.

Kerlinger, F. N. (1973). *Foundations of behavioral research*. New York: Holt, Rinehart and Winston.

Kirkpatrick, D. L. (1976). Evaluation of training. In R. L. Craig (Ed.), *Training and development handbook* (2nd ed., pp. 18.1–18.27). New York: McGraw-Hill.

Kirkpatrick, D. L. (1983). *A practical guide for supervisory training and development.* Reading, MA: Addison-Wesley.

Kirkpatrick, D. L. (1998). *Evaluating training programs: The four levels.* San Francisco: Berrett-Koeher.

Mager, R. F. (1962). *Preparing instructional objectives.* Palo Alto, CA: Fearon.

Meister, D. (1989). *Conceptual aspects of human factors.* Baltimore: Johns Hopkins University Press.

Muckler, F. A. (1977). Selecting performance measures: "Objective" versus "subjective" measurement. In L. T. Pope & D. Meister (Eds.), *Productivity enhancement: Personnel performance assessment in Navy systems* (pp. 169–178). San Diego, CA: Navy Personnel Research and Development Center.

Nullmeyer, R. T., & Rockway, M. R. (1984). Effectiveness of the C-130 weapon systems trainer for tactical aircrew training. In *Proceedings of the InterService Industry Training and Equipment Conference* (pp. 431–440). Arlington, VA: National Defense Industrial Association.

Peters, T. (1988). *Thriving on chaos.* New York: Harper & Row.

Roscoe, S. N., & Childs, J. M. (1980). Reliable, objective flight checks. In S. N. Roscoe (Ed.), *Aviation Psychology.* Ames, IA: Iowa State University Press.

Rossi, P. H., & Freeman, H. E. (1989). *Evaluation: A systematic approach.* Newbury Park, CA: Sage.

Saari, L. M., Johnson, T. R., McLaughlin, S. D., & Zimmerle, D. M. (1988). A survey of management training and education practices in U.S. companies. *Personnel Psychology, 41,* 731–743.

Sanders, N. M. (1989). Evaluation of training by trainers. In R. Brinkerhoff (Ed.), *Evaluating training programs in business and industry.* San Francisco, CA: Jossey-Bass.

Spears, W. D. (1986, March). *Design specification development for the C-130 model aircrew training system: Test and evaluation plan* (Seville TD 86-03). Irving, TX: Seville Training Systems.

Taylor, H. L., Lintern, G., Hulin, C. L., Talleur, D. A., Emanuel, T. W., Jr., & Phillips, S. I. (1999). Transfer of training effectiveness of a personal computer aviation training device. *The International Journal of Aviation Psychology, 9,* 319–335.

U.S. Department of Labor. (1999). Available: www.dol.gov

Vincente, K. J. (1999). *Cognitive work analysis.* Mahwah, NJ: Lawrence Erlbaum Associates.

Yin, R. K. (1984). *Case study research: Design and methods. Applied social research methods series (Vol. 5).* Newbury Park, CA: Sage.

Chapter 22

Epilogue:
Testing Technologies

Samuel G. Charlton
Waikato University
and Transport Engineering Research New Zealand Ltd.

When asked to prepare an epilogue to this book, I undertook the task with considerable trepidation. After all, what more could possibly remain to be said about HFTE that was not already said in the preceding chapters? One theme often adopted by chapters of this sort, and one which I was encouraged to pursue for this chapter, is to describe challenges for the future. The opportunity to prognosticate about future trends in HFTE, however, did little to alleviate my unease. Because technological change drives much of what we do in HFTE, forecasting the future of HFTE would require a forecast of technology itself, an assignment far beyond my expertise. Fortunately (from my immediate perspective), as HFTE takes some time to catch up with technological innovation, it is possible to look at existing technologies and see where HFTE is overdue in meeting the questions asked of it. This chapter will explore some of those shortcomings, briefly and from a somewhat more informal perspective than that adopted elsewhere in this book.

There is little question that technology shapes human culture every bit as much as our culture shapes technology. Every new technology has consequences for our lives, many of them unintended and unforeseen (Norman, 1993; Van Cott, 1990). It can be argued that HFTE's role in this relationship is to buffer or protect individual users from as many of the adverse consequences of new technology as possible. In this sense, the goal of HFTE is to catch technological side effects before they harm, frustrate, or impede users of the technology. By way of example, when the telephone was introduced some 120 years ago, it was not clear how the technology would be used. Some envisioned it as a broadcast medium, sending information and entertainment to groups of people gathered at a public or private locale to listen to the latest news or even musical and theatrical performances (Norman, 1993). Few people could foresee the impending changes brought by the introduction of widely available point-to-point communication. The availability of instant communication brought with it side effects related to overaccessibility. The frustration produced by hearing the telephone ring as one sits down to a meal, slips into a hot bath, or begins an important task has led some to adopt unlisted numbers, answering machines to screen calls, or yearn for telephone-free vacations. (With the advent of cellular telephony, hearing other people's phones ringing at inopportune times and places can induce even greater annoyance.)

More recently, the personal computer (PC) was developed with little anticipation of the changes that would be brought to our lives. Originally available for researchers and hobbyists, the PC was subsequently marketed for technologically savvy home users as a means of keeping household budgets, storing recipes, and playing games. The convergence of personal computers and telephony has brought us the Internet and the advent of today's wired society. Who could have foreseen the rise of the dot-com industry in the humble beginnings of the PC? At this stage, we can only guess at the ultimate impact of e-commerce, much less the implications of advances in other technologies, such as biotechnology and nanotechnology.

What of HFTE? Looking back at future challenges posed by authors over the years, the challenges were seen as arising from increasing automation and the demassification of information produced by new computing and communications technologies (Meister, 1986, 1996; Norman, 1993; Van Cott, 1990). In the case of automation, the need to test how well the systems could be monitored, diagnosed, and controlled by human supervisors was seen as a challenge for HFTE. In that regard, one can view the ascendance of research into situation awareness, cognitive workload, and naturalistic decision making as attempts to meet this challenge. But how well have those attempts succeeded? In large part, the issues surrounding the use of verbal reports as a measure of cognitive states (a problem as old as experimental psychology) have kept HFTE solutions in these areas somewhat disjointed and idiosyncratic. As well studied and important as they are, we have yet to come to agreement on measures, thresholds, and testable criteria for these cognitive states. To be sure, the sometimes dramatic differences in individual users and their activities makes the development of these sorts of psychological thresholds very difficult indeed. Until we do, however, we cannot identify requirements that will give designers and testers a tangible goal to work toward in developing effective and operable automated systems.

How much workload is too much? How much situation awareness is enough? How fatigued should we let operators become before we give them rest? For that matter, how much difficulty, discomfort, or pain is indicative of an unacceptable system? Unfortunately, our profession cannot provide answers to any of these questions (short of the rejoinder "well, it depends . . ."). In many cases we cannot even agree on the metrics to use. We can, of course, design our systems to fit an idealized user and then develop proficiency standards for selecting operators. In this way only the best and brightest would be selected to operate our air traffic control centers, power-generating stations, and telecommunications networks. But what of office automation equipment, our transportation systems, and consumer products? In these cases users' purchases will ultimately select the most effective and usable product, rather than vice versa. (Not to mention that relying on operator selection instead of emphasizing user-centered design would be a giant step backwards to where HF was 50 or 60 years ago.)

Although difficult, I would argue that establishing testable criteria for these cognitive states is not an impossible undertaking for HFTE. Given some latitude, I will even propose that an initial beachhead has already been made by calibrating psychological fatigue against alcohol intoxication and driving. In spite of the tremendous individual differences in drivers' alcohol tolerance and driving skills, lawmakers have specified a limit for driver impairment based on the quantification of blood alcohol levels (measurable by breath alcohol content). Across the full range of traffic conditions and vehicle types, we have set limits for drivers (0.08 or 0.05 BAC) that not only have the force of law but also now enjoy widespread popular support. In the area of driver fatigue, researchers are using this alcohol impairment benchmark to propose standards of driver impairment due to fatigue (Dawson, Lamond, Donkin, & Reid, 1999; Fairclough & Graham, 1999; Lenne, Triggs, & Redman, 1998). Stated simply, the idea is to quantify performance decrements associated with legally defined thresholds of alcohol intoxication and then calibrate them against equivalent performance decrements associated with fatigue. The result of this work may see the establishment of limits on the amount of sleep and rest needed by drivers by linking them to the societally acknowledged limit of driver impairment associated with alcohol consumption.

I am not suggesting (yet) that we set situation awareness and workload thresholds against the performance of intoxicated air traffic controllers, but wherever a performance threshold or modal performance level can be established for a task, that threshold could be linked to objective criteria for performance decrements due to excessive workload or poor situation awareness. Mathematically and logically, setting design requirements against cognitive state measures that are linked to operator performance criteria is straightforward and practical (Charlton, 1988). Further, development of criteria for these important HFTE measures, although they may lack the force of law, may at least have contractual force and improve our ability to ensure the effectiveness and suitability of new system designs.

Is it appropriate for the human factors profession to set these standards? In my view, the answer is unhesitatingly yes. If we do not, who will? Can we do a better job than we have done to date? Again, my answer is yes. We need to come to agreement on how to measure workload, fatigue, boredom, and so on and then identify minimum and maximum levels for those measures (by virtue of linkage to performance levels). Once we do, we will at last be able to tell system engineers and designers how much workload, fatigue, and discomfort is too much and how much situation awareness and ease of use is enough.

Turning to the second technological challenge posed earlier, the information explosion resulting from the convergence of telephones, computers, and the Internet, HFTE still has a long way to go in developing measures and defining requirements. To be sure, progress has been made in the areas of information display and software usability, but we have not addressed in any substantial way the problems associated with protracted sessions in front of computer screens (and talking on cell phones). The physical, cognitive, and social problems produced by occupational and recreational overuse of these technologies are with us now. Can HFTE buffer users from these sorts of technological side effects and, perhaps more to the point, should it?

To my way of thinking, the answer to both questions is, once again, yes. HFTE can and should include the wider implications of technology use in our measurement and analyses. Some would argue that extending HFTE to the voluntary overuse of technology by users smacks of guarding users against themselves and establishing a nanny-state mentality. In response, I argue that when technology use (or overuse) by even a willing user serves to harm other individuals, or society as a whole, the designers and testers are responsible for changing that technology such that it no longer affords user behavior that is harmful. As a case in point, setting limits on drivers' speeds is widely recognized as appropriate. Although designers are struggling with how to include users' respect for speed limits as an inherent product of the system design, some system solutions, such as speed bumps and lane markings, have been implemented. To address the transportation technology issues confronting us in the future, we will need to move beyond testing the individual driver's cognitive workload and driving speed. We will need to test the impact of new transportation design solutions (e.g., increased traffic densities, larger vehicles, highway bypasses, noise barriers, etc.) on the interactions between drivers (traffic) and the impact on surrounding communities. Similarly, the introduction of free flight in commercial aviation will have effects not only on individual pilots and aircrews but also on the air traffic in entire air corridors and on passengers waiting on the ground. As designers work toward technologies that do a better job at compelling users to behave responsibly and safely, HFTE must be ready with the methods and measures ready to evaluate their success. Moving from testing technologies' impact on a single user or operator to testing the effects on groups of users and nonusers affected by the technology is certainly a challenge but one deserving of our efforts.

It is often opined that human factors (and by implication, HFTE) is a young discipline. Reviewing the chapters in this book gave me the impression that HFTE is maturing and capable of more responsibility. Like a teenager coming of age but still unclear about the direction in which to exercise their capabilities, I believe HFTE is capable of much but promises much more still. Although I cannot imagine the new technologies that will drive our discipline in the future, there are a fair few challenges before us now. The two described in this chapter, the establishment of criteria for cognitive states and evaluation of the broader social effects of system design, are good places to begin. New technologies, new challenges, and new questions will make for fascinating times for our discipline in the years to come.

REFERENCES

Charlton, S. G. (1988). An epidemiological approach to the criteria gap in human factors engineering. *Human Factors Society Bulletin, 31,* 1–3.

Dawson, D., Lamond, N., Donkin, K., & Reid, K. (1999). *Quantitative similarity between the cognitive psychomotor performance decrement associated with sustained wakefulness and alcohol intoxication.* Limited circulation manuscript.

Fairclough, S. H., & Graham, R. (1999). Impairment of driving performance caused by sleep deprivation or alcohol: A comparative study. *Human Factors, 41,* 118–128.

Lenne, M., Triggs, T., & Redman, J. (1998). Sleep loss or alcohol: Which has the greater impact on driving? *Proceedings of the Road Safety Conference 1998.* Wellington, NZ: Land Transport Safety Authority and New Zealand Police.

Meister, D. (1986). *Human factors testing and evaluation.* Amsterdam: Elsevier.

Meister, D. (1996). Human factors test and evaluation in the twenty-first century. In T. G. O'Brien & S. G. Charlton (Eds.), *Handbook of human factors testing and evaluation* (pp. 313–322). Mahwah, NJ: Lawrence Erlbaum Associates.

Norman, D. (1993). *Things that make us smart: Defending human attributes in the age of the machine.* Reading, MA: Addison-Wesley.

Van Cott, H. P. (1990). From control systems to knowledge systems. In M. Venturino (Ed.), *Selected readings in human factors* (pp. 47–54). Santa Monica, CA: The Human Factors Society.

Author Index

Subject Index

A